Soils in Construction

Fifth Edition

W. L. Schroeder
Oregon State University, Emeritus

S. E. Dickenson
Oregon State University

Don C. Warrington
Pile Buck, Inc.

PEARSON

Prentice
Hall

Upper Saddle River, New Jersey
Columbus, Ohio

To my wife Judy.—Don

Library of Congress Cataloging-in-Publication Data

Schroeder, W.L. (Warren Lee)
  Soils in construction / W.L. Schroeder, S.E. Dickenson, D.C. Warrington.—5th ed.
    p. cm.
  Includes bibliographical references and index.
  ISBN 0-13-048917-4
  1. Soil mechanics. 2. Foundations. 3. Building. I. Dickenson, S.E. II. Warrington, D.C. III. Title.

TA710.S286 2004
624.1'5136—dc21   2003051758

Editor in Chief: Stephen Helba
Executive Editor: Ed Francis
Assistant Editor: Linda Cupp
Production Editor: Holly Shufeldt
Production Coordination: Preparé Inc.
Design Coordinator: Diane Ernsberger
Cover Designer: Mark Shumaker
Production Manager: Matt Ottenweller
Marketing Manager: Mark Marsden

Appendix B: Specifications are ©Copyright 2001 Pile Buck®, Inc., P. O. Box 64-3929, Vero Beach, FL, 32964, www.pilebuck.com and are reproduced with permission.

This book was set in Times by Preparé Inc. It was printed and bound by R.R. Donnelley & Sons Company. The cover was printed by The Lehigh Press, Inc.

Pearson Education Ltd.
Pearson Education Singapore Pte. Ltd.
Pearson Education Canada, Ltd.
Pearson Education—Japan

Pearson Education Australia Pty. Limited
Pearson Education North Asia Ltd.
Pearson Educación de Mexico, S.A. de C.V.
Pearson Education Malaysia Pte. Ltd.

10 9 8 7 6 5 4 3 2 1

ISBN 0-13-048917-4

Preface

For more than a quarter of a century, *Soils in Construction* has been an important resource for the teaching of the basics of soil mechanics, foundation design, and foundation construction to construction management students. The fifth edition continues this with extensive revisions to both the text and the accompanying graphics. The highlights of these changes are as follows:

- An expanded overview of the geology of soils and rocks (Chapters 1 and 2);
- Improved classification aids for the Unified system (Chapter 4);
- A more complete description of borrow and fill calculations (Chapter 8);
- Addition of a description of OSHA's requirements for trench safety and configuration (Chapter 10);
- A completely rewritten description of pile driving and driven pile design and testing, including new material on pile dynamics and a new pile hammer chart (Chapter 11);
- Breakout of all of the existing examples and new examples added to the text, so that students and instructors can more readily follow methods for solving problem;
- Addition of SI units to some of the calculations.

Many other details of the text have been enhanced and improved as well.

This book was originally prepared as a teaching aid for a course in the Construction Engineering Management program in the Civil Engineering Department at Oregon State University. The purpose of the course was to introduce students to the nature of soils and to illustrate how soil materials may influence certain construction operations. The course was not design-oriented. It was a terminal geotechnical course specifically arranged to deal with soils in construction for those who did not contemplate further study of soil mechanics or foundation engineering. The book is, therefore, suited for use in other similar programs.

The book begins with an introduction to soil materials. In conjunction with the testing methods in Appendix A, the material presented provides the basic background for understanding soil behavior and how construction specifications relate to it. From here the book turns to soils in the contruction contract. Specifications from example contracts influenced by soil materials are discussed, as is the application of principles of soil behavior to those specifications. The appendices contain both testing procedures designed to be used as instructional laboratory exercises (with references to similar standard testing methods) and a pile hammer chart.

Today persons with diverse responsibilities, such as contractors, owners, technicians, lawyers, and engineers from other specialties, interact with geotechnical design professionals. They often need to know and understand the designer's language and concerns, and they need to understand how to apply selected principles of the geotechnical disciplines to their own work. It is to such individuals that this book is directed.

In addition, particular thanks are due to Greg Ohrn, Northern Arizona University; Leslie M. Gioja, Parkland College; Timothy W. Zeigler, Southern Polytechnic State University; Constantine A. Ciesielski, East Carolina University; and Ho-Yeong (Julian) Kang, Texas A&M University, for their assistance with the fifth edition text review.

More than any other kind of book, a textbook is a "work in progress" and this one is no exception. Instructors, students and readers can send their questions, comments and suggestions by visiting http://www.vulcanhammer.net/soils/.

W.L. Schroeder
S.E. Dickenson
Don C. Warrington

Contents

PART II   EARTHWORK IN THE CONSTRUCTION CONTRACT   87

Chapter 9 Dewatering   163

Chapter 10 Excavations and
Excavation Supports   185

Chapter 11   Foundation Construction   211

Chapter 12   Construction Access and Haul Roads   260

PART ONE

Soil Materials

CHAPTER 1

Physical Character of Soil Constituents

Soil may be defined as an accumulation of solid particles produced by mechanical and chemical disintegration of rocks. It may contain organic constituents and water. This broad definition applies to a construction material that varies widely in its physical composition and behavior from location to location, and even on a particular site. This chapter describes the nature of various soil constituents and how they are developed from parent rock. Knowledge of the physical character of soil constituents is essential to an understanding of soil behavior during construction.

Studying this chapter should give the reader

1. A definition of soil material constituents and their origin.

2. Descriptive information concerning the size of soil particles of various types and other factors that control soil behavior.

3. An introduction to the fundamental reasons for the observed differences in behavior among various fine-grained soils.

1.1 ROCK, SOIL AND WEATHERING PROCESSES

The terms *rock* and *soil* are ones used by virtually everyone, yet a careful investigation of the nature of the earth's crust—let alone the complexities of a soil and rock profile for a specific job—make clear that the complex nature of the world around us makes generalizations both necessary and misleading. However, some type of categorization is necessary to quantify the properties of the material we have been given to build structures on. The first set of definitions are of rock and soil itself:

- *Rock* is the naturally occurring material composed of mineral particles so firmly bonded together that relatively great effort is required to separate the particles (i.e., blasting or heavy crushing forces).

- *Soil* is the conglomeration of naturally occurring mineral particles which are readily separated into relatively small pieces, and in which the mass may contain air, water, or organic materials (derived from decay of vegetation).

Considering rock first, rock can be divided into three categories of its own:

- *Igneous rocks* are rocks that are formed by the solidification of molten material, either by intrusion at depth in the earth's crust or by extrusion at the earth's surface.
- *Sedimentary rocks* are those that are formed by deposition, usually under water, of products derived by the disaggregation of preexisting rocks.
- *Metamorphic rocks* may be either igneous or sedimentary rocks that have been altered physically and sometimes chemically by the application of intense heat and pressure at some time in their geological history

Natural mechanical and abrasive forces degrade massive rock into smaller and smaller particles. These forces may be of thermal or gravitational origin. The action of these forces is referred to as weathering, although these phenomena can be from beneath the surface of the earth as much as from the action of atmospheric forces. Weathering can broadly be divided into two categories:

- Mechanical weathering of parent rock produces coarser soil particles such as gravel and sand, and silts which are finer. Soil particles produced by mechanical weathering have approximately three-dimensional shapes.
- Chemical weathering produces very small particles that are often in crystalline form. These particles are typically two-dimensional or flake shaped. Chemical weathering of rock results in clayey soils. The characteristics of clays depend on the nature of the parent rock, the environment in which the weathering takes place, and the length of time available for chemical alteration to develop.

Soils are broadly divided into *coarse-* and *fine-grained* soils, depending upon their grain size. The difference between these fractions can be seen in Figure 1.1. Generally speaking, a fine-grained soil is a soil where most of the particles pass the No. 200 sieve; that is, the particles are small enough to be able to pass through mesh openings which are 0.074 mm in size. Coarse-grained soils constitute all other soils. This distinction will be elaborated on in Chapter 4. Fine-grained soils are also referred to as *cohesive* soils and coarse-grained ones as *cohesionless* soils, for reasons that will be explained later.

Figure 1.1 A coarse soil and a fine soil. The coarse soil on the left is shown as it would appear wet or dry. The fine soil is shown in both states.

1.2   CHARACTER OF THE COARSE SOIL FRACTION

1.2.1   Particle Size and Sieve Analysis

Once a soil has been designated as either coarse or fine, it can be further differentiated by sub-division of the entire mass of solid material into size ranges. These are outlined in Table 1.1, including the various descriptive adjectives used in designating these fractions. Coarse soil constituents may be sand, gravel, cobbles, or boulders. Fine soil constituents (also referred to as "fines") consisting of silts and clays are discussed in the next section. An example of a laboratory sieve analysis used to separate soil into size fractions (ASTM D422) is shown in Figure 1.2. Appendix A.2 is a summary of this test.

TABLE 1.1   Definitions of Soil Components and Fractions (After NAVFAC DM 7.01)

1. Grain Size

Material	Fraction	Sieve Size
Boulders		12″+
Cobbles		3″–12″
Gravel	Coarse	3/4″–3″
	Fine	No. 4–3/4″
Sand	Coarse	No. 10–No. 4
	Medium	No. 40–No. 10
	Fine	No. 200–No. 40
Fines (Silt & Clay)		Passing–200

2. Coarse- and Fine-Grained Soils

Descriptive Adjective	Percentage Requirement
Trace	1–10%
Little	10–20%
Some	20–35%
And	35–50%

3. Fine-Grained Soils

Identify in accordance with plasticity characteristics, dry strength, and toughness as described in Figure 4.3.

Descriptive Term, Stratified Soils	Thickness
Alternating	
Thick	
Thin	
With	
Parting	0 to 1/16″ thickness
Seam	1/16 to 1/2″ thickness
Layer	1/2 to 12″ thickness
Stratum	greater than 12″ thickness
Varved clay	alternating seams or layers of sand, silt and clay
Pocket	small, erratic deposit, usually less than 1 foot
Lens	lenticular deposit
Occasional	one or less per foot of thickness
Frequent	more than one per foot of thickness

Figure 1.2 Soil fractions separated by sieving. Each fraction shown is the range of particle sizes passing the next larger sieve and retained on the sieve with which it is shown.

1.2.2 Particle Shape and Angularity

Particle shape, or angularity, for instance, greatly influences the ability of a granular or coarse-grained soil to form interparticle interlocks. Degrees of angularity are illustrated in Figure 1.3. Individual particles may be angular (sharp, distinct edges) or rounded. Subangular particles have distinct but rounded edges while subrounded particles are smooth and nonequidimensional. The greater the degree of sharpness and angularity of its particles, the greater ability a coarse soil has to form a stable, strong soil mass.

1.2.3 Specific Gravity

Specific gravity of soil particles can be defined in one of three ways:

- Absolute specific gravity is defined as

$$G_s = \frac{\gamma_s}{\gamma_w} \tag{1.1}$$

  where γ_s is the unit weight of a material and γ_w is the unit weight of water at the reference temperature, 4°C.

- Apparent specific gravity of a soil is the ratio of the weight in air of a given volume of the impermeable portion of a permeable material (i.e., the solid matter including its impermeable pores or voids) at a stated temperature.

- Bulk specific gravity of a soil is the ratio of the weight in air of a given volume of permeable material (including both permeable and impermeable voids normal to the material) at a stated temperature to the weight in air of an equal volume of distilled water at a stated temperature.

(a)

(b)

Figure 1.3 Angularity variations for sand particles. (*a*) Platte River sand, Colorado (magnification 50 ×). (*b*) Ottawa sand (magnification 50 ×).

Figure 1.4 indicates the distinction between these definitions of specific gravity. The specific gravity of solid particles usually reported is an apparent specific gravity. In determinations of specific gravity, the volume of solids and the weight of solids are determined in order that unit weight may be computed. The normal procedure (ASTM D854; see Appendix A.3) used to determine specific gravity involves volume determination by water displacement. Since impervious voids may not be filled by this process, the specific gravity normally reported is less than the absolute value. The importance of this distinction is usually minor.

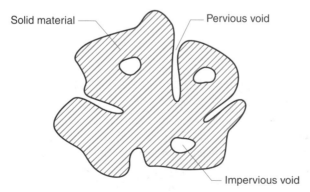

Solid material —
Pervious void —
Impervious void —

Figure 1.4 Specific gravity–volume relation. For bulk specific gravity, the particle volume includes solid material, impervious voids and pervious voids. For apparent specific gravity, the particle volume includes solid material and impervious voids. For absolute specific gravity, the particle volume includes solid material only.

Except for organic soils, the range of specific gravities for soil is small. Most soils (sands and clays) have a typical specific gravity of 2.65: soils such as dark-colored sand and sand–silt–clay mixtures have a specific gravity of 2.72. An example data sheet for laboratory specific gravity determinations is shown in Figure 1.5.

SPECIFIC GRAVITY TEST

TEST FOR   *Olequa Dam*

TEST BY   *EFS*   DATE   *4 March 1992*

SAMPLE DESCRIPTION   *Red-brown clay core material.*
test pit no. 1   depth - 12 feet

DETERMINATION NO.	1	2	3	4
BOTTLE NO.	5	5		
WT. BOTTLE + WATER + SOIL (W_1)	674.72	675.81		
TEST TEMPERATURE t, °C	22.8	23.2		
WT. BOTTLE + WATER (W_2)	643.13	643.09		
EVAPORATING DISH NO.	SM 2	SM 4		
WT. DISH + DRY SOIL	824.58	850.00		
WT. DISH	774.99	798.90		
WT. DRY SOIL (W_S)	49.59	51.10		
SPECIFIC GRAVITY OF WATER AT t,(G_t)	0.9976	0.9975		
SPECIFIC GRAVITY OF SOIL (G_S)				

REMARKS _____

_____

_____

$$G_S = \frac{G_t\, W_S}{W_S + W_2 - W_1}$$

1. $\dfrac{0.9976 \times 49.59}{49.59 + 643.13 - 674.72} = 2.72$

2. $\dfrac{0.9975 \times 51.10}{51.10 + 643.09 - 675.81} \approx 2.77$

Figure 1.5   Results and calculations of specific-gravity test.

1.2.4   Toughness and Durability

In addition to size and specific gravity, toughness and durability of coarse particles can be important factors. The quality of rock proposed for a specific use must often meet accepted standards for resistance to mechanical or chemical degradation. For these purposes, the Los Angeles abrasion test (ASTM C131) and the Soundness of Aggregates by Use of Sodium Sulfate or Magnesium Sulfate (ASTM C88) test are widely used. In these tests, representative samples are subjected to harsh mechanical or chemical action and their weight loss is determined. Performance is compared with performance of materials that have proven acceptable when construction specifications are written. Other characteristics of the coarse soil fraction such as surface texture and surface chemistry may control their acceptability. These are most important when the material is used for aggregate in Portland cement or asphalt concrete.

1.3   CHARACTER OF THE FINE SOIL FRACTION

The fine soil fraction consists of silts and clays (see Table 1.1). Silts are usually similar in shape to coarse fraction particles. They are very small. Silt is the result of mechanical weathering. Clays result from chemical weathering. The difference in origins of silts and clays produces very distinct differences in their behavior.

1.3.1   Chemical Structure of Clays

Clays are typically hydrous aluminum silicates composed of distinct structural units with unique relationships to one another. The fundamental structural units are tetrahedral and octahedral ionic arrangements such as those shown in Figure 1.6. These units are assembled in sheets parallel to their basal planes. The silica units form a sheet with common oxygen ions at the corners. The basal oxygen ions share an electron with adjacent silica. The resulting electrostatic charge per unit is 21, since, of the available 28 due to the oxygens, three are balanced by

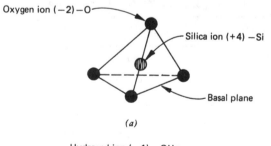

(a)

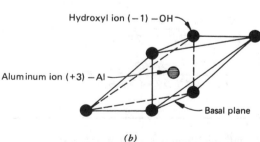

(b)

Figure 1.6   Fundamental clay building blocks. (Reprinted, by permission, from Grim, R.E., *Clay Mineralogy*, McGraw-Hill, New York, 1953.) (*a*) The silica tetrahedron. (*b*) The octahedral unit.

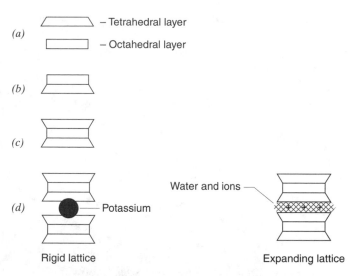

Figure 1.7 Schematic clay mineral structures. (*a*) The basic units. (*b*) 1 : 1 clay mineral structure. (*c*) 2 : 1 clay mineral structure. (*d*) Typical 2 : 1 clay lattice structures.

sharing with adjacent silica and four are balanced by the central silica. The octahedral unit assembles with each hydroxyl common to three units. The resulting charge per unit is 11, since the available six negatives are divided by three.

The octahedral or tetrahedral sheets formed as described above are mutually attractive and unite in various ways to form the basic two-dimensional structure of clay minerals. The more common forms are shown schematically in Figure 1.7. These structures tend to exhibit excess negative charges on their flat sides and positive charges at their edges.

1.3.2 Types of Clays

The structure of *kaolinite* is schematically represented by Figure 1.7*b*. Adjacent units tend to stack, one on the other (see Figure 1.8), as the oxygens in the basal layer of the tetrahedral unit are attracted to the hydrogens in the octahedral unit. A comparatively weak hydrogen bond results. Kaolins are very stable minerals and are among the last formed in the weathering process. They exhibit very little physicochemical activity compared to other clay minerals, because of a comparatively large grain size and low surface charge.

Illite exhibits intermediate physicochemical activity. Interlayer potassium ions form a weak ionic bond with the oxygen ions in the silica sheets. A deficiency of potassium results in greater unsatisfied valence charge and correspondingly greater activity. Illite is a very commonly occurring clay mineral.

Montmorillonite is a third characteristic clay mineral. In the absence of the interlayer bonding noted in illite's structure, the interlayer space is expandable. Typically, it is filled with water. Depending on the abundance or deficiency of water available to occupy this space, a montmorillonitic soil may exhibit marked volume changes upon drying or wetting. These materials are particularly troublesome when abundant water is present and are the most claylike

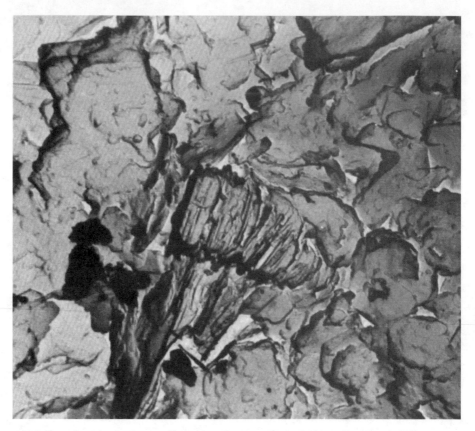

Figure 1.8   Parallel particle stacking in a clay soil—a ped (magnification 25,000 ×).

(active) of the clay minerals. They are among the poorest of available soils for most structural support purposes. Montmorillonites are very unstable chemically and are among the initially formed chemical-weathering products. They derive principally from volcanic parent materials in a moist environment.

Other clay minerals have special properties that may assume major importance in construction works. *Halloysite*, for instance, is similar in structure to kaolinite, except that the plately particles tend to assume a tubelike shape. The shape is thought to result from the presence of interlayer adsorbed water, which weakens the hydrogen bond between the tetrahedral and octahedral sheets. A halloysite develops an irreversible change upon drying and becomes less claylike than in its natural state. Laboratory procedures that require drying before testing may, thus, misrepresent the true nature of a natural material, often to the detriment of the contractor. Drying in preparation for testing results in tests that may show that the soil is less plastic and, therefore, more workable than it actually may be in the field where drying has not occurred.

Amorphous (noncrystalline) materials occur widely. These soils are claylike, yet do not have the simple structure described for crystalline clays. Very finely divided amorphous forms typically occur as gels and, in sufficient quantity, coat other mineral particles. In this manner, they serve to control soil behavior. While amorphous materials are an interesting phenomenon to some, to distinguish them from other clay soils may be, for the construction engineer, large-ly academic.

1.3.3 Cohesive and Cohesionless Soils

As we said earlier, clays are referred to as cohesive soils. The chemical bonding that takes place between the atoms gives the soils the ability to hold together without the need of external pres-sure. This ability is dependent upon both the type of clay and the water content. Although in theory it is better to have soil cohesion, in reality the variables of that cohesion can make both building and supporting structures in cohesive soil difficult in certain cases.

Coarse-grained soils, on the other hand, are termed cohesionless because they do not have this kind of bonding. Any strength they have is derived from the mutual friction of the par-ticles, which is mobilized by external pressure. These effects, which are very important in de-sign and construction, are discussed in Chapter 5.

1.4 SUMMARY

This chapter has shown that soils occur as coarse-grained or fine-grained materials. Coarse soils include sands, gravels, cobbles, and boulders; fine soils are silts and clays. If a soil parti-cle, say, a coarse sand grain, is subdivided, its volume remains the same, while its surface area increases. Further subdivision of the resulting particles increases the surface area geometrical-ly while total volume and weight are constant. The sand particle, if subdivided often enough, is eventually reduced to a large number of particles in the clay-size range. If the specific surface of a soil is defined as the area of the particle surfaces per unit of weight, an index of the effect of particle size on surface area is established. Consider the information in Table 1.2.

If it is recalled that the surfaces of clay minerals possess an electrochemical charge, it is evident from Table 1.2 that the presence of this charge, as a factor in determining soil behavior, becomes increasingly important as particle size decreases. Many of the observed differences in soil behavior for coarse and fine materials may be explained on the basis of this information. Coarse soil behavior is controlled largely by the weight and denseness of packing of the soil particles, while fine soils, clays in particular, owe much of their often troublesome nature to in-terparticle physicochemical forces that arise from their natural electrostatic charges.

The reader should now

1. Be able to distinguish among soil constituents based on particle size.

2. Know how to explain the origin of surface charge on fine soil particles.

TABLE 1.2   Specific Surfaces of Soil Materials

Material	Specific Surface (m^2/g)
Sand (0.1 mm)	0.03
Kaolinite	10
Illite	100
Montmorillonite	1000

Source: T. W. Lambe and R. V. Whitman, *Soil Mechanics*, New York; Wiley, 1969.

REFERENCES

American Society for Testing and Materials, *1988 Annual Book of ASTM Standards*, Philadelphia.

Coduto, Donald P., *Geotechnical Engineering: Principles and Practices*. Upper Saddle River, NJ: Prentice Hall, 1999.

Department of the Army, *Laboratory Soils Testing*. EM 1110-2-1906. Washington, DC: U.S. Army Corps of Engineers, 1986.

Department of the Navy, *Soil Mechanics*. NAVFAC DM 7.01. Norfolk, VA: Naval Facilities Engineering Command, 1986.

Grim, R.E., *Clay Mineralogy*. New York: McGraw-Hill, 1953.

Lambe, T.W., and Whitman, R.V., *Soil Mechanics*. New York: Wiley, 1969.

Scott, Ronald F., *Principles of Soil Mechanics*. Reading, Mass: Addison-Wesley, 1963.

PROBLEMS

1. Describe the difference between rock and soil.

2. What are the three types of rocks that are found in the earth? Describe each of them.

3. All three types of specific gravity are measured for the solids of a certain soil. Which would you expect to be the greatest? Why?

4. What type of clay experiences the most volume changes? What causes these changes? What difficulty does this pose for the foundation?

CHAPTER 2

Natural Soil Deposits

Knowledge of the nature of soil constituents is of little value unless supplemented by information concerning the properties of assemblies of particles in a soil deposit. It has been said that to fully understand soil behavior, it is essential that one have at least a basic understanding of geologic processes. Some of these concepts and their effect on the nature of natural soil deposits are discussed in this chapter. Additional study of the geologic sciences will prove valuable to engineering and construction students. The purposes of this chapter are

1. To introduce the concept of soil structure, and how individual soil particle arrangements arise.

2. To explain how differences in particle arrangements may affect mass soil behavior.

3. To distinguish between residual and transported soils.

4. To illustrate typical examples of various natural soil deposits.

2.1 SOIL STRUCTURE

2.1.1 Sands

The primary structure of a coarse soil is typically single grained. Individual particles may assume relatively stable or unstable positions according to their mode of deposition. The dense configuration shown in Figure 2.1 typically occurs in deposits built in an active water environment. Beach sands and river gravels are representative examples. In the absence of moving water or some other agent to effect the necessary dense arrangement, coarse soil deposits may be loose. For example, loose deposits are typically formed in quiet water or may result when a dense deposit is disturbed (e.g., by a landslide).

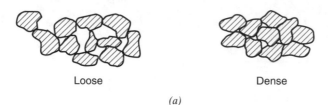

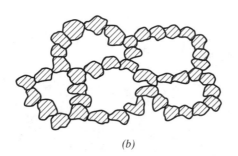

Loose

Dense

(a)

Figure 2.1   Some primary soil structures. (Reprinted, by permission, from Taylor, D.W., *Fundamentals of Soil Mechanics*, Wiley, New York, 1948) (*a*) Single grain. (*b*) Honeycomb.

(b)

Very fine sands or silts may assume a honeycomb configuration similar to that in Figure 2.1*b*. In some instances, the gravitational forces during deposition of such materials are not sufficient to overcome interparticle attractive forces, and a very open structure results. The term *metastable* is sometimes used to describe this condition because of its inherent sensitivity to even a minor disturbance. Soil deposits of this nature often appear to be firm and strong but become wet and unworkable as the excavation process breaks down the primary structure.

2.1.2   Clays

Clay soil exhibits a structure very strongly influenced by the chemical environment existing during deposition and by stress history thereafter. In the case of clay soils resulting from weathering of rock in place, the relic structure of the parent rock material may be evident in the resulting soil.

Clay soil particles deposited as sediment in freshwater repel one another as they settle because of the like electrostatic surface charge they carry. In the resulting sediment, the particles align themselves in a face-to-face arrangement as the repulsive forces from their surface charges come into equilibrium. Forces due to the soil weight are resisted by electrostatic repulsion until a condition similar to Figure 2.2 is reached. If the clay, on the other hand, is deposited in saltwater, the excess of available free ions present in the water effectively neutralizes the particles' surface charges. While suspended they are, thus, free to attract each other gravitationally and form flocs or aggregates of particles. The single grains in these flocs are haphazardly arranged as illustrated by Figure 2.2*b*. A very porous soil deposit results, with a structure that is characteristically weak and compressible.

Microscopy techniques suggest that some clays may exist in randomly arranged packets or peds, like those shown in Figures 1.8 and 2.2*c*, which individually are made up of highly oriented individual particles. This concept has been used to explain observed soil behavior inconsistent with previous concepts of structural arrangements.

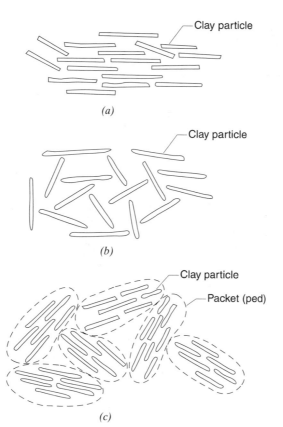

Figure 2.2 Structure or fabric of clay soils. (*a*) Dispersed. (Reprinted, by permission, from Lambe, T.W., *The Structure of Compacted Clay*, Journal, Soil Mechanics and Foundations Division, American Society of Civil Engineers, Vol. 84, SM2, 1958.) (*b*) Flocculated. (*c*) Packet or ped.

Some soils develop a secondary structure during or after formation or deposition. A clay deposit, for example, may be highly dispersed (Figure 2.2*a*) by its own weight and subsequent drying as it is exposed to the atmosphere. As the drying continues or the surface is eroded, a well-developed set of minute discontinuities is sometimes produced. Similarly, slickensides (surfaces with a soil mass which have been smoothed and striated by shear movements on these surfaces) may be produced if a sediment is disturbed by landsliding or tectonic deformation. Discontinuities from any source may dominate the overall deposit's behavior.

2.2 SOIL DEPOSIT ORIGINS AND TYPES

Accumulations of soil particles form soil deposits in a well-defined sequence of geologic processes. Figure 2.3 illustrates the progression of mechanical and chemical weathering that converts all types of rock to residual soils, these soils' transport and deposition to form sediments, and the eventual conversion of some of the sediments to sedimentary rock through a process of induration. While mechanical and chemical weathering begin anew, other geologic processes such as metamorphism and volcanic activity may alter the nature of the parent rock.

In describing the soils below, general statements will be made about the suitabilities of various types of soil for construction and support of foundations. As they appear in the earth, soils vary greatly; exceptions to rules are almost as common as the rules themselves. Both the engineer and the contractor must rely on a thorough knowledge of the actual soil conditions for

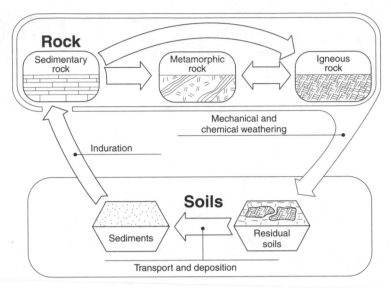

Figure 2.3 Geologic processes affecting origin of soil deposits.

a particular site for proper construction and foundation performance to take place. This is the result of detailed geotechnical investigations which are a necessary prerequisite to any successful foundation installation. These general statements cannot be taken to be a substitute for this type of investigation.

2.2.1 Residual Soils

Residual soils consist of material formed by the disintegration of underlying parent rock or partially indurated material. Typical residual soil profiles are shown in Figure 2.4. The soil developed usually reflects the chemical and mineralogical makeup of the parent material. The soil layer may be further subdivided into *horizons* determined by their current chemical composition and state of weathering. Extensive residual soil deposits occur throughout the United States.

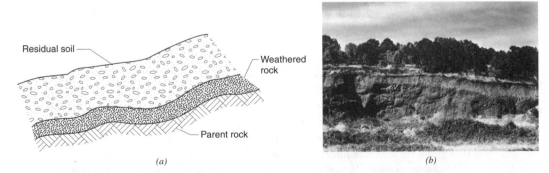

Figure 2.4 Residual soil profiles. (*a*) Schematic residual soil profile showing approximate parallelism in weathering zones. (*b*) Residual soil profile, derived from volcanic rock in western Oregon.

Residual sands and fragments of gravel size are formed by solution and leaching of cementing material, leaving the more resistant particles; commonly quartz. Generally, these present favorable foundation conditions. *Residual clays* are formed by decomposition of silicate rocks, disintegration of shales, and solution of carbonates in limestone. These soils have variable properties requiring detailed investigation. These deposits in many cases present favorable foundation conditions except in humid and tropical climates, where lateritic soils present difficulties when dried and rewetted.

Organic soils consist of an accumulation of highly organic material formed in place by the growth and subsequent decay of plant life. These include *peat*, a somewhat fibrous aggregate of decayed and decaying vegetation matter having a dark color and odor of decay; and *muck*, peat deposits which have advanced in stage of decomposition to such extent that the botanical character is no longer evident. These are very compressible and entirely unsuitable for supporting building foundations.

2.2.2 Transported Soils

Where topographical features are unstable, we may have transported soils or no soils at all. A section through an area of deposition is shown in Figure 2.5. A thin weathered zone of the underlying rock may or may not exist, depending on the geologic history of the region. Unlike the residual profile, wherein the soil thickness is more or less constant, a transported soil deposit's thickness bears no relationship to bedrock contours whatsoever. The degree of horizontal and vertical homogeneity within a transported deposit ranges from none to high, depending on the transporting agent and material source. Transported soils may be grouped according to the agents of transportation. Examples are given in Table 2.1.

Alluvial soils consist of material transported and deposited by running water. Alluvial deposits may vary markedly both horizontally and vertically, owing to the nature of the current patterns in the parent stream. These include a wide variety of soils. An alluvial soil deposit is shown in Figure 2.5b. An alluvial sand deposit is shown in Figure 2.6.

Floodplain deposits are laid down by a stream within that portion of its valley subject to inundation by floodwaters. *Point bars* are alternating deposits of arcuate ridges and swales

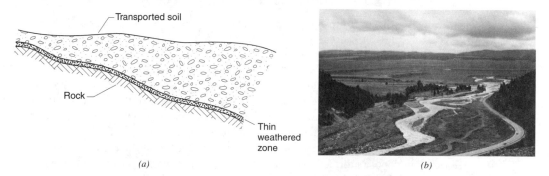

(a)   *(b)*

Figure 2.5 Transported soil profiles. (*a*) Schematic transported soil profile. Note that soil depth bears no particular relationship to rock surface topography. The soil thickness varies according to physical conditions during deposition. (*b*) Alluvial deposits at the mouth of the Madison River Canyon, Montana. Several levels of older terrace deposits are shown in the background.

TABLE 2.1   Transporting Agents and Soil Deposits

Agent	Deposit Name	Depositional Environment
Water	Alluvium (Figure 2.6)	Flowing water
	Marine	Quiet brackish water
	Lacustrine	Quiet fresh water
Ice	Till (Figure 2.7)	Glacial ice contact zones
Wind	Loess (Figure 2.8)	Variable
	Dune sand	Arid or coastal lands
Gravity	Colluvium (Figure 2.9)	Below slide area
	Talus	Base of cliff

Figure 2.6   Cross-bedding in a natural sand deposit. Some transported soils, unlike this deposit, show a regular sequence of materials of uniform thickness.

(lows formed on the inside or convex bank of mitigating river bends.) *Channel fill* consists of deposits laid down in abandoned meander loops isolated when rivers shorten their courses. *Backswamp* is the prolonged accumulation of floodwater sediments in flood basins bordering a river. Generally, these types of soils present favorable foundation conditions, but require detailed investigations for nonhomogeneous conditions.

Alluvial terrace deposits are relatively narrow, flat-surfaced, river-flanking remnants of flood plain deposits formed by entrenchment of rivers and associated processes. They are usually drained and oxidized. Generally, they present favorable foundation conditions. *Estuarine deposits* are mixed deposits of marine and alluvial origin laid down in widened channels at mouths of rivers and influenced by tide of the body of water into which they are deposited. They are generally fine grained and compressible, experiencing long-term consolidation and the settlement problems that go along with that. There are many local variations in soil conditions.

Alluvial-lacustrine deposits consists of material deposited within lakes (other than those associated with glaciation by waves, currents, and organochemical processes. The clays are frequently varved (i.e., layered by the annual deposition of material). They are usually uniform in horizontal direction. These soils are fine-grained soils, and are generally compressible. *Piedmont deposits* are alluvial deposits at foot of hills or mountains. They consist of extensive plains or alluvial fans. Generally, they present favorable foundation conditions.

Deltaic deposits have been formed at the mouths of rivers that result in extension of the shoreline. They are generally fine-grained and compressible. There are many local variations in soil conditions.

Glacial soils are made up of material transported and deposited by glaciers, or by meltwater from a glacier. An important type of glacial soil is the *glacial till*, which is an accumulation of debris, deposited beneath, at the side (lateral moraines), or at the lower limit of a glacier (terminal moraine). Material lowered to ground surface in an irregular sheet by a melting glacier is known as a ground moraine. Glacial till consists of material of all sizes in various proportions from boulder and gravel to clay. The deposits are unstratified. Generally, they present favorable foundation conditions; however, rapid changes in conditions are common. An example of glacial till is shown in Figure 2.7.

Glacio-fluvial deposits consist of coarse and fine-grained material deposited by streams of meltwater from glaciers. Material deposited on ground surface beyond terminal of glacier is known as an outwash plain. Gravel ridges known as kames and eskers are formed in the process. There are many local variations, but generally these present favorable foundation conditions. *Glacio-lacustrine* deposits come about when material is deposited within lakes by meltwater

Figure 2.7   Coarse glacial gravel deposit (till) in the State of Washington. Till may vary in particle size composition according to its source. Many midwestern U.S. tills consist principally of clays.

Figure 2.8   Excavation in loessial soil deposits near Vicksburg, Mississippi. True loess banks will normally stand with a vertical face because of cohesive strength resulting from natural cementation.

from glaciers. They consist of clay in central portions of lake and alternate layers of silty clay or silt and clay (varved clay in peripheral zones.) They are uniform in a horizontal direction.

Aeolian or wind-laid deposits are typically well sorted with a uniform particle size. They include *loess*, which is a calcareous, unstratified deposit of silts or sandy or clayey silt traversed by a network of tubes formed by root fibres now decayed. These deposits are relatively uniform and characterised by ability to stand in vertical cuts. They have a collapsible structure. Deep weathering or saturation can modify characteristics. An example of loess is shown in Figure 2.8. Aeolian soils also include *dune sands*, which are mounds, ridges, and hills of uniform fine sand characteristically exhibiting rounded grains. They have a uniform grain size and may exist in relatively loose condition.

Marine soils are made up of material transported and deposited by ocean waves and currents in shore and offshore areas. They include *shore deposits* of sands or gravels formed by the transporting, destructive, and sorting action of waves on the shoreline. They are relatively uniform and of moderate to high density. *Marine clays* consist of organic and inorganic deposits of fine-grained material. They are generally very uniform in composition, are compressible, and usually are very sensitive to remolding.

Colluvial soils have been transported and deposited by gravity. They include *talus* (deposits created by gradual accumulation of unsorted rock fragments and debris at base of cliffs) and *hillwash* (fine colluvium consisting of clayey sand, sand silt, or clay.) The previous move-

Figure 2.9   Colluvial debris accumulated above residual soil profile in Oregon.

ment indicates possible future difficulties. They present generally unstable foundation conditions. A colluvial debris is shown in Figure 2.9.

Pyroclastic soils were ejected from volcanoes and transported by gravity, wind and air. *Ejecta* are loose deposits of volcanic ash, lapilli, bombs, etc. *Pumice* is frequently associated with lava flows and mudflows, or may be mixed with nonvolcanic sediments. Pyroclastic soils have typically shardlike particles of silt size with larger volcanic debris. Weathering and redeposition produce highly plastic, compressible clay. They present unusual and difficult foundation conditions.

2.3 SUMMARY

The subsurface conditions in a natural deposit may often be anticipated with great accuracy if one understands the rudiments of the relations between landform and origin (geomorphology). Unanticipated subsurface conditions are one of the most, if not the most, frequent cause of disputes in execution of a construction contract. If all parties to the contract have adequate information regarding the nature of the subsurface materials and the sequence of their occurrence, and in addition are able to ascertain the effects these conditions will have on their respective operations, such disputes need not arise.

This chapter was intended to illustrate that the physical formation of a natural soil deposit may produce different soil structures. These different structural arrangements of soil particles result in observable differences in soil mass behavior. The reader should

1. Understand how and why these different structural arrangements come about.

2. Be able to identify and visualize different soil structures, and describe their behavior in general terms.

3. Know the difference between residual and transported soil deposit origins and be able to identify types of transported soils.

REFERENCES

Deere, D.U., and Patton, F.D., "Slope Stability in Residual Soils," *Proceedings, 4th Pan American Conference on Soil Mechanics and Foundation Engineering*, 1, 1971.

Lambe, T.W., "The Structure of Compacted Clay," *Journal, Soil Mechanics and Foundations Division, American Society of Civil Engineers*, 84, SM2, 1958.

Department of the Navy, *Soil Mechanics*. NAVFAC DM 7.01. Norfolk, VA: Naval Facilities Engineering Command, 1986. (Much of the discussion on soil types is adapted from this document.)

Spangler, M.G., *Soil Engineering*. Scranton, PA: International Textbook Company, 1969.

Taylor, D.W., *Fundamentals of Soil Mechanics*. New York: Wiley, 1948.

Terzaghi, K., and Peck, R.B., *Soil Mechanics in Engineering Practice*. New York: Wiley, 1967.

PROBLEMS

1. What is the difference between the interaction of fresh water and salt water on clays?

2. Why are organic soils unsuitable for foundation support? Do you think they would also pose problems for construction as well? How do these problems differ from pyroclastic soils?

3. Define the following terms:
   a. Glacial till
   b. Colluvial soils
   c. Deltaic deposits
   d. Alluvial-lacustrine deposits

CHAPTER 3

Soil Index Properties

In order to relate soil behavior to physical properties, it is convenient to have standard procedures for the testing and reporting of results. Test methods are used to obtain index properties that relate to the engineering behavior of soil materials. Some test results may permit judgments concerning the nature of individual soil particles, while others may be useful in assessing properties of an entire soil mass composed of individual particles. The behavior of sands and gravels may be inferred from the shape, size, and density of packing of the constituent particles. The behavior of silts and clays is more nearly controlled by surface activity of the particles and, in particular, interaction of the particles with water.

This chapter introduces concepts and definitions for index properties of soil constituents and for the soil mass. The reader will learn about

1. The particle size distribution curve for coarse-grained soils.

2. Plasticity characteristics for fine-grained soils and their relationship to natural water content.

3. Phase relationships (air, water, and solid) for the soil mass.

3.1 OVERVIEW OF SOILS TESTING

The ability to quantify soil properties is a key to both properly configuring foundations on the soil and in the process of construction itself. With soils, the soil properties we use cannot be separated from the tests themselves. It is therefore important for these tests to be properly executed. There are some important principles in the execution that the contractor needs to be aware of, both in the use of these properties in bid preparation and in the actual construction of a particular structure. Some of these principles are as follows:

- Test procedures must be properly applicable to the soil in question. For example, Atterberg limits and hydrometer tests are applicable to fine-grained soils, while sieve analysis is used for coarse-grained soils. In soils where the grain size is widely variable (e.g., material coarser than 3/4 inch for compaction testing), special variations in the procedures are available, and these must be noted by the laboratory when they are performed and by the contractor when reviewing the results of the tests.

- Test procedures must be followed as they are written. Variances from these procedures, be they ASTM, AASHTO, or EM 1110-2-1906, will produce results that can be misleading. In performing these tests, common errors should be understood and avoided. Personnel who perform laboratory tests must be properly trained and exercise good laboratory practice in the execution of these tests.

- If there are variations in the procedures, they should be noted on the reports. This is especially important in field testing, where variations in procedure are very common. An example of this is with standard penetration tests; the type of hammer, for example, is important to note, as its efficiency can significantly affect the results and how they are used.

- Samples for field or laboratory testing must be representative of the soil in the ground at the site. They must be taken from a wide variety of locations on the site, both in plan and elevation. Unrepresentative samples will give a misleading profile of the soil on a jobsite, which will result in unrealistic bids or construction problems that otherwise could have been foreseen.

- Samples should be handled carefully, both in the field and in the laboratory. One phrase that is commonly heard in soils testing is "undisturbed samples." This means that the sample being tested in the laboratory is identical in soil structure, moisture content, etc., to its state in the earth. Samples must be labeled properly when they are removed in the field so that when laboratory tests are complete, an accurate soil profile of the jobsite can be put together. They should be tested in the laboratory as soon as possible; if it is necessary to store them, storage must be done to insure that the soil's properties do not change with time. Most test procedures have recommendations that deal with the proper procedure to store soil samples. Disturbed samples should never be used for any tests other than soil classification, specific gravity, or water content.

- Data from laboratory tests must be properly recorded, documented, and analyzed.

3.2 COARSE-GRAINED CONSTITUENTS—GRAIN SIZE DISTRIBUTION

Sands and gravels are most usefully described by reference to the distribution of particle sizes, the shape of particles, the relative density (see Chapter 8), and silt and clay content. In some cases, the toughness and durability of particles is of major importance. All of these factors considered together determine the acceptability of a coarse soil for a particular use. Grains size distribution and fines content are usually represented on a semilogarithmic plot such as that shown in Figure 3.1. The distribution curves shown are determined by mechanical analysis as indicated in Chapter 1. The percentage of a sample greater than a given size is determined, for coarse soils, by passing the entire sample through a nest of standard sieves, and for fine soils, by hydrometer analysis (ASTM D422; see Appendix A.2). The relation between percentage composition and size is thus determined. The maximum particle size in the finer nth percentage of a sample is indicated by the notation D_n. Curves a and b on Figure 3.1 represent two greatly

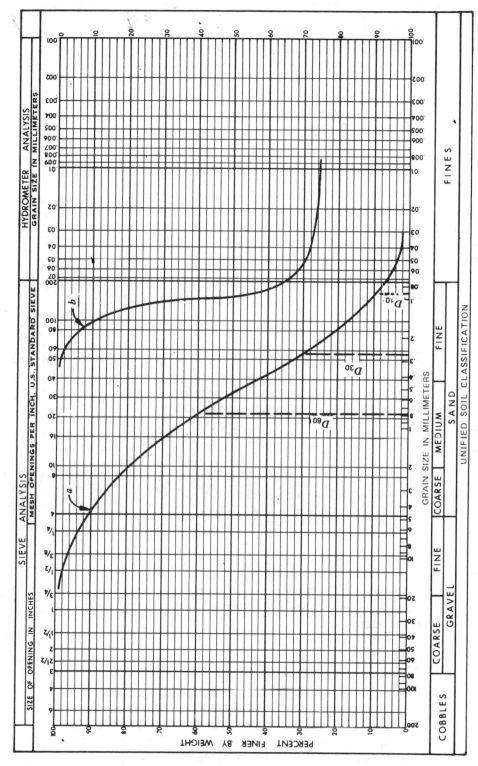

Figure 3.1 Particle size distribution curves.

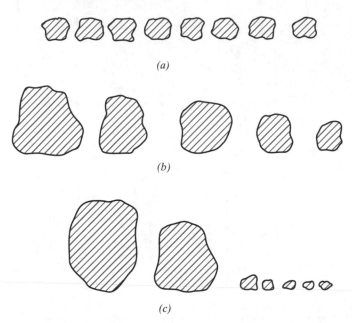

Figure 3.2 Particle size distributions.
(*a*) Uniform size distribution.
(*b*) Nonuniform size distribution.
(*c*) Gap gradation.

different soil materials. The former contains a small percentage of fines and has a broad distribution of particle sizes. The latter is almost one-third silt and clay and two-thirds fine sand. It will be shown in Chapter 4 that indices from a particle size distribution curve are useful for soil identification and classification. The uniformity coefficient

$$C_u = \frac{D_{60}}{D_{10}} \tag{3.1}$$

indicates the degree to which the particles are the same size. If particles are the same diameter (see Figure 3.2), C_u is 1. C_u for a real soil may, therefore, be any number greater than 1, the general rule being that increasing values represent an increasingly wide range of particle size differences. The coefficient of curvature,

$$C_c = \frac{D_{30}^2}{D_{60}D_{10}}, \tag{3.2}$$

describes the smoothness and shape of the gradation curve. Very high or very low values indicate that the curve is irregular. Acceptable values of each depend upon the soil type and are given in Figure 4.2.

EXAMPLE   Sieve Analysis

The data shown in Figure 3.3 illustrate a test in which the particle size distribution, including percentage of fines, but not the fine particle size distribution, is to be determined. The procedure used to determine the size distribution is based on ASTM D1140. The sample was first sieved dry. It was determined that 26.45 g of soil passed the No. 200 sieve. The washing and sieving procedure described in ASTM D1140 was then followed for that portion of the sample

SIEVE ANALYSIS DATA AND GRAIN SIZE CURVE

TEST FOR _Backfill for footings_
DATE _12 June 1993_
TEST BY _REF_
SAMPLE DESCRIPTION _Sandy gravel - collected from borrow pit_
Sample No. 7268 (12)

U.S. STANDARD SIEVE SIZE	WEIGHT RETAINED	TOTAL WEIGHT RETAINED	TOTAL WEIGHT PASSING	PERCENT FINER
3"	0	0	956.5 g	100.0
1½"	212.3 g	212.3 g	744.2 g	77.8
¾"	142.6 g	354.9 g	601.6 g	62.8
NO. 4	209.1 g	564.0 g	392.5 g	41.0
NO. 10	152.2 g	716.2 g	240.3 g	25.1
NO. 40	93.6 g	809.8 g	146.7 g	15.3
NO. 100	52.8 g	862.6 g	93.9 g	9.8
NO. 200	61.6 g	924.2 g	32.3 g	3.3
PAN TOTAL	32.3 g	956.5 g	0	0

FINE FRACTION	
UNWASHED WEIGHT	366.1 g
WASHED WEIGHT	360.2 g
FINES BY WASHING	5.9 g
FINES BY SIEVING	26.4 g
FINES TOTAL	32.3 g

NOTES

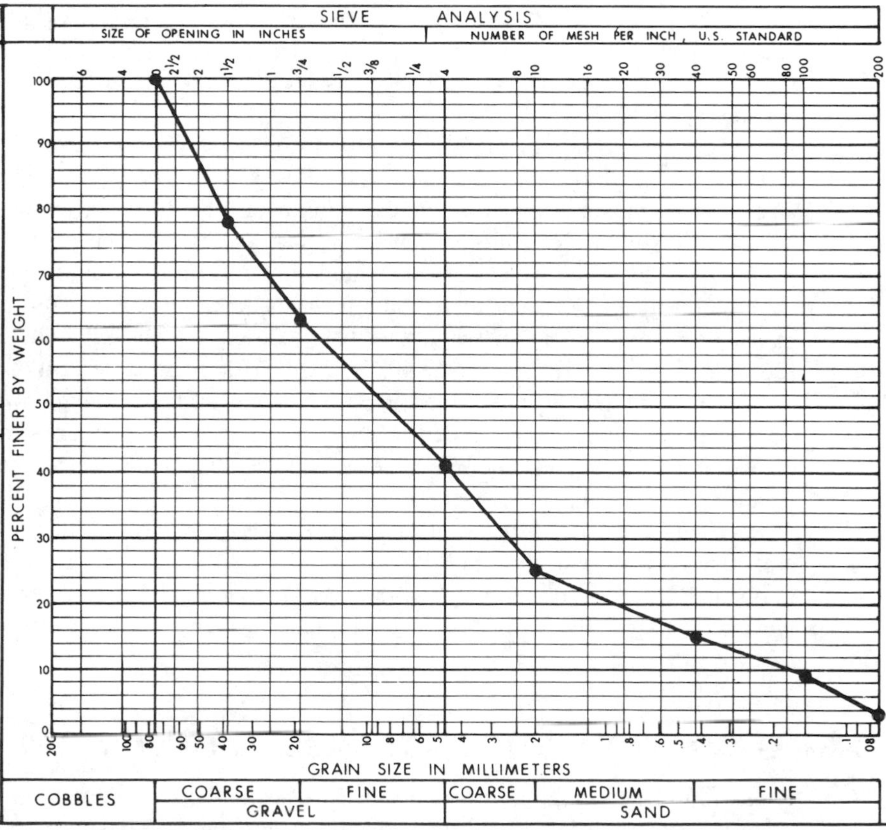

Figure 3.3 Data and results from laboratory sieve analysis.

27

passing the No. 4 sieve. The tabulated weights retained are those determined by both dry siev-ing and washing. About 5.9 g of fines were removed from the minus No. 4 fraction by washing. The combined weight of fines in the sample, 32.3 g, is represented by the sum of the weights from sieving and washing. Weights retained on each sieve and the pan are tabulated. From these numbers, the cumulative weights retained on each sieve are obtained by summing the weights retained on it and all coarser sieves. The total weight of the sample, less the total weight re-tained on any sieve, gives the total weight passing, which can then be expressed as a percent-age. Distribution of particle sizes finer than the No. 200 sieve is developed from hydrometer test results (ASTM D422).

In some cases, large samples must be divided to obtain a representative portion for test-ing which can be accommodated by a particular set of sieves and the requirements of ASTM D422 concerning minimum sample size. The total weight of the sample required for a sieve analysis depends on maximum particle size.

EXAMPLE   Combined Sieve Analysis

The lower right-hand corner of Figure 3.4 illustrates a procedure for dividing a large sample for sieve analysis. The original sample in this case was divided with a sample splitter. A split por-tion was subsequently divided three more times, with the result of the fourth split being the 6100 g sample indicated. This sample is considered identical to the original larger sample be-cause the splitter divides a given sample into practically identical halves. It is the 6100-g sam-ple in this case that will be tested to produce a particle size curve like those shown in Figures 3.1 and 3.2.

The split portion of the sample is divided on the No. 10 sieve. The coarser portion is sieved directly, while the finer portion is split twice before sieving to reduce its total weight to an acceptable value; that is, a weight that will not result in too great a weight on any one of the finer sieves. Weights retained for both sievings are indicated in the example.

The two sieve analyses diagrammed in Figure 3.4 must be combined to produce a single analysis for the particle size curve that represents the original 6100-g sample. This is done by means of the calculations shown in the table in Figure 3.4. The KEY and NOTES in the figure illustrate the procedure used. Some 5000 g of the 6100 g, or 82 percent of the total sample, are coarser than the No. 10 sieve. Of the 5000 g, 3500 g ($1000 g + 800 g + 700 g + 1000 g$), or 70 percent, pass the 3-in. sieve, Percentages passing each sieve size are similarly computed and entered in the lower portion of the table as indicated by (1) in the key. The process is repeated for the portion passing the No. 10 sieve.

The percentages of the total sample (3) for both the coarser than No. 10 and finer than No. 10 fractions are then multiplied by the percentage for each sieve size for each split portion to obtain the percent of combined sample for the fraction. The numbers for both fractions are then added to give the percent of the total sample, which corresponds to a point on the desired particle size distribution curve.

3.3   FINE-GRAINED CONSTITUENTS: SOIL PLASTICITY

Because of the controlling importance of the effect of surface activity on the behavior of fine-grained soils, description of these materials by reference to their particle sizes is practically meaningless. The practical distinction between silt and clay is made, not on the basis of an

COMBINED SIEVE ANALYSIS

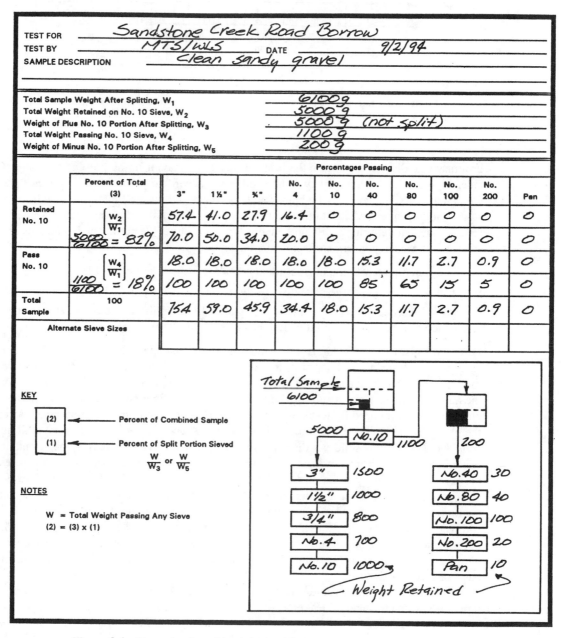

TEST FOR _Sandstone Creek Road Borrow_

TEST BY _MTS/WLS_   DATE _9/2/94_

SAMPLE DESCRIPTION _Clean sandy gravel_

Total Sample Weight After Splitting, W_1						_6100 g_						
Total Weight Retained on No. 10 Sieve, W_2						_5000 g_						
Weight of Plus No. 10 Portion After Splitting, W_3						_5000 g (not split)_						
Total Weight Passing No. 10 Sieve, W_4						_1100 g_						
Weight of Minus No. 10 Portion After Splitting, W_5						_200 g_						

Percentages Passing

	Percent of Total (3)	3"	1½"	¾"	No. 4	No. 10	No. 40	No. 80	No. 100	No. 200	Pan
Retained No. 10	$\left[\dfrac{W_2}{W_1}\right]$	57.4	41.0	27.9	16.4	0	0	0	0	0	0
	$\dfrac{5000}{6100} = 82\%$	70.0	50.0	34.0	20.0	0	0	0	0	0	0
Pass No. 10	$\left[\dfrac{W_4}{W_1}\right]$	18.0	18.0	18.0	18.0	18.0	15.3	11.7	2.7	0.9	0
	$\dfrac{1100}{6100} = 18\%$	100	100	100	100	100	85	65	15	5	0
Total Sample	100	75.4	59.0	45.9	34.4	18.0	15.3	11.7	2.7	0.9	0
Alternate Sieve Sizes											

KEY

(2)	◄── Percent of Combined Sample
(1)	◄── Percent of Split Portion Sieved

$$\frac{W}{W_3} \text{ or } \frac{W}{W_5}$$

NOTES

W = Total Weight Passing Any Sieve

(2) = (3) × (1)

Total Sample 6100

5000 → No.10 → 1100   200

3"	1500
1½"	1000
3/4"	800
No.4	700
No.10	1000

No.40	30
No.80	40
No.100	100
No.200	20
Pan	10

Weight Retained

Figure 3.4   Example of combined sieve analysis.

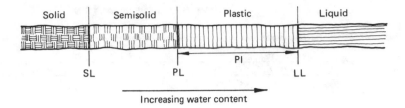

Figure 3.5   States of fine soil consistency. As water content increases, the soil becomes increasingly fluid.

arbitrary size distinction, but on the basis of material behavior in the presence of water. Consider Figure 3.5. The consistency of fine soil varies according to the amount of water present. Completely dry, the soil may be hard (solid), while at high water contents it may be almost a slurry (liquid). Intermediate states of consistency are semisolid and plastic states. A plastic material is one that deforms readily without cracking or rupture. We define the boundaries of these states of consistency in terms of soil water content:

- The water content at the plastic-liquid boundary is the liquid limit (LL).
- The water content at the plastic-semisolid boundary is the plastic limit (PL).
- The difference between the liquid and plastic limits is the range of water contents over which a soil is plastic, and it is designated as the plasticity index (PI).

Different soils may be distinguished by their plasticity characteristics because these characteristics vary with surface activity of the constituent particles. The more active soils (claylike) are more plastic than the inactive soils (silts). This phenomenon may be explained by examining the nature of the water near the surface of a clay particle. Figure 3.6 is a conceptual view of a clay particle surrounded by water. Since water's molecular structure is dipolar, the water near the clay particle is effectively immobilized by the surface charge. It is adsorbed and may be considered essentially solid. As distance from the particle surface increases, the orientation of water is reduced in degree until, at the boundary of the particle's influence (limit of diffuse double layer), the viscosity is that of free water. With an abundance of water, soil particles would be separated by free water and the mixture would be fluid. As the amount of water decreases, the particles are separated by increasingly stiffer water. The mixture becomes like a solid. The soil water system, therefore, has a range of water contents over which it may be plastic. Changing the chemical composition of the constituent phases (soil type or fluid) causes a change in the plasticity range or the plasticity index. This permits us to distinguish among soils on the basis of their plasticity.

The plasticity index values may be used to identify or classify fine soils. They are determined according to standard test procedures (ASTM D4318); an overview of these procedures is given in Appendix A.4. The tests themselves are illustrated in Figure 3.7. An example set of test data is shown in Figure 3.8. The shrinkage limit (SL) may be determined (ASTM D427), but is not widely used as an index property. It represents the solid–semisolid interface and is the water content at which the soil volume is a minimum. High plasticity indices or high liquid limits are characteristic of clay soils. The use of these indices in identification and classification is described in Chapter 4.

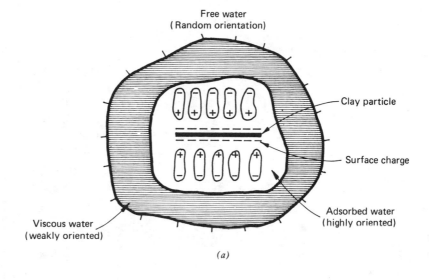

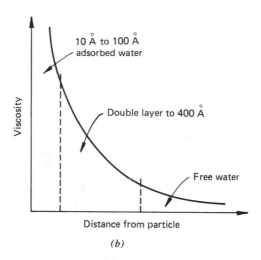

Figure 3.6 The clay–water system. (*a*) Conceptual clay–water system. (*b*) Viscosity in the double layer.

The variations in water content of clays are important for a wide variety of reasons. They affect the engineering properties of the soil and thus the building design. For the contractor, the effect can be more prosaic but very important. For example, the jobsite that is dry and easy to move construction equipment about can become a quagmire if enough rain raises the water content of the soil towards and beyond the liquid limit. This may necessitate matting in order to move equipment around, or even affect the type of equipment that is brought on site. The contractor must be prepared for this type of change if conditions are not the same as when the job was bid or began.

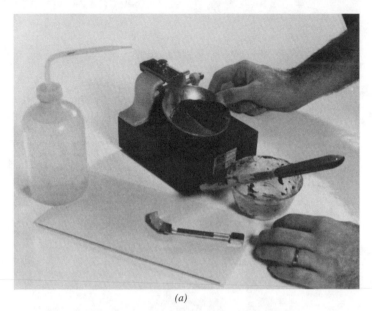

(a)

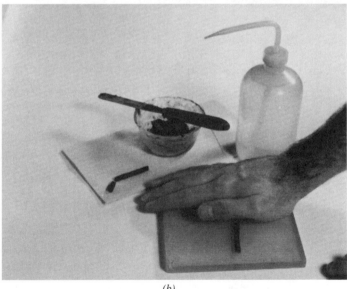

(b)

Figure 3.7 Atterberg limits tests for fine soil index properties. (a) The liquid limit test. (b) The plastic limit test.

3.4 INDICES FOR THE SOIL MASS

Index values for coarse and fine constituents are descriptive of the physical nature of the constituents. They are not always sufficient to permit conclusions regarding soil performance for a given purpose. To supplement these indices, we must also know soil unit weight or density, water content, and other relations among the air–water–solid phases of our material.

ATTERBERG LIMITS AND WATER CONTENT

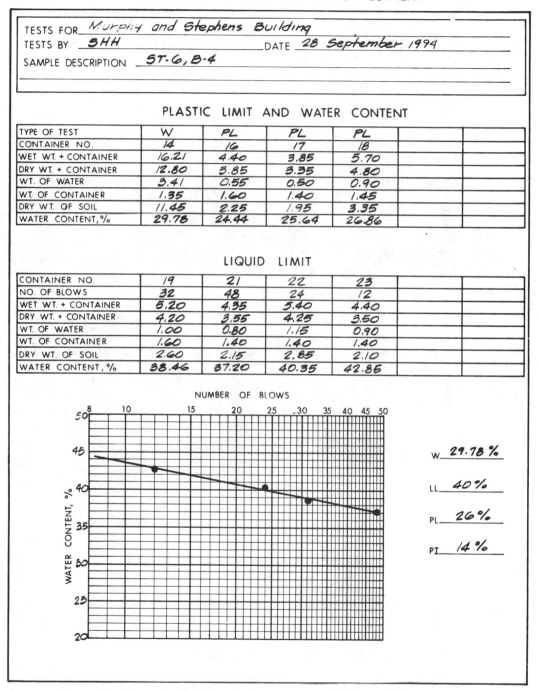

TESTS FOR _Murphy and Stephens Building_

TESTS BY _SHH_   DATE _28 September 1994_

SAMPLE DESCRIPTION _ST-6, B-4_

PLASTIC LIMIT AND WATER CONTENT

TYPE OF TEST	W	PL	PL	PL		
CONTAINER NO.	14	16	17	18		
WET WT. + CONTAINER	16.21	4.40	3.85	5.70		
DRY WT. + CONTAINER	12.80	3.85	3.35	4.80		
WT. OF WATER	3.41	0.55	0.50	0.90		
WT. OF CONTAINER	1.35	1.60	1.40	1.45		
DRY WT. OF SOIL	11.45	2.25	1.95	3.35		
WATER CONTENT, %	29.78	24.44	25.64	26.86		

LIQUID LIMIT

CONTAINER NO.	19	21	22	23		
NO. OF BLOWS	32	48	24	12		
WET WT. + CONTAINER	5.20	4.35	5.40	4.40		
DRY WT. + CONTAINER	4.20	3.55	4.25	3.50		
WT. OF WATER	1.00	0.80	1.15	0.90		
WT. OF CONTAINER	1.60	1.40	1.40	1.40		
DRY WT. OF SOIL	2.60	2.15	2.85	2.10		
WATER CONTENT, %	38.46	37.20	40.35	42.85		

W _29.78 %_

LL _40 %_

PL _26 %_

PI _14 %_

Figure 3.8 Data and calculations for water content and Atterberg limits test.

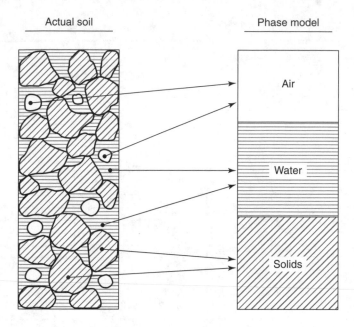

Figure 3.9 Substitutions for phase calculations.

Soil is a mixture of air, solid particles, and water or other fluids. These are mixed together in actual soil; for computational and conceptual purposes, it is more convenient to treat them as if they were gathered separately in the soil. This is shown in Figure 3.9. From this, the various basic volume and weight variables can be defined as in Figure 3.10. The basic relationships of these variables are

$$V = V_A + V_W + V_S, \tag{3.3}$$

$$V_V = V_A + V_W, \tag{3.4}$$

and

$$W = W_W + W_S, \tag{3.5}$$

where the variables are as shown in Figure 3.10. Air is assumed to be weightless for soil index calculations. By definition, saturated soils are those soils for which $V_A = 0$, in which case $V_V = V_W$.

All of the commonly used weight and volume relationships can be derived from Equations (3.3), (3.4), and (3.5), in conjunction with unit weight and specific-gravity definitions. However, most of these relationships are shown already defined in Table 3.1 for both saturated and unsaturated soils. The next example shows the application of these relationships.

EXAMPLE   Volume and Weight Calculations

Consider a natural soil sample taken from a proposed borrow pit. The volume of the hole from which the sample was taken was 1.1 ft^3. The total sample weighed 130 lb and, after drying, weighed 119 lb.

Volumes   Phase model   Weights

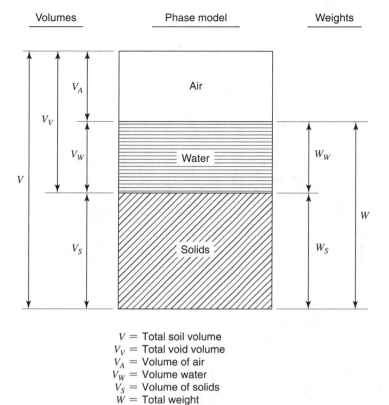

V = Total soil volume
V_V = Total void volume
V_A = Volume of air
V_W = Volume water
V_S = Volume of solids
W = Total weight
W_W = Weight of water
W_S = Weight of solids

Figure 3.10   Air–water–solid soil phases.

The water, or moisture, content (ASTM D2216) of the sample is thus

$$w = \frac{W_w}{W_s} = \frac{130 - 119}{119} = 0.092 = 9.2\%.$$

The field, or total, unit weight or density is

$$\gamma_T = \frac{W_s + W_w}{V} = \frac{119 + 11}{1.1} = 118 \text{ pcf.}$$

The dry-unit weight is

$$\gamma_D = \frac{W_s}{V} = \frac{119}{1.1} = 108 \text{ pcf.}$$

TABLE 3.1   Soil Mass Index Relations

Diagram labels (schematic of soil phase relations):

Weights for unit volume of soil — γ_{SAT}, γ, γ_D

Volume Components: Volume of air or gas; Volume of voids V_V (V_0, V_w, Vol of H_2O); Volume of solids V_S; V Total volume of sample

Phase column: Air or Gas / H_2O (Assumed Weightless, Wt. of H_2O); SOLIDS (Weight of solids); Soil Sample

Weight Components: W_w; V_S; V_T Total weight of sample

Properties		Saturated Sample (W_S, W_w, G, Are Known)	Unsaturated Sample (W_S, W_w, G, V, Are Known)	Weight Components	Supplementary Formulas Relating Measured and Computed Factors		
V_S	Volume of Solids	$\dfrac{W_S}{G\gamma_w}$		$V-(V_a+V_w)$	$V(1-n)$	$\dfrac{V}{(1+e)}$	$\dfrac{V_v}{e}$
V_w	Volume of Water	$\dfrac{W_w}{\gamma_w}$	$\dfrac{W_w}{\gamma_w}$	V_V-V_a	SV_V	$\dfrac{SVe}{(1+e)}$	SV_Se
V_a	Volume of Air or Gas	Zero	$V-(V_S+V_w)$	V_V-V_w	$(1-S)V_V$	$\dfrac{(1-S)Ve}{(1+e)}$	$(1-S)V_Se$
V_V	Volume of Voids	V_S+V_w	$V-\dfrac{W_S}{G\gamma_w}$	$V-V_S$	$\dfrac{V_Sn}{1-n}$	$\dfrac{V_e}{(1+e)}$	V_Se
V	Total Volume of Sample	V_S+V_w	Measured	$V_s+V_a+V_w$	$\dfrac{V_S}{1-n}$	$V_S(1+e)$	$\dfrac{V_V(1+e)}{e}$
n	Porosity	$\dfrac{V_V}{V}$	$\dfrac{V_V}{V}$	$1-\dfrac{V_S}{V}$	$1-\dfrac{W_S}{GV\gamma_w}$	$\dfrac{e}{1+e}$	$\dfrac{wG}{S}$
e	Void Ratio	$\dfrac{V_V}{V_S}$	$\dfrac{V_V}{V_S}$	$\dfrac{V}{V_S}-1$	$\dfrac{GV\gamma_w}{W_S}-1$	$\dfrac{W_wG}{W_SS}$	$\dfrac{n}{1-n}$

VOLUME COMPONENTS

Properties	Saturated Sample (W_S, W_W, G, Are Known)	Unsaturated Sample (W_S, W_W, G, V, Are Known)	Supplementary Formulas Relating Measured and Computed Factors		
WEIGHTS FOR SPECIFIC SAMPLE					
W_S Weight of Solids	Measured	Measured	$\dfrac{W_T}{(1+W)}$	$GV\gamma_w(1-n)$	$\dfrac{W_wG}{eS}$
W_W Weight of Water	Measured	Measured	wW_S	$S\gamma_w V_v$	$\dfrac{eW_sS}{G}$
W_t Total Weight of Sample	$W_S + W_w$	$W_S + W_w$	$W_S(1+w)$		
WEIGHTS FOR SAMPLE OF UNIT VOLUME					
γ_D Dry Unit Weight	$\dfrac{W_S}{V_S + V_W}$	$\dfrac{W_S}{V}$	$\dfrac{W_t}{V(1+w)}$	$\dfrac{G\gamma_w}{(1+e)}$	$\dfrac{G\gamma_w}{1+wG/S}$
γ_T Wet Unit Weight	$\dfrac{W_S + W_W}{V_S + V_W}$	$\dfrac{W_S + W_W}{V}$	$\dfrac{W_T}{V}$	$\dfrac{(G+Se)\gamma_w}{(1+e)}$	$\dfrac{(1+w)\gamma_w}{w/S+1/G}$
γ_{SAT} Saturated Unit Weight	$\dfrac{W_S + W_W}{V_S + V_W}$	$\dfrac{W_S + V\gamma_w}{V}$	$\dfrac{W_S}{V} + \left(\dfrac{e}{1+e}\right)\gamma_w$	$\dfrac{(G+e)\gamma_w}{(1+e)}$	$\dfrac{(1+w)\gamma_w}{w+1/G}$
γ_{SUB} Submerged (Buoyant) Unit Weight	$\gamma_{SAT} - \gamma_w$		$\dfrac{W_S}{V} - \left(\dfrac{1}{1+s}\right)\gamma_w$	$\left(\dfrac{G+e}{1+e} - 1\right)\gamma_w$	$\left(\dfrac{1-1/G}{w+1/G}\right)\gamma_w$
COMBINED RELATIONS					
w Moisture Content	$\dfrac{W_w}{W_s}$		$\dfrac{W_t}{W_s} - 1$	$\dfrac{Se}{G}$	$S\left[\dfrac{\gamma_w}{\gamma_D} - \dfrac{1}{G}\right]$
S Degree of Saturation	1.00	$\dfrac{V_w}{V_V}$	$\dfrac{W_w}{V_V\gamma_w}$	$\dfrac{wG}{e}$	$\dfrac{w}{\left[\dfrac{\gamma_w}{\gamma_D} - \dfrac{1}{G}\right]}$
G Specific Gravity	$\dfrac{W_S}{V_S\gamma_w}$		$\dfrac{Se}{w}$		

Source: (From NAVFAC DM 7.01)

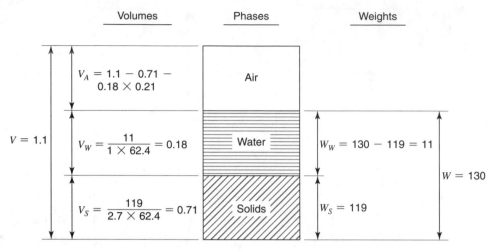

Figure 3.11 Example of air–water–solid phase calculations.

To determine another index property, it is convenient to represent what is known about the sample phases in a sketch similar to that of Figure 3.11. Since specific gravity may be closely estimated (the actual specific gravity could, of course, be determined by testing), the volume of the solid phase may be computed:

$$V_S = \frac{W_S}{G_S \gamma_w} = \frac{119}{2.7 \times 62.4} = 0.71 \text{ cu.ft.}$$

Similarly, the volume of water is

$$V_W = \frac{W_W}{G_W \gamma_w} = \frac{11}{1 \times 62.4} = 0.18 \text{ cu.ft.}$$

By subtraction,

$$V_A = V_V - V_W = V - V_S - V_W = 1.1 - 0.71 - 0.18 = 0.21 \text{ cu.ft.}$$

The weights and volumes of all phases are now known. Other defined indices may now be computed. For instance, the degree of saturation is

$$S = \frac{V_W}{V_V} = \frac{0.18}{0.39} = 0.46 = 46\%.$$

EXAMPLE   Volume and Weight Calculations, Known Volume

A wet soil sample has a volume of 0.0005 m^3 and a weight of 9.5 N. After the sample is dried, its weight is found to be 8 N. The solids have a specific gravity of 2.65. Determine the unit weight of the soil, void ratio, degree of saturation, and water content.

The formulas needed to determine these quantities can be found in Table 3.1. The unit weight is first found by dividing the weight of the sample by the volume of the sample; thus,

$$\text{Unit Weight: } \gamma = \frac{W}{V} = \frac{9.5}{.0005} = 19 \text{ kN/m}^3.$$

The rest of the quantities can be computed as follows:

Void Ratio:
$$e = \frac{GV\gamma_W}{W_S} - 1 = \frac{2.65 \times .0005 \times 9.81}{9.5} - 1 = 0.62;$$

Water or Moisture Content:
$$w = \frac{W_t}{W_s} - 1 = \frac{9.5}{8} - 1 = .18 = 18\%;$$

Degree of Saturation:
$$S = \frac{wG}{e} = \frac{.18 \times 2.65}{.62} = .79 = 79\%.$$

3.5 SUMMARY

The successful anticipation of soil behavior during construction may depend on availability of laboratory testing information for soils from the site. This chapter has presented certain indices of soil behavior that may be computed from laboratory test results. These same indices are often readily determined indicators of the engineering properties of the soil that have been presumed in design and must be obtained during construction. As such, index properties are specified as standards to meet in the execution of a contract. A thorough understanding of their significance is, therefore, essential.

Following study of this chapter and solution of the sample problems, the reader should

1. Be able to define soil constituent and soil mass index properties.

2. Know how to estimate consistency of fine soils given Atterberg limits and water content.

3. Thoroughly understand phase relation calculations.

REFERENCES

American Society for Testing and Materials, *1988 Annual Book of ASTM Standards,* Vol. 04.08, Soil and Rock, Building Stones; Geotextiles, Philadelphia.

Department of the Army. *Laboratory Soils Testing.* EM 1110-2-1906. Washington, DC: U.S. Army Corps of Engineers, 1986.

Department of the Navy. *Soil Mechanics.* NAVFAC DM 7.01. Norfolk, VA: Naval Facilities Engineering Command, 1986.

Lambe, T.W., and Whitman, R.V., *Soil Mechanics.* New York: Wiley, 1969.

Means, R.E., and Parcher, J.V., *Physical Properties of Soils.* Columbus, OH: Merrill, 1963.

Scott, Ronald F., *Principles of Soil Mechanics.* Reading, MA: Addison-Wesley, 1963.

PROBLEMS

1. A 5-lb soil sample was taken from a fill. The volume of the hole from which the sample was taken was 0.044 ft^3. The sample weighed 4.4 lb after oven drying. It was determined that the specific gravity of the soil was 2.65. Compute

   a. The total density or unit weight of the fill.

   b. The dry density or unit weight of the fill.

   c. The fill void ratio.

    d. The fill porosity.

    e. The degree of saturation of the fill.

    f. The total unit weight of the fill if its voids were to be filled with water (saturated density).

2. A saturated clay soil has a water content of 54 percent and a specific gravity of solids of 2.7. What are its total unit weight and its dry unit weight?

3. A stockpile of gravel that has a water content of 6 percent is rained on so that its water content increases to 9 percent. Compute the difference in the weights of dry gravel in a 10-ton truckload taken from the pile before and after the rain.

4. A sandy soil has a total unit weight of 120 pcf, a specific gravity of solids of 2.64, and a water content of 16 percent. Compute

    a. Its dry unit weight.

    b. Its porosity.

    c. Its void ratio.

    d. Its degree of saturation.

5. If 1 ft^3 of soil with a void ratio of 0.9 is disturbed and then replaced at a void ratio of 1.16, what is its final volume? Express the final volume as a percentage of the original volume.

6. A moist soil has a total unit weight of 110 pcf and a water content of 6 percent. Assuming that the dry unit weight is to remain constant, determine how much water, in gallons per cubic yard, must be added to increase the water content to 12 percent.

7. A soil sample is obtained by pushing a 16-in.-long tube into the side of a pit. The tube has an inside diameter of 3 in. and weighs 1.8 lb. The soil in the tube is trimmed so that its ends are flush with the ends of the tube. If the total weight of the tube and sample is 8.62 lb and the water content of the soil is 43.6 percent, what are the unit weight and dry unit weight of the sample? If the soil is saturated, what is the average specific gravity of its solid constituents?

CHAPTER 4

Soil Classification

Various systems have been devised for classifying soils in order to assign them descriptive names or symbols. Such systems are designed to group soils according to the physical characteristics of their particles or according to the performance they may exhibit when subjected to certain tests or conditions of service. It is hoped that, using classification symbols, engineers and contractors will be able to improve communications concerning site conditions. Soil classification systems described in this chapter are examples of systems available for engineering use and, in the case of the Unified and AASHTO systems, those most widely used.

The objectives of this chapter are to

1. Familiarize the reader with soil classification systems in general.
2. Show how popular soil classification systems are used.
3. Provide practice in the use of the Unified and AASHTO soil classification systems.

4.1 TEXTURAL CLASSIFICATION

Natural soils may be assigned descriptive names, such as *silty clay, clayey sand,* and *sandy gravel*, if the amounts of various constituent sizes are known. If the soil contains mostly sand and some clay, then clayey sand would seem appropriate. To establish such a system of classification, one must assign size limits to the soil fractions and establish percentage compositions corresponding to descriptive names. The result will be a chart similar to that shown in Figure 4.1. This particular chart is the basis for soil descriptions published in soil maps by the U.S. Department of Agriculture. These maps, widely available, are sometimes useful references for

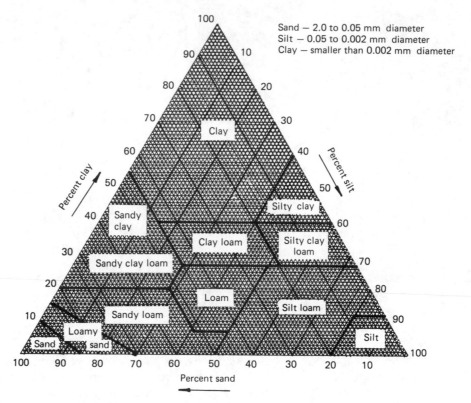

Figure 4.1 Textural soil classification chart, U.S. Department of Agriculture.

engineering purposes, particularly where characteristics of only superficial soils are of principal concern, which for construction purposes they frequently are.

On the chart of Figure 4.1, the index lines corresponding to the percentages of each soil fraction are simply intersected. A soil with 50 percent clay, 20 percent sand, and 30 percent silt would be classified, for example, as *clay*, since the intersection of the index lines is in the clay region. Index lines for clay are horizontal, index lines for silt are down and to the left, and index lines for sand are up and to the left, from the corresponding sides of the chart. If the soil contains 20 percent or more gravel, a *gravelly* prefix is added to the classification, and the percentages of sand, silt, and clay are taken as percentages of the soil fraction with the gravel excluded.

Textural classification of soils is of use to the builder primarily for coarse-grained soils (sands and gravels). It is these materials for which performance depends principally on the relative amounts and sizes of particles. For fine soils (silts and clays), textural classification provides little information of practical use for engineering purposes. The behavior of these soils is controlled by factors other than particle size. In particular, plasticity characteristics are of great importance.

4.2 UNIFIED SOIL CLASSIFICATION SYSTEM

4.2.1 Laboratory Classification

The Unified Soil Classification System (ASTM D2487, D2488) is used by the U.S. Bureau of Reclamation, the U.S. Army Corps of Engineers, the U.S. Forest Service, and many private consulting engineers. Arthur Casagrande developed the system for the Corps of Engineers during the 1940's. Its original purpose was to classify soils for use in roads and airfields; however, it since has been revised to include use in embankments and foundations as well. The Unified System permits classification of soils using either laboratory or field procedures. Field procedures are of particular value to the field engineer.

For a laboratory-based classification in the Unified System, index values of the soil are needed. The most important distinction the Unified System makes is whether a soil is coarse or fine grained. As mentioned in Chapter 1, coarse-grained soils are those satisfying the condition that 50% or more of the grains will not pass the No. 200 sieve; put another way, these are soils in which more than half of the soil particles are larger than 0.074 mm (0.0029"). Fine-grained soils are just the opposite (i.e., those in which more than half of the soil particles pass through the No. 200 sieve and are smaller than 0.074 mm (0.0029")). If the soil is coarse, complete classification requires a grain-size distribution curve. From this curve, percentage composition of the constituents and the coefficients of uniformity and curvature are available. For the special case of a coarse soil containing more than 12 percent fines (silt and clay), the liquid limit and plasticity index of the fine fraction are also needed. With these indices the soil will be assigned a descriptive symbol consisting of two or more letters. These letters and their meanings are given in Table 4.1.

The Unified System for coarse-grained soils is shown in Figure 4.2. This chart and the classifications for fine grained soils (Figure 4.3) are both used for defining the various classifications and actually determining the classification of a given soil. Materials with 5–12% of the grains passing through the No. 200 sieve are borderline cases and can be designated with a pair of classifications, such as GW-GM, SW-SC, etc. If the soil is fine grained, classification

TABLE 4.1   Unified Classification Symbols

Symbol	Description
G	Gravel
S	Sand
M	Silt
C	Clay
O	Organic
Pt	Peat
W	Well graded
P	Poorly graded
L	Low liquid limit compressibility; lean (clays)
	Low liquid limit; (silts); plasticity
H	High liquid limit, compressibility; fat (clays)
	High liquid limit; elastic (silts)
NP	Nonplastic

Primary Divisions for Field and Laboratory Identification			Group Symbol	Typical Names	Laboratory Classification Criteria		Supplementary Criteria For Visual Identification
Coarse-grained soils. (More than half of material finer than 3-inch sieve is larger than No. 200 sieve size.)	Gravel. (More than half of the coarse fraction is larger than No. 4 sieve size, about 1/4 inch.)	Clean gravels. (Less than 5% of material is smaller than No. 200 sieve size.)	GW	Well graded gravels, gravel–sand mixtures, little or no fines.	$C_u = \dfrac{D_{60}}{D_{10}}$ greater than 4. $C_z = \dfrac{(D_{30})^2}{D_{10} \times D_{60}}$ between 1 and 3.		Wide range in grain size and substantial amounts of all intermediate particle size.
			GP	Poorly graded gravels, gravel-sand mixtures, little or no fines.	Not meeting both criteria for GW.		Predominantly one size (uniformly graded) or a range of sizes with some intermediate sizes missing (gap graded).
… do …	… do …	Gravels with fines. (More than 12% of material is smaller than No. 200 sieve size.)	GM	Silty gravels, and gravel–sand–silt mixtures.	Atterberg limits below "A" line, or PI less than 4.	Atterberg limits above "A" line with PI between 4 and 7 is borderline case GM–GC.	Nonplastic fines or fines of low plasticity.
			GC	Clayey gravels, and gravel–sand–clay mixtures.	Atterberg limits above "A" line, and PI greater than 4.		Plastic fines.
… do …	Sands. (More than half of the coarse fraction is smaller than No. 4 sieve size.)	Clean sands. (Less than 5% of material is smaller than No. 200 sieve size.)	SW	Well graded sands, gravelly sands, little or no fines.	$C_u = \dfrac{D_{60}}{D_{10}}$ greater than 6. $C_z = \dfrac{(D_{30})^2}{D_{10} \times D_{60}}$ between 1 and 3.		Wide range in grain sizes and substantial amounts of all intermediate particle sizes.
			SP	Poorly graded sands and gravelly sands, little or no fines.	Not meeting both criteria for SW.		Predominantly one size (uniformly graded) or a range of sizes with some intermediate sizes missing (gap graded).

Figure 4.2 Unified Classification System, Coarse-Grained Soils (after NAVFAC DM 7.01)

Primary Divisions for Field and Laboratory Identification			Group Symbol	Typical Names	Laboratory Classification Criteria		Supplementary Criteria For Visual Identification
… do …	… do …	Sands with fines. (More than 12% of material is smaller than No. 200 sieve size.)	SM	Silty sands, sand–silt mixtures.	Atterberg limits below "A" line, or PI less than 4.	Atterberg limits above "A" line with PI between 4 and 7 is borderline case SM-SC.	Nonplastic fines or fines of low plasticity.
			SC	Clayey sands, sand–clay mixtures.	Atterberg limits above "A" line with PI greater than 7.		Plastic fines.

Figure 4.2 (*continued*)

Primary Divisions for Field and Laboratory Identification		Group Symbol	Typical Names	Laboratory Classification Criteria		Supplementary Criteria For Visual Identification		
						Dry Strength	Reaction to Shaking	Toughness Near Plastic Limit
Fine-grained soils. (More than half of material is smaller than No. 200 sieve size.) (Visual: more than half of particles are so fine that they cannot be seen by naked eye.)	Silts and clays. (Liquid limit less than 50.)	ML	Inorganic silts, very fine sands, rock flour, silty or clayey fine sands.	Atterberg limits below "A" line, or PI less than 4.	Atterberg limits above "A" line with PI between 4 and 7 is borderline case ML-CL.	None to slight	Quick to slow	None
	… do …	CL	Inorganic clays of low to medium plasticity; gravelly clays, silty clays, sandy clays, lean clays.	Atterberg limits above "A" line, with PI greater than 7.		Medium to high	None to very slow	Medium
	… do …	OL	Organic silts and organic silt-clays of low plasticity.	Atterberg limits below "A" line.		Slight to medium	Slow	Slight

Figure 4.3   Unified Classification System, Fine-Grained Soils (after NAVFAC DM)

Primary Divisions for Field and Laboratory Identification		Group Symbol	Typical Names	Laboratory Classification Criteria	Supplementary Criteria For Visual Identification		
					Dry Strength	Reaction to Shaking	Tough- ness Near Plastic Limit
... do ...	Silts and clays. (Liquid limit greater than 50.)	MH	Inorganic silts, micaceous or diatomaceous fine sands or silts, elastic silts.	Atterberg limits below "A" line.	Slight to medium	Slow to none	Slight to medium
	... do ...	CH	Inorganic clays of high plasticity, fat clays.	Atterberg limits above "A" line.	High to very high	None	High
	... do ...	OH	Organic clays of medium to high plasticity.	Atterberg limit below "A" line.	Medium to high	None to very slow	Slight to medium
... do ...	Highly organic soils	Pt	Peat, muck and other highly organic soils.	High ignition loss, LL and PI decrease after drying.	Organic color and odor, spongy feel, frequently fibrous texture.		

Figure 4.3 (*continued*)

requires that we know both the liquid limit and plasticity index (the "Atterberg Limits" referred to on the chart). These limits are then used to determine the distinction between silts and clays and high and low compressibility using Figure 4.4.

EXAMPLE Unified Classification System

To illustrate the use of these charts, assume that Soil *A* with the gradation curve shown in Figure 4.5 is to be classified. In addition to the sieve tests, laboratory tests on the Atterberg limits show that LL = 63 and PL = 42; thus, PI = LL − PL = 21. We must classify the soil according to the Unified System.

Since only about 10 percent passes through the No. 200 sieve, the soil is coarse grained, and so we would use Figure 4.2 for soil classification. Also, only 10 percent of the soil (about 11 percent of the coarse fraction) is gravel. The soil is, therefore, sand. Finally, since the fine fraction is between 5 and 12 percent of the sample weight, the soil will have a dual classification. By examining the gradation curve, we determine that

$$C_u = \frac{D_{60}}{D_{10}} = \frac{0.6 \text{ mm}}{0.074 \text{ mm}} = 8.1$$

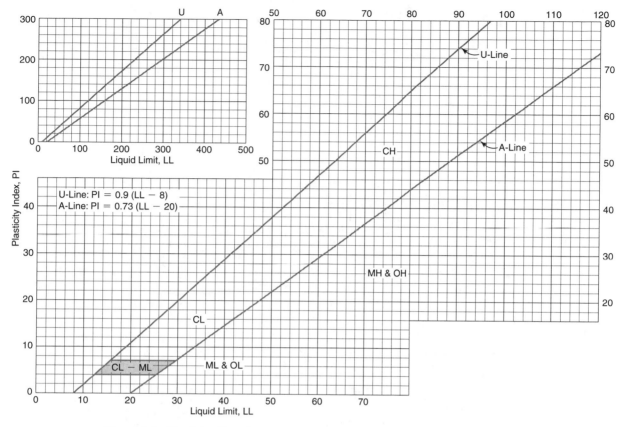

Figure 4.4 Plasticity Chart for the Unified Soil Classification System (From EM 1110-2-1906)

and

$$C_c = \frac{D_{30}^2}{D_{60}D_{10}} = \frac{(0.2)^2}{0.6 \times 0.074} = 0.9.$$

The value of C_c eliminates GW and SW classifications from consideration. The other gravel classifications (GP, GM, and GC) are also eliminated, as 50% gravel is required, but our soil only has 10% gravel. This leaves SP, SM, and/or SC. We say "and/or" because the fine fraction dictates a dual classification that falls somewhere between the SP on the one hand and the SM or SC on the other. So one of the dual classifications will be SP. Since the plasticity index PI is greater than 7, we might make the second classification SC. However, SC also requires that the PI/LL combination fall above the "A" line in Figure 5.5, which it does not. We thus have a silt (as opposed to a clay), and this dictates that the second classification be SM. The soil is accordingly classified as SP–SM.

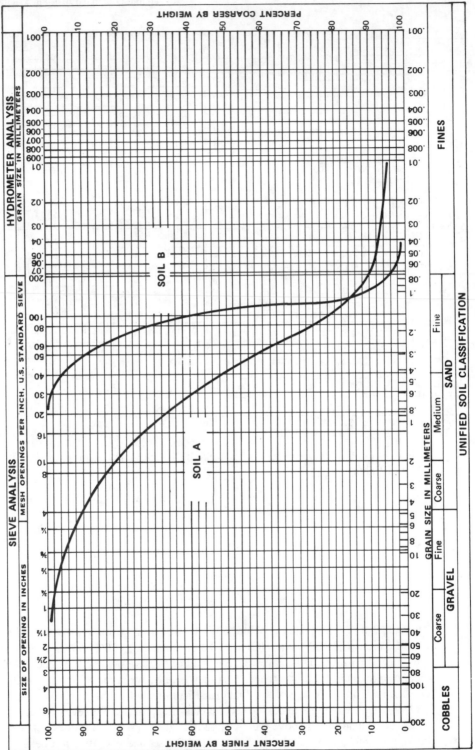

Figure 4.5 Gradation curves for soil classification examples.

48

Turning to soil B, we classify the soil in the following sequence:

1. Approximately 4% of the soil passes the No. 200 sieve; we are still dealing with a coarse-grained soil and use Figure 4.2 for classification.
2. All of the soil passed through the No. 4 sieve, so this soil is sand.
3. Since less than 12% of the material is smaller than the No. 200 sieve, it is a clean sand.
4. The coefficients of uniformity and curvature are, respectively,

$$C_u = \frac{D_{60}}{D_{10}} = \frac{0.15 \text{ mm}}{0.1 \text{ mm}} = 1.5$$

and

$$C_c = \frac{D_{30}^2}{D_{60}D_{10}} = \frac{0.13^2}{0.15 \times 0.1} = 1.13$$

5. The curvature coefficient C_c is between 1 and 3; thus, it meets that criterion for SW soils. However, the uniformity coefficient C_u is less than 6; therefore, this soil is classified as an SP soil.

It is interesting to note that, for "clean" soils (soils that have less than 5% fines,) the Atterberg limits do not enter into the soil classification.

The previous example for soil A illustrated the use of the plasticity chart to classify fine-grained soils. The coordinates of liquid limit and plasticity index are plotted and the soil is assigned the symbol that represents the area in which the point falls. It should be noted that organic soils generally are distinguished from inorganic soils by color and smell. Dark soils with a distinctly objectionable smell (usually hydrogen sulfide) are organic. Positive identification of the presence of organics can be made in most cases by running liquid limit tests on the sample after air drying and again after oven drying. If the liquid limit is less than three-fourths the standard value because of oven drying, the soil is considered organic.

4.2.2 Field Classification

Field classification using the Unified System should lead to the same results (classification symbols) as laboratory classification.

For coarse soils, the percentages of constituent sizes are simply estimated for field classification purposes. It is easiest to work by spreading the soil on a flat surface as shown in Figure 4.6. Coarse soils include particles that may be seen with the unaided eye. Individual fine soil particles cannot be seen without magnification. Therefore, a fine soil is identified in the field as one that contains more than 50 percent by weight of visually indistinguishable particles. It is important to point out that often aggregates of fine particles are mistaken for sand or gravel. Care should be taken in field classification to ensure that this difficulty does not arise. Having decided if a soil is coarse or fine, one then proceeds to complete the classification, as described in the paragraphs that follow.

For coarse soils the distinction between sand and gravel is based on passage through the No. 4 sieve. For field classification, 1/4-inch particle size is generally used. For clean sands or gravels (those with less than 12 percent fines), the gradation designation W or P is assigned, depending on whether a wide range of particle sizes is represented or only one or two sizes predominate. Dirty

Figure 4.6 Visualizing the particle size distribution in a coarse soil sample.

sands and gravels have a second designator *M* or *C*, according to whether the fines are silty or clayey. This distinction is based on tests that are also appropriate for fine-grained soils.

With fine-grained soils, the progress through the classification process is supplemented by simpler tests for dilitancy, dry strength and toughness that are accomplished with little or no equipment. Responses typical of the various soil groups are also given and are further described in Figures 4.2 and 4.3.

Dilitancy is a soil's reaction to shaking. To test, after removing particles larger than the No. 40 sieve size, prepare a pat of moist soil with a volume of about 1/2 cubic inch. Place the pat in the open palm of one hand and shake horizontally, striking vigorously against the other hand several times. A positive reaction consists of the appearance of water on the surface of the pat which changes to a livery consistency and becomes glossy. When the sample is squeezed between the fingers, the water and glass disappear from the surface, the pat stiffens, and finally it cracks or crumbles. The rapidity of appearance of water during shaking and of its disappearance during squeezing assist in identifying the character of the fines in a soil. Typical results are shown in Figure 4.7. Very fine clean sands give the quickest and most distinct reaction whereas a plastic clay has no reaction. Inorganic silts, such as a typical rock flour, show a moderately quick reaction.

Dry Strength refers to the crushing characteristics of a soil. To test, after removing particles larger than the No. 40 sieve size, mold a pat of soil to the consistency of putty, adding water if necessary. Allow the pat to completely dry by oven, sun or air drying, and then test its strength by breaking and crumbling between the fingers. This strength is a measure of the character and quantity of the colloidal fraction contained in the soil. Typical results are shown in

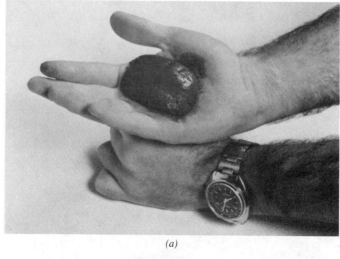

(a)

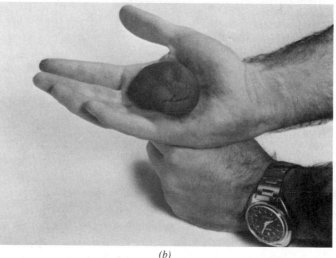

(b)

Figure 4.7   Response of a nonplastic silt to the dilatancy test. (*a*) After jarring the sample by tapping the hand to densify the sample (note shiny surface). (*b*) After opening the hand to allow the sample to expand (note dull surface).

Figure 4.8. High dry strength is characteristic of clays in the CH group. A typical inorganic silt possesses very slight dry strength. Silty fine sands and silts have about the same slight dry strength, but can be distinguished by the feel when powdering the dried specimen. Fine sand feels gritty, whereas a typical silt has the smooth feel of flour.

Toughness refers to the consistency of a soil near the plastic limit. To test, after removing particles larger than the No. 40 sieve size, a specimen of soil about 1/2 cubic inch in size is molded to the consistency of putty. If it is too dry, water must be added, and if it is sticky, the specimen should be spread out in a thin layer and allowed to lose some moisture by evaporation.

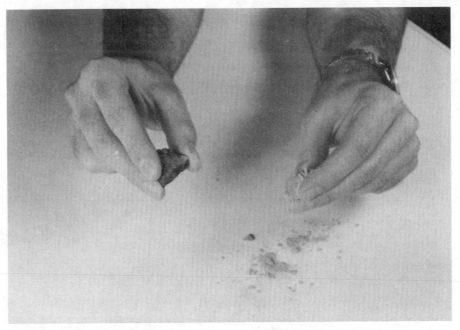

Figure 4.8 Demonstration of dry strength of a clay (left) and a silt (right).

Then the specimen is rolled out by hand on a smooth surface or between the palms into a thread about 1/8″ in diameter. The thread is then folded and rerolled immediately. During this manipulation, the moisture content is gradually reduced and the specimen stiffens, finally loses its plasticity, and crumbles when the plastic limit is reached. After the thread crumbles, the pieces should be lumped together and a slight kneading action continued until the lump crumbles. The tougher the thread near the plastic limit and the stiffer the lump when it finally crumbles, the more potent is the colloidal clay fraction of the soil. Weakness of the thread at the plastic limit and quick loss of coherence of the lump below the plastic limit indicate either inorganic clay of low plasticity or materials such as kaolin-type clays and organic clays that occur below the A-line. Highly organic clays have a very weak and spongy feel at the plastic limit.

Generally, the distinction between silts and clays is easily made based on dilatancy (ability to change volume) or dry strength. For moist soils the toughness test is useful. It is more difficult, especially for silt, to distinguish between high and low elasticity or plasticity. Probably the best method of learning to do so is to observe the reactions to these simple tests of soils that have been classified using laboratory procedures. It is also useful, especially for beginners, to summarize observations on a data sheet, such as that shown in Figure 4.9, before assigning a classification.

4.3 AASHTO CLASSIFICATION SYSTEM

The Unified Classification System groups soils according to their behavior in a wide variety of uses. The American Association of State Highway and Transportation Officials (AASHTO) and the Federal Highway Administration (FHWA) use a classification system (AASHTO 3282)

VISUAL SOIL IDENTIFICATION

TEST BY __TKN__   FOR __Capital Corporation__   DATE __2 February 1993__

SAMPLE NO.	MAX. SIZE	ESTIMATED GRADATION			COLOR (Wet)	DESCRIPTION AND IDENTIFICATION	GROUP SYMBOL
		GRAVEL (%)	SAND (%)	FINES (%)		Consider particle size and shape, gradation, fines, plasticity, dry strength, reaction to shaking test, odor, undisturbed consistency.	
B6, ST 1	No. 40	0	10	90	dark brown	dark brown inorganic stiff silty clay	CL
B6, ST 2	No. 40	0	10	90	light brown	light brown inorganic soft clayey silt	MH
B6, ST 3	No. 40	0	10	90	light brown	light brown inorganic soft clayey silt	MH
B6, SS 1	No. 4	0	75	25	grey-brown	grey brown angular silty sand	SM
B6, SS 2	No. 4	0	75	25	grey-brown	grey brown angular silty sand	SM

Figure 4.9 Summary of observations for visual soil identification and classification.

53

intended to indicate the behavior of materials used as highway subgrades. There are seven basic soil groups designated A-1 through A-7. Soils are placed in these groups according to performance characteristics, A-1 being the best and A-7 the worst. An A-8 designation is reserved for peat or highly organic soils. A-1 to A-3 soils are sands and gravels, whereas A-4 to A-7 soils are silts and clays.

Some basic differences between the AASHTO System and the Unified System are as follows:

- Gravel in the AASHTO System is that material passing the 3-in. sieve and retained on the No. 10 sieve. In the Unified System, the lower particle size bound is defined by the No. 4 sieve.

- The AASHTO System recognizes only fine and coarse sand. Fine sand has the same particle size range in both classification systems, while coarse sand in the AASHTO System is that material passing the No. 10 sieve and retained on the No. 40. Coarse sand in the AASHTO System is thus the same as medium sand in the Unified System. The smallest gravel-size material in the AASHTO System is included as coarse sand in the Unified System.

- The definition of fine and coarse soils is different in the two systems. In the AASHTO System, fine soils (A-4 to A-7) are those containing 35 percent or greater fines. The corresponding fines content in the Unified System is 50 percent.

Material quality within the basic soil groups is further indicated by an additional number or letter designation, according to the results of sieve analysis and Atterberg limits tests. A summary of the classification system is given in Figure 4.10. This is used by proceeding from left to right with the use of the available test data. The first soil group encountered in this manner that matches the test data is the correct classification. The plasticity chart shown in Figure 4.11 is of considerable aid in the selection of a classification for fine soils. It incorporates the same information as that shown in Figure 4.10 and may be used in place of it.

Because soil behavior is influenced not only by the amount of fine material present, but also by its plasticity, AASHTO classifications include a parenthesized number that supplements the basic group symbols. As with the group symbols, increasing size of number indicates decreasing quality of material. The group index is determined from the nomograph shown in Figure 4.12, by adding two partial group indices, one based on plasticity index and one based on liquid limit. The partial group indices are determined by extending a straight line from the percent fines scale through the liquid limit scale and the plasticity index scale to intersect the partial group index axis.

The use of the AASHTO System may be illustrated by reclassifying the soils in the previous section. Soil A (Figure 4.5) is classified as an A-2-7, while soil B is an A-2-4. Soil A is considered a poorer material because of its moderately high fines content.

Sandy silt with 60 percent fines, a liquid limit of 67, and a plasticity index of 15 would classify as an MH in the Unified System and as an A-7-5 (11) in the AASHTO System. The group index (11) consists of partial group indices 8.5 (based on liquid limit) and 2.5 (based on plasticity index).

4.4   SUMMARY

Soil classification is an aid in predetermining the behavior of soils under various conditions of service and during construction. Such characterizations can provide very useful preliminary information on a job site. Classifications in the systems discussed are often found in the

GENERAL CLASSIFICATION

	GRANULAR MATERIALS (35 percent or less of total sample passing No. 200)							SILT-CLAY MATERIALS (More than 35 percent of total sample passing No. 200)			
	A-1		A-3	A-2				A-4	A-5	A-6	A-7
GROUP CLASSIFICATION	A-1-a	A-1-b	A-3	A-2-4	A-2-5	A-2-6	A-2-7	A-4	A-5	A-6	A-7-5, A-7-6
Sieve analysis, percent passing:											
No. 10	50 max.										
No. 40	30 max.	50 max.	51 min.								
No. 200	15 max.	25 max.	10 max.	35 max.	35 max.	35 max.	35 max.	36 min.	36 min.	36 min.	36 min.
Characteristics of fraction passing No. 40:											
Liquid limit				40 max.	41 min.	40 max.	41 min.	40 max.	41 min.	40 max.	41 min.
Plasticity index	6 max.		NP	10 max.	10 max.	11 min.	11 min.	10 max.	10 max.	11 max.	11 min.*
Group Index**	0		0	0	0	4 max.		8 max.	12 max.	16 max.	20 max.

Classification procedure: With required test data available, proceed from left to right on chart; correct group will be found by process of elimination. The first group from the left into which the test data will fit is the correct classification.

*P.I. of A-7-5 subgroup is equal to or less than L.L. minus 30. P.I. of A-7-6 subgroup is greater than L.L. minus 30 (see Fig. 4.13)

**See group index formula or Fig. 4.14 for method of calculation. Group index should be shown in parentheses after group symbol as: A-2-6(3), A-4(5), A-6(12), A-7-5(17), etc.

Figure 4.10 AASHTO classification for highway subgrade material.(Courtesy of American Association of State Highway and Transportation Officials.)

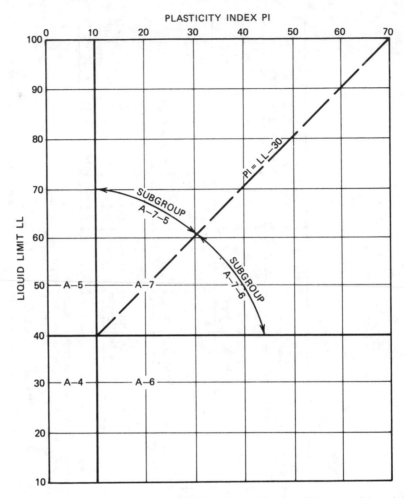

Figure 4.11 Plasticity chart for AASHTO soil classification. (Courtesy of American Association of State Highway and Transportation Officials.)

subsurface information section contained in contract documents for earthwork. It helps the bidder and the contractor during construction to understand the meaning of these various terms and symbols. They are particularly useful when supplemented by other information from investigative reports.

After having studied this chapter and solving the problems at the end, the reader should

1. Have a reading knowledge of available engineering soil classification systems.
2. Be able to correctly classify soils in the AASHTO and Unified systems, given appropriate laboratory test data.
3. Be able to classify soils in the Unified System given only the soil to work with.
4. Know how to describe soils classified in the AASHTO System and the Unified System given soil group symbols.

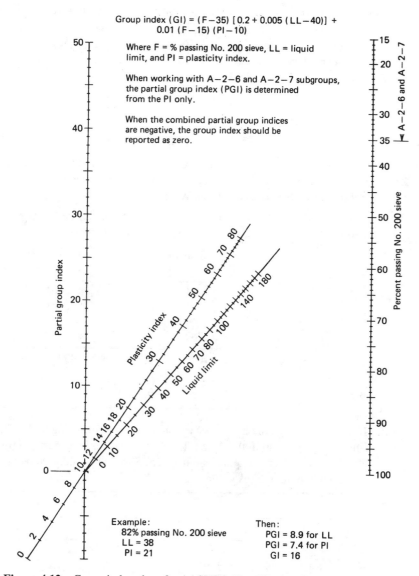

Figure 4.12 Group index chart for AASHTO Classification System. (Courtesy of American Association of State Highway and Transportation Officials.)

REFERENCES

American Association of State Highway and Transportation Officials, *Highway Materials, Part I, Specifications,* Washington, D.C., 1978.

American Society for Testing and Materials, *1988 Annual Book of ASTM Standards,* Vol. 04.08, Soil and Rock, Building Stones; Geotextiles, Philadelphia.

Casagrande, A., "Classification and Identification of Soils," *Transactions, American Society of Civil Engineers,* 113, 1948.

Department of the Army. *Laboratory Soils Testing*. EM 1110-2-1906. Washington, DC: U.S. Army Corps of Engineers, 1986.

Department of the Navy. *Soil Mechanics*. NAVFAC DM 7.01. Norfolk, VA: Naval Facilities Engineering Command, 1986.

U.S. Department of Agriculture, *Soil Survey Manual,* Handbook No. 18, 1951.

U.S. Department of the Interior, Bureau of Reclamation, *Earth Manual,* Denver, 1963.

PROBLEMS

1. A dark gray soil is described as having a rapid reaction to the shaking or dilatancy test, no dry strength, and an offensive odor. Describe the soil and assign it a tentative classification in the Unified Soil Classification System.

2. Given the test results that follow, classify the soil in the Unified Classification System. Outline the steps you take in arriving at your result.

U.S. Standard Sieve No.	Percent Finer
4	100
10	93
40	71
200	48

Atterberg Limits of Minus-No. 40 Fraction

$$LL = 33$$
$$PL = 30$$

3. Test results for a number of soils are shown in the following table. Classify each soil in the Unified System and in the AASHTO System, using your best judgment when specific information you might require is not given. Show any calculations you may find necessary and list any assumptions you make.

Soil No.	1	2	3	4	5	6	7	8	9
Sieve No.					Percent Finer				
4	73								67
10	50	92	100		86			88	43
40	40	52	69	100	61	92		78	19
100	25	41	52		30		100	19	16
200	15	28	46	62	19	90	69	7	7
					Alterberg Limits				
LL	20	29	37	72	39	38	NP	NP	42
PL	15	15	30	45	30	19	NP	NP	21

4. Describe a soil classified as SW and a soil classified as CH in the Unified Soil Classification System as you would to a person having no knowledge of soil classification procedures or equipment.

CHAPTER 5

Stress Analysis and Engineering Properties

The construction contractor is concerned with analyses using engineering properties of soils in those instances where construction of temporary work such as shoring for an excavation, false-work for a structure, or a dewatering system requires some engineering design using soil parameters. In some circumstances, the contractor's staff does the analysis and design work. In other cases, outside consultants are employed. In either event, the design requires some knowledge of engineering soil properties and the principles governing soil behavior. This chapter is an introduction to the subject. After studying this chapter, the reader should have developed an understanding of the effective stress concept and learned the principles of analysis for gravity stresses in soils. The engineering properties, soil strength, permeability, and compressibility will be related to soil index properties discussed earlier. It will be shown, for instance, that strength varies with the changing water content of clay soils and that permeability varies with the grain size of coarse soils.

The reader should learn

1. The effective stress principle.

2. How to make static stress calculations.

3. About capillarity.

4. The relative permeabilities of various soils and how permeability values may be determined.

5. What is meant by compressibility.

6. The mechanism leading to consolidation of soils.

7. The factors controlling soil strength.

8. Relative strengths of troublesome soils and how they may be inferred from index properties.

5.1   THE EFFECTIVE STRESS PRINCIPLE

Soil engineering properties and behavior are strongly influenced by stresses and stress history. For this reason, it is important to understand the principles on which stress determinations are based, and how to make fundamental stress calculations.

Figure 5.1a shows a sphere of weight W_s and volume V_s resting inside a container with a square base of width B. If we assume that the container is weightless, then the average stress on the surface on which the container rests is W_s/B^2. This is the stress due to the weight of the sphere. The sphere contacts the inside of the container at a point, but we may think of its weight being distributed over the inner container base for purposes of illustration. The average stress on the base in these circumstances would be W_s/B^2, the same as on the supporting surface. If, as in Figure 5.1b, we fill the container with water to depth z, the stress on the base becomes

$$p = \frac{W_s + (zb^2\gamma_w - V_s\gamma_w)}{B^2}. \tag{5.1}$$

We may think of Equation (5.1) as representing the total stress due to the weight of the sphere and water. The stress on the inner container base has two components. That due to water is obviously

$$u = z\gamma_w. \tag{5.2}$$

By subtracting Equation (5.2) from Equation (5.1), we obtain the stress due to the sphere:

$$p' = \frac{W_s - V_s\gamma_w}{B^2}. \tag{5.3}$$

The difference between the stresses due to the sphere in Figures 5.1a and 5.1b results from the buoyancy effect of the water. Implicit in the preceding discussion is the fact that

$$p = p' + u. \tag{5.4}$$

To illustrate how this observation relates to soils, we may consider many spheres, representing soil particles, as shown in Figure 5.2. At some depth z, the average total stress due to all of the spheres and the water will be the sum of their respective weights divided by the area they

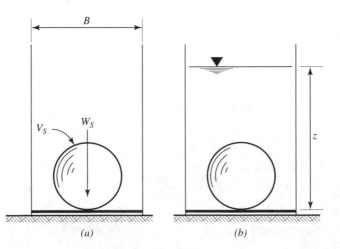

Figure 5.1   Illustration for effective stress-principle discussion.

(a)   (b)

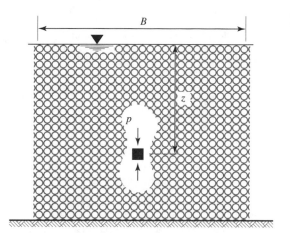

Figure 5.2 Idealized soil stratum.

occupy. In fact, if the sum of the weights is W_s and the sum of the volumes is V_s, the total stress for an area $B \times B$ is represented by Equation (5.1). Following the same logic as in the foregoing illustration, it is obvious that Equation (5.4) remains valid. This equation represents the effective stress principle, a most important principle in describing soil behavior in engineering applications. It may be stated as follows: *Total stress* (p) *is equal to the sum of effective stress* (p') *and pore water pressure* (u) or, alternatively, *effective stress is equal to total stress minus pore water pressure.* Pore water pressure is also sometimes referred to as neutral stress.

5.2   VERTICAL EARTH PRESSURE CALCULATIONS

In theory, vertical earth pressures may be directly calculated using Equation (5.4) and appropriate material properties. The reality is that real soil profiles are stratified with varying unit weights; this stratification is further complicated by the water table, which reduces the effective stress below its surface. Equation (5.4) must thus be applied layer by layer. Keeping track of the effects of these varying layers can be difficult without an illustration; such an illustration is a "P_o Diagram[1]", and an example of this is shown in Figure 5.3. P_o diagrams are enormously useful in both visualizing the progression of effective stress with depth and in reducing the possibility of error in calculations. With modification they can also be used for lateral earth pressures as well.

EXAMPLE   Vertical Earth Pressure Calculation Using P_o Diagrams

In Figure 5.3, a simple stratigraphic column is shown and the unit weights of the soil materials involved are given. This column is representative of a 40-ft sand layer overlying rock. The water table is at a depth of 10 ft. Analysis of the layers is as follows:

- Layer 1 (0–10 ft): As it lies above the water table, the unit weight in this region may be considered as the total unit weight of the material. If we consider a 1-ft^2 column from the

---

[1]In this text we would properly call this a "p' diagram" for effective stress; P_o is used in other texts and gives its name to this diagram.

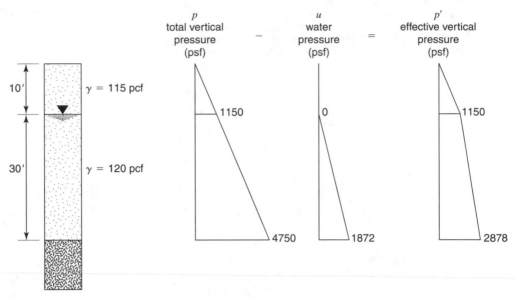

Figure 5.3 Example of vertical earth pressure analysis.

surface down, then the vertical pressure at a point will be equal to the total unit weight of the material above that point times the vertical distance to the point. At 10 ft in depth, therefore, the total vertical pressure is $(10 \text{ ft})(115 \text{ pcf}) = 1150$ psf.

- Layer 2 (10 ft–40 ft): The unit weight shown for the lower 30 ft is also the total unit weight of the material but, in this case, the material is submerged below the water table. At 40 ft in depth, the total vertical pressure is the sum of the vertical pressure at 10 ft and the change in vertical pressure for the lower 30 ft. The change in vertical pressure considering the unit weight of the soil only is $(30 \text{ ft})(120 \text{ pcf}) = 3600$ pcf. Thus, the total vertical pressure is 1150 psf + 3600 psf = 4750 psf. However, since this layer is below the water table, we must subtract the effect of the unit weight of the water, which is $(30 \text{ ft})(62.4 \text{ pcf}) = 1872$ psf. Hence, the effective stress is 4750 psf − 1872 psf = 2878 psf at a depth of 40 ft. An alternative way of calculating submerged effective stress changes is to subtract the unit weight of the water γ_w from the wet unit weight of the soil and multiply it by the depth of the layer, or $(30 \text{ ft})(120 \text{ pcf} - 62.4 \text{ pcf}) = 1728$ psf. The effective stress at a depth of 40 ft is thus 1150 psf + 1728 psf = 2878 psf, which is the same as before.

Both the total and effective stress states are illustrated in Figure 5.3. It is important to note that this effective vertical pressure is the pressure due to the effective weight of the soil particles. To obtain the total pressure, it is necessary to add the water pressure to the effective pressure.

5.3 CAPILLARY STRESSES

The water surface in a brim-full glass stands above the lip because of surface tension at the air-water interface. When the glass is less than full the interface turns up at its edges, pulling up on the water. The magnitude of the surface tension T_s at an air-water interface is nominally

0.0075 g/mm. This taut interface, or meniscus, can cause water to rise above a free water surface in a small tube and in continuous small soil voids. This effect is referred to as capillary rise and can be significant in building design and construction.

5.3.1 Computation of Capillary Rise

Figure 5.4 illustrates a hypothetical soil column in a deposit with the free water surface at a–a. A single capillary above a–a is shown shaded to indicate it is filled with water. Other continuous voids will also be filled to some height. The contact between the interface meniscus and soil particles is at some angle α, and the surface tension force acts along that angle. This force acts to pull water above the free surface, a–a. If the effective diameter of the capillary is d, then the total force exerted by the meniscus on the water column is $\pi d T_s \cos \alpha$. Static equilibrium requires that the meniscus support a water column of height z_c, with the result that

$$\pi d T_s \cos \alpha = \frac{\pi z_c d^2 \gamma_w}{4}, \tag{5.5}$$

and if $\alpha = 0$, then

$$z_c = \frac{4 T_s}{d \gamma_w}. \tag{5.6}$$

Equation (5.6) allows us to estimate the height of capillary rise in soils if we can estimate d, which is related to particle size. We know that d may be approximated by $D_{10}/5$ if D_{10} is taken as the effective size of particles.

Figure 5.4 Capillary rise in soils.

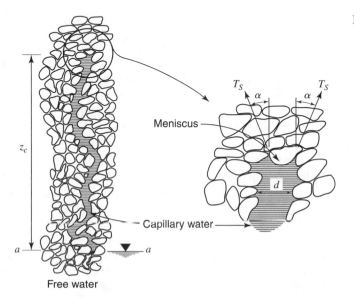

EXAMPLE   Computation of Capillary Rise

Since fine sand or silt soils are by definition those which pass the opening of the No.-200 sieve, and this opening is 0.074 mm, D_{10} could be estimated as this opening size. Then, from Equation (5.6),

$$z_c = \frac{4 \times 0.0075 \text{ g/mm}}{\dfrac{0.074 \text{ mm}}{5} \times .001 \text{ g/mm}^3} \cong 2000 \text{ mm} \cong 2 \text{ m}. \tag{5.7}$$

Similar calculations will show that the height of capillary rise is very large in very fine soils and almost negligible in clean, coarse sands and gravels.

Capillary action also increases effective stress above the free water surface. For instance, force equilibrium at the meniscus in Figure 5.4 requires that

$$\frac{\pi d^2 u}{4} + \frac{\pi z_c d^2 \gamma_w}{4} = 0, \tag{5.8}$$

where u is the pore water pressure. From Equation (5.8),

$$u = -z_c \gamma_w. \tag{5.9}$$

Thus, in the illustration of Figure 5.3, if the soil had been such that z_c was greater than or equal to 10 ft, the pore water pressure at the ground surface would have been

$$u = -10 \times 62.4 = -624 \text{ psf}. \tag{5.10}$$

Since the total stress at the ground surface is zero, the effective stress principle requires that, at the ground surface, the effective stress be 624 psf. With similar calculations, it can be shown that there is a linear increase in effective stress from the free water surface to the height of capillary rise, caused by capillary action. This increment in effective stress is in addition to that due to soil weight. Its importance will be discussed later.

5.3.2   Frost Heave

In finer soils, capillary action can result in a high degree of saturation above the water table. This results in low strength, which is usually undesirable in engineered works. In cold regions, where low temperatures persist for sufficiently long periods, the combination of near surface groundwater, freezing temperatures and high rapid capillary rise produce frost heaving. Silty soils, which have a high capillary rise and are relatively permeable (see Section 5.4), are referred to as *frost-susceptible soils*. Clayey soils with relatively low permeability do not allow rapid enough accumulation of water to produce frost heaving by capillary action, and coarse soils with low capillary rise are not susceptible to frost.

Frost heaving results from the formation and growth of ice lenses in surface soils during freezing periods. Loss of strength from the high degree of saturation is produced after thawing. Problems that arise include pavement break-up, foundation heaving and settlement. Pavement break-up occurs when heavy traffic runs on roads where ice lenses have formed in the subgrade, then thawed, a phenomenon all too familiar to highway departments in the northern United States and Canada. Foundation heaving may occur in improperly designed or operated cold storage facilities in any climate on sites with a high water table and silty subsoils. In one such situation, in a temperate climate, a warming system in the gravel fill below a cold storage plant malfunctioned, and ice lenses which formed in the subgrade caused the floor slab to heave more than four feet before the cause was determined. Where perennially frozen ground (*permafrost*) exists, improper design or construction of buildings results in warming of the subsoils, thawing of the ice lenses, and settlement.

There must be three factors present for problems like those described above to occur: frost-susceptible soils (typically silts), a source of water (high groundwater table), and long periods of freezing temperatures. The elimination of any one of these factors, either by design or because it may not be naturally present, eliminates the frost heave problem.

5.4 PERMEABILITY

The permeability of a soil is a measure of the ease with which a particular fluid flows through its voids. Usually, we are concerned with the flow of water through soils. A highly pervious material may be either a headache or a blessing for the contractor. If an excavation below the groundwater level is to be dewatered, a highly pervious soil will require a carefully planned and executed pumping system with comparatively large capacity. On the other hand, highly pervious material from a borrow source may often be placed and compacted immediately after it is excavated. High permeability permits the excess water to escape almost instantly. Excavation below water in an impervious soil may require no pumping at all. However, saturated impervious borrow may require days of drying before placement. The extra handling involved is an added expense.

5.4.1 Coefficient of Permeability and Darcy's Law

The permeability coefficient k may be considered as the apparent velocity of seepage through a soil under a unit hydraulic gradient. It is numerically equal to the seepage volume per unit of time through a unit area. Analysis of a seepage problem to determine pumping capacity for a dewatering system thus requires not only a representative k for the soil but also a determination of the subsurface flow pattern (area) and associated hydraulic gradients. Such analyses may also provide estimates of uplift pressures under hydraulic structures.

A physical feel for the concept of permeability may be obtained by considering a simple test for its determination. (See Figure 5.5, for example.) If water is flowing through the device (through the soil) at a constant rate q, we may say that the amount collected in a given period t is

$$Q = qt. \tag{5.11}$$

The apparent velocity of flow is

$$v = \frac{q}{A}. \tag{5.12}$$

Darcy's law for flow through porous media may be stated as

$$v = ki, \tag{5.13}$$

where v is the apparent velocity, k is a constant for the material, the coefficient of permeability, and i is the hydraulic gradient. The latter quantity, defined in terms of Figure 5.5, is

$$i = \frac{H}{L}, \tag{5.14}$$

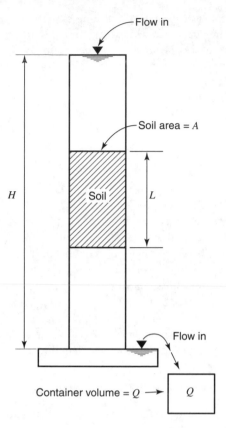

Flow in

Soil area = A

H

Soil

L

Flow in

Container volume = Q → | Q |

Figure 5.5   A simple constant head permeameter. The inlet and outlet are arranged to maintain constant water surface elevations.

where H is the head causing flow over the distance L. Combining Equations (5.11) through (5.14) and solving for k, we have

$$k = \frac{QL}{HAt}. \qquad (5.15)$$

Typical values of permeability for various soils are given in Table 5.1. It is evident that soils vary markedly in their permeability, with gravels being about 1 million times more pervious than clays. In some instances, it is necessary to determine the permeability coefficient k more accurately than is possible using general descriptions such as shown in Table 5.1.

TABLE 5.1   Typical Soil Permeabilities

Soil	Permeability Coefficient, k (cm/sec)	Relative Permeability
Coarse gravel	$\geq 10^{-1}$	High
Sand, clean	$10^{-1} - 10^{-3}$	Medium
Sand, dirty	$10^{-3} - 10^{-5}$	Low
Silt	$10^{-5} - 10^{-7}$	Very low
Clay	$\leq 10^{-7}$	Impervious

5.4.2 Laboratory Tests

Equation (5.15) and the apparatus shown in Figure 5.5 are the direct basis for the constant head permeability test (ASTM D2434), which is used to determine the coefficient of permeability for coarse soils. For soils with values of k less than 10^{-3} cm/sec, the water flows through the apparatus and thus the test is too slow, and so falling head tests are used. Falling head tests are also based on Darcy's law, but use it in an indirect way. Both of these tests are outlined in Appendix A.7. Laboratory permeability tests are generally the least accurate laboratory soil tests, as sample disturbance can affect the results significantly.

5.4.3 Field Tests

Recognizing the difficulty of laboratory sampling such materials and the inherent variability of the deposits from which they come, however, it is considered more reliable to determine these values in the field, when possible. These tests require considerable planning and expense, but offer by far the most reliable information.

The layout of a field-pumping test is shown in Figure 5.6 for the case of a well penetrating an open soil aquifer. It can be shown that the pumping rate from the well is

$$q = \frac{\pi k (H_2^2 - H_1^2)}{\ln\left(\dfrac{R_2}{R_1}\right)}.$$ (5.16)

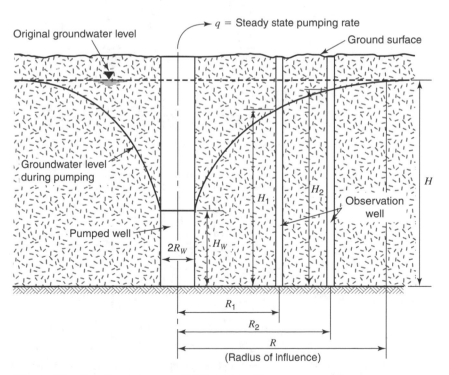

Figure 5.6 Pumping test on a well fully penetrating an open aquifer.

If the aquifer is confined between impervious layers and if H exceeds the aquifer thickness, Equation (5.16) does not apply. By solving Equation (5.16) for k, we obtain

$$k = \frac{q}{\pi(H_2^2 - H_1^2)} \ln\left(\frac{R_2}{R_1}\right).$$

(5.17)

Thus, by measuring the pumping rate q, and knowing the groundwater depth in two observation wells, we can establish a representative value of k for field conditions. Equation (5.17) may be used to estimate k if no observation wells are available and if it is modified to the form

$$k = \frac{q}{\pi(H^2 - H_w^2)} \ln\left(\frac{R_2}{R_w}\right).$$

(5.18)

This approach is approximate, in that the radius of influence must be estimated and the water depth in the well is not a true point on the draw-down curve.

Some representative values of k from pumping tests are shown in Figure 5.7. For example, from Figure 5.7, for $D_{10} = 0.3$ mm, $k = 0.2$ cm/sec. It is important to note the magnitude

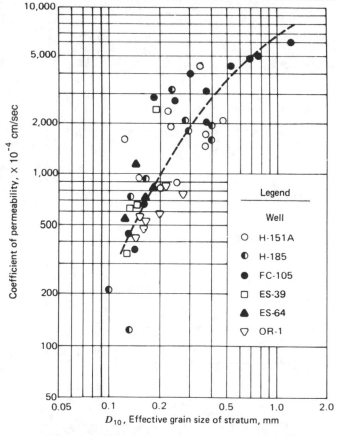

Figure 5.7   Coefficient of permeability of sands determined from pumping tests. (Courtesy of the Corps of Engineers, U.S. Army.)

of differences in data point values where the effective sizes of the soil strata are the same. The coefficient of permeability may vary five times or more for a given grain size. This simply illustrates that site-specific pump tests always provide the best information.

5.4.4 Permeability in Fine Soils

The discussion of permeability here has concerned mostly clean sand and gravelly soils, since these materials are those that are usually amenable to dewatering on construction projects. It is important to note that the permeability of coarse soils is influenced dramatically by gradation and fines content. For instance, well-graded gravelly sand may have a coefficient of permeability ten times that of uniform fine sand with the same effective particle size. Thus, the use of general correlations, such as that shown in Figure 5.7, involves considerable risk of error. In addition, the permeability of coarse soils is very much dependent on their fines content. The presence of as little as five percent fines in a coarse soil can reduce permeability as much as three orders of magnitude from that of the same soil with no fines. Fifteen percent fines reduce permeability three to five orders of magnitude. Additional discussion of permeability determination is provided in Chapter 9 on dewatering.

5.5 COMPRESSIBILITY

Soils subjected to increased effective stress decrease in volume. The stress increase may come about from structure foundations, embankments, or even lowering of the groundwater table. The decrease in volume results in surface settlement. If the soil supports a structure and the settlement is large enough, damage may result. Some soils are very compressible; others are not. If the soil compressibility and loads to be imposed are known, these settlements may be estimated. Good estimates can aid in construction planning where settlements created by construction activities are potentially damaging. This possibility arises where long-term heavy surface loads are applied, both in and around the area loaded by the structure and where groundwater lowering is required.

5.5.1 Consolidation Settlement in Soils

Consider Figure 5.8. The actual soil stratum is represented by column a. It consists of a solid and a liquid phase. This stratum is equivalent to stratum c, wherein the phases have been separated. As the load Δp is applied, the stratum compresses an amount Δh. Since the solid phase is incompressible, the fluid must be expelled from the voids for the settlement Δh to occur. This is indicated by stratum d. If we remember that the void ratio is

$$e = \frac{V_v}{V_s},$$ (5.19)

then, before the load is applied,

$$e_1 = \frac{h_{v1}}{h_s}$$

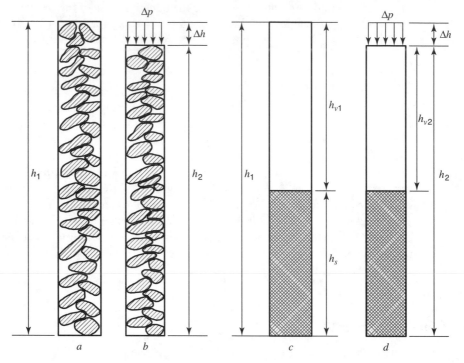

Figure 5.8 Idealization of loaded, compressible soil stratum.

if it is assumed that the area of the column is unity. After settlement is complete,

$$e_2 = \frac{h_{v2}}{h_s}$$

and

$$\Delta e = e_1 - e_2 = \frac{h_{v1}}{h_s} - \frac{h_{v2}}{h_s} = \frac{h_{v1} - h_{v2}}{h_s}.$$

Setting

$$\Delta h = h_{v1} - h_{v2},$$

and then combining and rearranging, we have

$$\frac{\Delta h}{h_1} = \frac{\Delta e h_s}{h_s + h_{v1}} = \frac{\Delta e h_s}{h_s + e_1 h_s} = \frac{\Delta e}{1 + e_1}.$$

Therefore,

$$\Delta h = \frac{h_1 \Delta e}{1 + e_1} = h_1 \frac{e_1 - e_2}{e_1 + 1}. \tag{5.20}$$

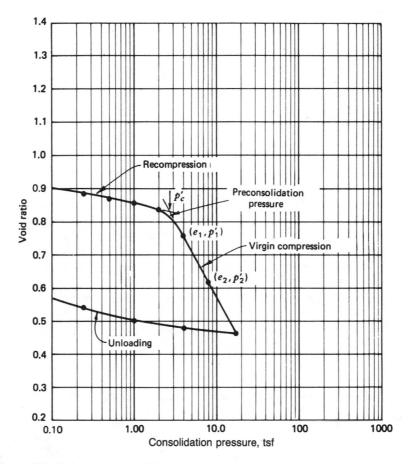

Figure 5.9 Laboratory consolidation test results.

Equation (5.20) states mathematically that, given the original height and void ratio of a soil stratum and the change in void ratio that might result from a load, we could predict surface settlement. The height in a field situation may be determined by subsurface exploration, and the void ratio may be obtained from tests on samples obtained in the process. The change in void ratio is related to the applied load and is determined for a range of loads in a laboratory consolidation test (ASTM D2435; see Appendix A.11). In the tests, a sample, usually about 1 in. thick and 1 1/2 in. in diameter, is compressed between permeable porous discs. As load is applied, deformation dial readings are taken to record the change in thickness of the sample as water flows from it, through the porous discs. Typical results are shown in Figure 5.9. The slope of the virgin compression curve in Figure 5.9 is

$$C_c = \frac{e_1 - e_2}{\log_{10} p_2' - \log_{10} p_1'} = \frac{e_1 - e_2}{\log_{10}\left(\dfrac{p_2'}{p_1'}\right)}.$$

The slope of the recompression curve can be computed in a similar manner. Substituting into Equation (5.20), we get

$$\Delta h = \frac{h_1 C_c}{1 + e_1} \log_{10}\left(\frac{p_2'}{p_1'}\right). \tag{5.21}$$

This relationship is used to predict settlement. The change in effective stress on a soil element, the stratum height, and C_c (the compression index) must be known. The compression index may be considered as a measure of soil compressibility. Most clays in nature are precompressed; that is, they have been loaded in the past to some greater effective stress than they presently carry. For this reason, when they are reloaded, such loading takes place along the recompression curve shown in Figure 5.9. These clays and also most sands exhibit comparatively low compressibility. Soils that have not been precompressed are much more compressible. Subjected to a change in load, these materials compress along the steeper virgin compression curve. To determine whether a soil has been precompressed, the preconsolidation pressure p_c' (see Figure 5.9) is compared to the vertical effective pressure p_o' at the depths from which the sample was taken. The preconsolidation pressure may be estimated by intersecting tangents to the recompression and virgin compression segments of the pressure-void ratio curve. Other methods are available. If p_c' and p_o' are approximately equal, the soil has not been precompressed and is said to be *normally consolidated*. For the case where p_c' exceeds p_o', the soil has been precompressed and is said to be *overconsolidated*.

5.5.2 Time Rate of Consolidation

It has been noted that compression or consolidation involves expulsion of water from soil voids. It is logical, therefore, that the process should be time dependent. The process is not instantaneous. The rate at which consolidation occurs is governed by soil permeability and the efficiency of subsurface drainage provided by the more pervious soil layers. Sands generally are so pervious that any consolidation that occurs does so very rapidly upon application of load. For clays, the consolidation process may take many months or years to complete. Time rate of settlement estimates may be made from information determined in each load increment (each data point on the curve in Figure 5.9) for the consolidation tests mentioned above. Two ways of presenting the same laboratory data for a load increment are shown in Figures 5.10 and 5.11. In both methods the time scale is compressed.

A constant c_v, the coefficient of consolidation, controls the rate of compression of the soil. It is directly proportional to the coefficient of permeability and is defined as

$$c_v = \frac{T_u H^2}{t_u}, \tag{5.22}$$

where

H = the longest path taken by water seeping from the soils as a result of application of the consolidating pressure increment Δp. *For one-way drainage, it represents the entire thickness of the consolidating layer. For two-way drainage (i.e., with pervious layers above and below the consolidating layer), it represents half of the layer thickness.*

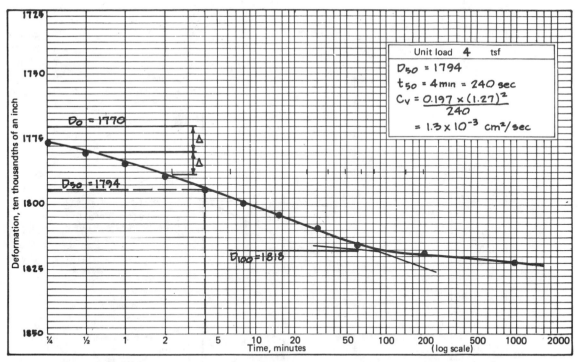

Figure 5.10   Laboratory consolidation test data. Deformation versus logarithm of time.

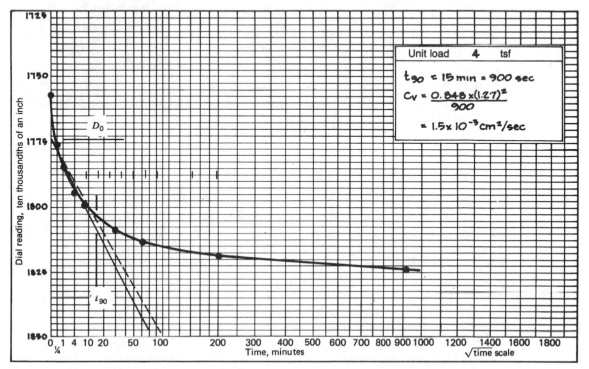

Figure 5.11   Laboratory consolidation test data. Deformation versus square root of time.

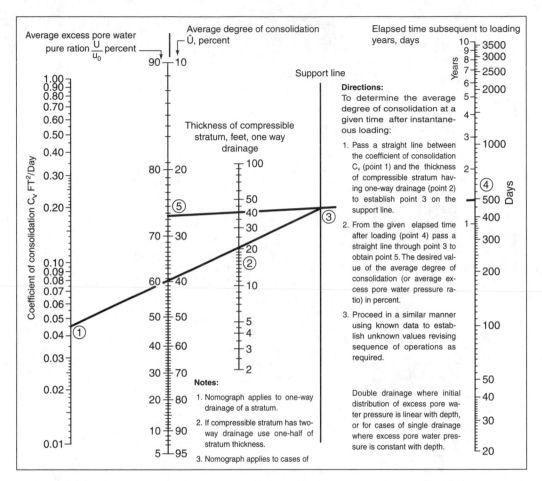

Figure 5.12 Nomograph for determining consolidation with vertical drainage (from NAVFAC DM 7.01)

t_u = time for a given percentage of consolidation to occur

T_u = a theoretical constant (see Table 5.2)

The value of c_v for a given soil may be determined from the relations for that soil represented by either Figure 5.10 or Figure 5.11 and Equation (5.22). The test data are used to determine the time to a given degree of consolidation. In addition to using Equation (5.22) and Table 5.2, the nomograph in Figure 5.13 can be used to determine the degree of consolidation.

In Figure 5.10, the semilogarithmic plot of time versus the dial readings during a load increment of the test shows a typical shape. The deformation dial reading D_{100}, at which primary consolidation is complete (degree of consolidation equals 100 percent) is found by intersecting tangents to the lower segments of the curve. Because the horizontal scale of the plot is logarithmic, there is no zero on the scale at which to determine D_0. It may be located, however, by determining the vertical distance between two points on the initial part of the curve whose times

TABLE 5.2   Theoretical Time Factors for Consolidation Rate Analysis, One W Initial Distribution of Pore Water Pressure

Degree of Consolidation, $u\ (\%)$	T_u
0	0
10	0.008
20	0.031
30	0.071
40	0.126
50	0.197
60	0.287
70	0.403
80	0.567
90	0.848
100	∞

Figure 5.13   Triaxial (cylindrical compression) shear test in progress.

are in the ratio of four to one (Δ as shown), and then extending a horizontal line to the vertical scale, the same distance above the earlier time. As shown in Figure 5.10, the points selected were 1/2 and 2 minutes. With D_0 and D_{100} established, the deformation dial reading at 50 percent consolidation is located midway between them, and the corresponding time t_{50} determined. The value of c_v is computed from Equation (5.22), with H equal to the half-sample thickness if the test sample can drain towards both faces, as is common.

An alternative procedure that employs the same deformation data plotted against the square root of time as shown in Figure 5.11 requires location of the time for 90-percent consolidation. A straight line is drawn through the longest segment of the early part of the curve. A second straight line (the dashed line) is then drawn so that it extends 1.15 times the distance of the first line from the vertical axis. It intersects the curve at t_{90}. Again Equation (5.22) and Table 5.2 are employed to compute c_v. The results should not be greatly different by this method from those obtained by the previous procedure, but exact agreement would not be expected because of the individual judgments that must be made in each.

The values of c_v obtained from laboratory tests are used in time-settlement analyses. There are many limitations on the accuracy of time-settlement analyses, and they are generally considered as estimates only. Predictions of the rate of surface settlement in a given situation may be made, but usually, when it is important to know, field measurements will be made to evaluate these predictions.

5.6   SOIL STRENGTH

The strength of soil is a variable and elusive property. Strength characteristics of some materials are represented by concepts such as yield point or tensile strength. A given grade of steel is an example. Other materials, such as concrete, are considered in a similar fashion. Concrete strength, though, is very much subject to the quality of its components, their proportions, and the workmanship that goes into its manufacture. The end product is more variable for concrete than it is for steel, even though the strength of both materials is expressed as a simple single number, compressive or tensile strength.

Except for compacted fills, we deal with soils as they are in nature. Because they are not manufactured products, variations in properties are the rule, not the exception. They are natural materials, and in most cases we use what is available, adjusting our designs and working methods to accommodate conditions. Soil strength is not described in a simple single value of tensile strength or compressive strength. There are many kinds of strength tests, some of which apply only to a limited number of soils. In the general case, soil strength may be determined in tests that permit the control of stresses on the test specimen. The triaxial shear test shown in Figure 5.13 is an example.

In this case, the soil specimen is enclosed in an impervious membrane to separate it from the fluid in the plastic cell that surrounds it. Drainage is controlled by access tubes to the base of the sample. Stresses on the sample are determined by the pressure of the fluid in the cell, which can be set at any value, and by a loading piston that extends through a bushing in the top of the cell. In the usual test, the pressure in the cell fluid is held constant, and the vertical pressure on the sample is increased by loading the piston at the top.

The objective of any soil strength test is to determine the shearing strength of the soil for a given set of conditions. When a soil is loaded to failure in a compression test, as illustrated in Figure 5.14, the stresses at failure are plotted on a Mohr diagram for interpretation. The corre-

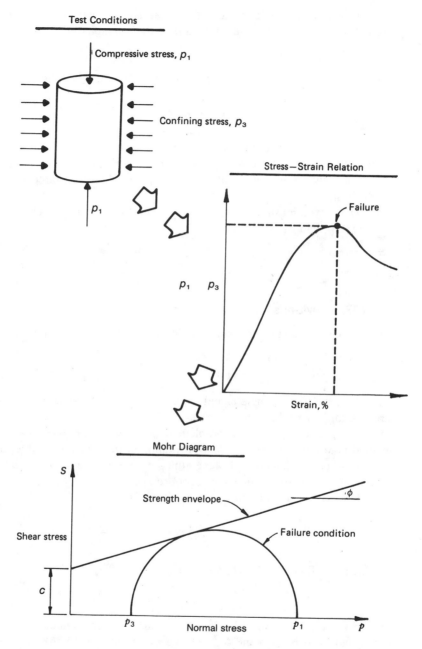

Figure 5.14   Development of soil strength parameters from triaxial test data.

sponding Mohr circle is plotted and the strength envelope located. In Figure 5.14 only one Mohr circle is shown, though two or more circles and, therefore, two or more tests, are needed to establish the position of the strength envelope. The strength parameters that describe the location of this line are its slope angle ϕ and the intercept on the shear stress axis c. The angle ϕ

is designated as the angle of internal friction, and c is designated as cohesion. The shearing strength s at any pressure is determined from the equation of the envelope:

$$s = c + p \tan \phi. \tag{5.23}$$

For a given confining pressure p_3, the failure stress p_1 can be found by the equation

$$p_1 = p_3 \tan^2\left(45° + \frac{\phi}{2}\right) + 2c \tan\left(45° + \frac{\phi}{2}\right). \tag{5.24}$$

For two specimens, we would have two sets of p_1 and p_3, and thus could solve simultaneously for ϕ and c. More pairs of specimens would allow us to compute more values of ϕ and c and thus increase our confidence in the data; however, three specimens are customary for this test.

The maximum shear stress is found at the top of Mohr's circle, which takes place at the average of p_1 and p_3. This stress is

$$s_{max} = \frac{p_1 - p_3}{2}. \tag{5.25}$$

EXAMPLE   Mohr's Circle

Two triaxial tests on the same soil sample have the following results:

1. $p_1 = 295$ kPa, $p_3 = 20$ kPa
2. $p_1 = 329$ kPa, $p_3 = 40$ kPa

Find the friction angle ϕ and the cohesion c for the specimen. Also, compute the shear strength for both of these cases.

It is possible to solve this problem graphically using Figure 5.14; however, an alternative method is to use Equation (5.24). Substitution of the two values of p_1 and p_3 will yield two equations in two unknowns. However, the presence of both the squared terms and trigonometric functions will complicate solving for ϕ and c.

One way to simplify this solution is to define

$$N_\phi = \tan\left(45° + \frac{\phi}{2}\right)^2, \tag{5.26}$$

in which case Equation (5.24) reduces to

$$p_1 = p_3 N_\phi + 2c\sqrt{N_\phi}. \tag{5.27}$$

Substituting the variables into Equation (5.27) and simultaneously solving these equations yields $c = 100$ kPa and $N_\phi = 1.7$. Solving Equation (5.26) for ϕ yields $\phi = 15°$.

These equations can be used for two pairs of results at a time. For three triaxial tests on a sample, three pairs of results should yield three very similar values of c and ϕ. Variations with this can indicate either irregularities with the test or a curved failure envelope.

Concerning the maximum shear stress, using Equation (5.25), the maximum shear stress for specimen (1) is $(295 - 20)/2 = 138$ kPa and specimen (2) is $(329 - 40))/2 = 145$ kPa. It can be seen, therefore, that as the confining stress increases, the shear strength of the soil does also. Figure 5.3 shows that, as the depth increases, the effective stress—which is the main source of confining stress in soils in the earth—increases with depth, and with it the shear strength for soils where $\phi > 0$.

5.6.1 Strength of Sands and Gravels

Uncemented sands and gravels are considered to be cohesionless materials. Their strength envelope passes through the origin of the Mohr diagram; hence, $c = 0$, and Equation (5.24) reduces to

$$p_1 = p_3 \tan^2\left(45° + \frac{\phi}{2}\right).\tag{5.28}$$

If the sand is unconfined (i.e., $p_3 = 0$), then any value of $p_1 = 0$, and from Equation (5.23), $s = 0$. From our experience, we know this is true. Dry or saturated sand on the ground surface has no strength whatever. At some depth, where the material is confined by pressure due to the soil surrounding it, it can support considerable load.

   Moist sand (partially saturated) has some cohesive strength or, more correctly, apparent cohesion. This strength is imparted by meniscus forces arising from surface tension (see Section 5.3) in the pore water. Figure 5.15 is a hypothetical section through moist sand showing the three constituent phases—air, water, and solids. The air–water interface is curved. This curvature and the surface tension on the interface produce intergranular forces between individual particles. The soil, therefore, has strength since the intergranular normal forces produce frictional resistance. When the soil is dry or saturated the air-water interface is not present, and the apparent cohesion is zero.

   The water in a partially saturated mass is in tension. It has been shown that this produces an apparent strength. The water pressure may also be positive. This reduces strength since it

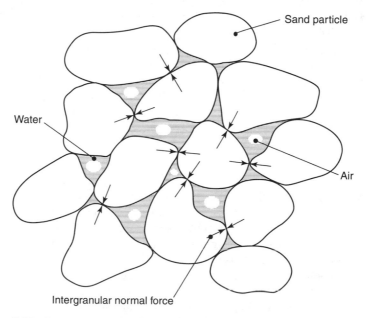

Figure 5.15 Source of strength for moist sands.

reduces intergranular normal forces or effective stresses. When a very loose sand is shaken it tends to densify. If the voids are filled with water, the water must be expelled before densification occurs. When very loose sands are disturbed, the pore water is therefore forced into compression and a positive pressure develops as it attempts to escape. Since the pressure in the fluid increases, the pressure between grains decreases and soil strength is reduced. When the reduction is sufficiently large, the soil has no strength at all and is said to have liquefied. This phenomenon is known as *liquefaction* and it is that condition that exists when a sand or silt is said to be "quick." The same condition can be produced by upward flow of water through a sand bed. This condition can occur in dewatering operations; it is discussed in Chapter 9.

The strength of a cohesionless soil depends on density or unit weight, grain size, shape, mineralogy, fines content, and a number of other factors. Primarily, it is controlled by density, which determines both the arrangement of particles and pressure at depth (see Section 5.2). Strength arises from intergranular friction and interlocking controlled by these factors. Variations in strength are indicated by variations in the angle of internal friction. It may vary from 28° to about 45° for a range of conditions from very loose to very dense. Equation (12.8) provides a general relationship between angle of internal friction and relative density (see Section 8.2). Table 7.1 correlates angle of internal friction and standard penetration resistance. Generally, cohesionless soils are the most desirable of earth materials. In most cases there is little concern for their strength capabilities.

5.6.2   Strength of Clays

Clay soils are cohesive in that they possess some strength at zero normal pressure. Consequently, their strength envelope looks like that shown in Figure 5.14. The slope of the strength envelope depends greatly on drainage conditions as the clay is being loaded. For most short-term loading conditions, the usual situation during construction, the slope of the strength envelope is near zero if the clay is saturated. The strength of the clay is, therefore, expressed by the cohesion term c. For the case in which $\phi = 0$, Equation (5.24) reduces to

$$p_1 = p_3 + 2c, \tag{5.29}$$

and so the pressure difference $p_1 - p_3$ at failure is numerically equal to $2c$ and is most conveniently determined in the unconfined compression test in which $p_3 = $ zero (see Figure 5.14). In this case, we can solve Equation (5.29) for the cohesion c:

$$c = \frac{p_1}{2} = \frac{q_u}{2}. \tag{5.30}$$

Unconfined compression test (ASTM D2166; see Appendix A.9) results are the usual measure of clay strength. A number of devices are on the market which expedite the testing process, and which can be used in the field. These are discussed in more detail in Chapter 7.

For a given clay, unconfined strength or cohesive strength will depend strongly on the water content and its relationship to the plastic and liquid limit. Table 7.2 indicates typical ranges of strength corresponding to standard nomenclature for soil consistency. Typical unconfined compression test data are shown in Figure 5.16.

Soils of stiff to hard consistency seldom present problems when encountered in construction. Very soft to medium clays are very troublesome when they must be supported or excavated and often are unsatisfactory materials for temporary foundations in shoring systems.

UNCONFINED COMPRESSION TEST DATA

TEST FOR _Northwest Cement_

BY____ _BJM_

DATE ____ _8/13/92_

SAMPLE IDENTIFICATION ____ _Light Brown Silty Clay (CL)_

PROVING RING NO. AND CONSTANT ____ _542J - 0.34 lb./0.0001 in._

SPECIMEN WEIGHT _140.89g_ SPECIMEN HEIGHT _2.80 in._

SPECIMEN DIAMETER

    TOP ____

    CENTER ____ _1.40 in._

    BOTTOM ____

    AVERAGE ____ _1.40 in._

AVERAGE SPECIMEN AREA, A_0 ____ _1.539 in.²_

AVERAGE LOADING RATE ____

SPECIMEN PROTECTION ____ _Specimen sheared with no moisture protection_

STRAIN DIAL (IN.)	LOAD DIAL (IN.×10⁻⁴)	LOAD, P (LB.)	AXIAL STRAIN, ϵ	AREA, A $A_0/(1-\epsilon)$ (IN.²)	P/A (LB/IN.²)	WATER CONTENT DETERMINATION
0	0	0	0	1.539	0	
0.02	30	10.20	.0071	1.550	6.58	Can no. ___ 43
0.04	86	29.24	.0143	1.561	18.73	Wt. wet _10.58 g._
0.06	127	43.18	.0214	1.572	27.46	Wt. dry _9.33 g._
0.08	152	51.68	.0286	1.584	32.62	Wt. water _1.25 g._
0.10	167	56.78	.0357	1.595	35.59	Tare _3.57 g._
0.12	177	60.18	.0428	1.607	37.44	Wt. solids _5.76 g._
0.14	186	63.24	.0500	1.620	39.03	w% _21.7%_
0.16	191.5	65.11	.0571	1.632	39.89	
0.18	197	66.98	.0643	1.644	40.74	Sketch of Failure
0.20	202	68.68	.0714	1.657	41.44	
0.22	205	69.70	.0786	1.670	41.73	
0.24	211	71.74	.0857	1.683	42.62	
0.26	215	73.10	.0929	1.696	43.10	
0.28	218	74.12	.1000	1.710	43.34	
0.30	220	74.80	.1071	1.723	43.41	
0.32	222	75.48	.1143	1.737	43.45	

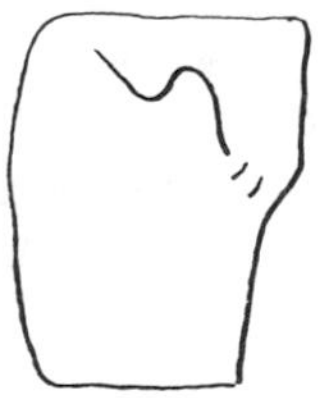

REMARKS:

     Maximum axial stress = 6390 psf = q_u

Figure 5.16 Unconfined compression test data.

5.7   SUMMARY

Soil engineering properties are of concern to contractors only when needed for analysis of their operations. Analysis of problems that involve selection of soil parameters is best done by someone thoroughly familiar with soil behavior and the purpose of the proposed work. Often such an individual is available within the contractor's organization and will usually be a civil engineer. When no person so qualified is available, the contractor should seek outside help. Some classes of problems where engineering properties are important considerations are discussed in later chapters.

The reader should now have obtained an introductory-level understanding of the fundamental engineering properties of soil, permeability, compressibility, and strength. Specifically, the reader should be able to

1. Make static stress calculations, including capillary effects.

2. Define the terms *permeability, compressibility*, and *strength* used in connection with analysis of soil behavior.

3. Describe ways in which soil engineering properties are determined.

4. Compute engineering properties of soils from standard test results.

5. Explain how index properties and engineering properties of soils are related.

REFERENCES

American Society for Testing and Materials, *1988 Annual Book of ASTM Standards*, Vol. 04.08, Soil and Rock, Building Stones; Geotextiles, Philadelphia.

Andersland, O.B. and Anderson, D.M., *Geotechnical Engineering for Cold Regions*. New York: McGraw-Hill, 1978.

Cheney, R.S., and Chassie, R.G., *Soils and Foundations Workshop Manual*. Washington, DC: U.S. Department of Transportation, FHWA, 1982.

Department of the Navy, Naval Facilities Engineering Command, "Soil Mechanics", *Design Manual 7.1*, Alexandria, VA, 1982.

Peck, R.B., Hanson, W.E., and Thornburn, T.H., *Foundation Engineering*. New York: Wiley, 1973.

Terzaghi, K., and Peck, R.B., *Soil Mechanics in Engineering Practice*. New York: Wiley, 1967.

U.S. Army Corps of Engineers, *Investigation of Underseepage and Its Control, Lower Mississippi River Levees*, Waterways Experiment Station TM 3-424, Vicksburg, MS: 1956.

PROBLEMS

1. The sequence of soils at a site is shown in Figure 5.17. Calculate the total pressure, water pressure, and effective pressure on horizontal planes at each change of soil type, and at the groundwater table. Show your results on a sketch, indicating the variations of each of these pressures with depth.

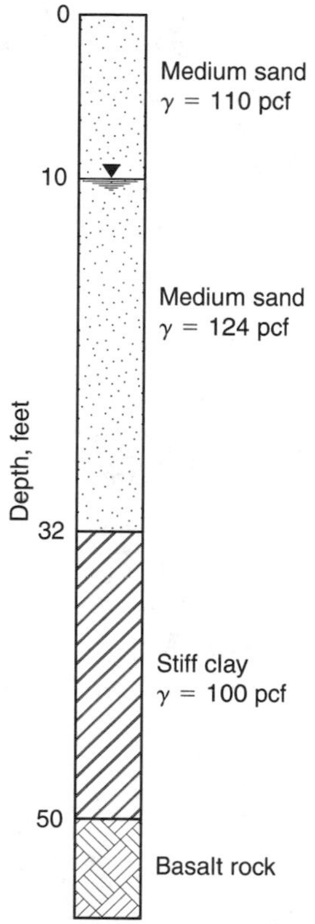

Figure 5.17 Soil profile for Problem 1.

2. Solve Problem 1 if the sands overlying the clays were silts with the same densities. State any assumptions you make.

3. Estimate the height of capillary rise in a clay soil. State your assumptions.

4. A series of three experiments was run on a single sample using apparatus like that shown in Figure 5.2. Q was 60 cm^3, L was 12 cm, and A was 25 cm^2. Paired values of H and t are in the following table:

H (cm)	t (sec)
20	140
40	73
60	50

Demonstrate the validity of Darcy's law (Equation 5.3) in this instance. Estimate the coefficient of permeability of the sample.

5. A 6-in.-diameter well that fully penetrates a 40-ft-thick open aquifer is pumped at a rate of 100 gpm. The drawdown in the well is 22 ft when a steady water level is established. The drawdown in a well 152 ft distant is 12 ft. Estimate the coefficient of permeability of the soil deposit.

6. Discuss the validity of the answer obtained in Problem 5. Suggest how you would improve the test data obtained to produce a better value of k. Why?

7. Data from a one-dimensional consolidation test are presented in the following table:

Consolidating Pressure (tsf)	Void Ratio
1/8	1.87
1/4	1.86
1/2	1.84
1	1.81
2	1.76
4	1.66
8	1.51
16	1.32
4	1.36
1	1.44
1/4	1.49

Plot the void ratio–pressure relationship and estimate the compression index for both the virgin compression curve and the recompression curve.

8. Data from a single load increment during a consolidation test on a 3/4-in.-thick sample with drainage on both ends follow:

Elapsed Time (minutes)	1/4	1	4	8	16	30	60	120	240	480	1200
Deformation Dial (0.0001 in.)	1600	1582	1544	1510	1465	1410	1330	1269	1232	1210	1200

Determine the coefficient of consolidation using both the logarithm of time and square root of time methods.

9. Data from unconfined compression tests on a sample (one test on the undisturbed sample, one on the same sample remolded) are shown in the following table:

Axial Strain (%)	Undisturbed Axial Stress (psi)	Remolded Axial Stress (psi)
0	0	0
2	6.1	1.4
4	6.8	2.5
6	6.9	3.2
8	6.9	3.7
10	6.8	4.2
12	6.7	4.5
14	6.6	4.7
16	6.5	4.8
18	—	5.0
20	—	5.1

Plot the axial stress versus strain curves. Estimate the unconfined compressive strength and soil cohesion in each case, in tons per square foot. What is the effect of remolding? Why?

10. The major principal stress at failure for a triaxial compression test on dense sand was 5.15 tsf. The minor principal stress, or lateral pressure, was 1.45 tsf. Plot the Mohr circle at failure and, assuming the sand to be cohesionless, determine the angle of internal friction. Check your results using Equation (5.24).

11. Calculate the unconfined compressive strength of a moist, fine sand. Carefully state any assumptions you might make. Illustrate your calculations with sketches if appropriate.

PART TWO

Earthwork in the Construction Contract

CHAPTER 6

The Contract and Contract Documents

In any construction work under contract, the agreement between the owner and the contractor regarding the work to be done is contained in the contract documents. These documents also include certain other information that establishes the basis for the agreement and aids in the administration of the contract. The documents are prepared by the owner's representative, who is usually an engineer or architect, unless the owner serves in this capacity. The responsibilities of all parties are explicitly put forth. Since any contract is practically limited in format and scope only by the imagination of the parties involved, no discussion of these matters is ever all encompassing. The sections that follow do, however, contain information related to standard features of most contract documents. The intent of this chapter is to call attention to those features specifically treating provisions affected by soil conditions. Figure 6.1 shows that the number of such items may be large.

The chapter includes

1. A discussion of the relationships and responsibilities among parties to contracts.

2. Enumeration of contents of typical construction contract documents.

3. An example special specification for execution of a particular portion of the earthwork in a project.

4. A section on settling disputes.

5. Coverage of the "changed condition" situation where soil conditions encountered in construction are different from those anticipated.

Figure 6.1 A major construction project involving several operations related to soil behavior. These include excavation and backfilling, dewatering, excavation supports, shallow foundation and pile foundation construction, and soil stabilization.

6.1 PARTIES TO THE CONTRACT

The administration of a construction contract principally involves three parties:

- The owner, who needs the work done.
- The contractor, who, executes the work according to the plans and specifications.
- The engineer, who prepared the plans and specifications of the work.

The engineer is usually best suited to see that the owner's interests are served during construction and, therefore, is present on the job during construction as the owner's representative. The contract, however, is between the owner and the contractor.

6.1.1 Relationships of the Parties

The relationships among contracting parties in typical, competitively bid private work are shown in Figure 6.2. The three parties' responsibilities are divided according to their benefits. The owner needs the work done and, therefore, must pay the contractor to do it and pay the engineer to assist with planning and construction. The owner's staff will generally oversee the progress of the work of both the contractor and the engineer. Some owners become closely involved in the work while others prefer the engineer to administer their interests in all phases of the project.

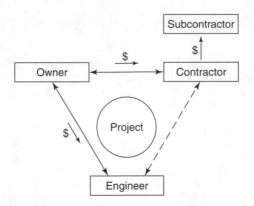

Figure 6.2   Contractual relationships in private contracts.

The contractor is obligated to the owner, but deals mostly through the engineer. The contractor employs subcontractors to assist with specialty work and is responsible for their work as if they were part of his or her own forces. Subcontractors, in principle, do not deal directly with the owner or the engineer. Their contract, if they have one, is directly with the contractor.

The engineer is not a party to the construction contract. He or she has no financial stake in its outcome, but, as the designer is in the best position to administer the contract in a situation where the owner wants to minimize expense and the contractor wants to maximize profit. The principal disadvantage of the relationships shown in Figure 6.2 lies with the fact that the contractor is sometimes asked to deal with two parties rather than one. The owner and engineer generally should speak with one voice. Unfortunately, this is not always the case.

Figure 6.3 illustrates the contractual relationships in which the engineer and owner are one and the same. Such situations are common when the owner maintains a large engineering staff. For very large organizations, where design and administrative functions may be rigidly compartmentalized, the relationships may be more nearly represented by Figure 6.2. Technical (engineering) and administrative (ownership) groups may be virtually independent. Government projects are a good example to which Figure 6.3 applies. The arrangement in Figure 6.3 offers the possible advantage of being streamlined administratively. But, in this instance, the engineer's interests are more closely tied to those of the owner than in the previous case. In the three-party relationship the contractor can appeal to the owner in the event of a dispute with the engineer. In the two-party relationship there is no one else to talk to.

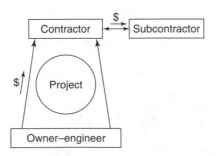

Figure 6.3   Contractual relationships with an owner–engineer.

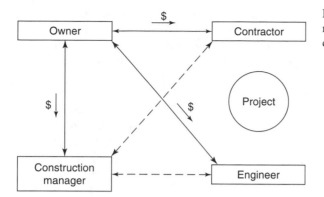

Figure 6.4 Contractual relationships established by the construction management method.

6.1.2 Construction Management

Construction work of some types is being increasingly handled through the construction management process. The owner's representative in this situation is the construction manager. Frequently, the construction manager may be the contractor, although that is not a requirement. He or she may also be the engineer, in which case the relationships are essentially as shown in Figure 6.3. Whoever serves as construction manager must be knowledgeable enough of both the design and construction functions to offer the owner the advantage of this method of doing business. If we presume an independent construction manager, the relationships among the parties are as shown in Figure 6.4. Proponents of this method will argue that it speeds construction because the manager may initiate some phases of work before completion of design, since the contractor will have been selected. Because the construction expertise of the contractor is available, his or her knowledge may also have an economizing effect on design.

6.2 DOCUMENT CONTENTS

In arranging for the selection of a contractor under the usual method (see Figure 6.2), the owner for whom the proposed work is to be done will advertise his or her intention to accept bids. This advertisement will contain a very general description of the work and direct prospective bidders to obtain further information regarding the project from the designated source, usually the engineer. Prospective bidders will be provided with a set of contract documents upon application. These documents specify the work to be done and administrative procedures and arrangements among the parties involved, and include also the contract agreement and bidding forms and requirements. The agreement and work will be governed entirely by this document. The usual essential features of the documents are included in the following sections:

- Advertisement for bids
- Proposal (bid form)
- Contract
- Information for bidders
- General conditions
- Special specifications
- Drawings

The work to be accomplished is detailed in the drawings and further clarified by the special specifications. The quantities and prices agreed on are listed in the contractor's completed proposal. The general conditions outline the procedures by means of which the project will be administered and define the duties and responsibilities of the parties concerned. Also contained are bonding and insurance requirements for the contractor. In many cases, the special specifications and drawings will include information that may be available concerning site subsurface conditions. Often, however, this information will not be included in the contract documents. It will be referenced therein as available in the engineer's office for inspection. This type of arrangement is intended to preclude the possibility that subsurface conditions anticipated by the designer become part of the contract by inference. Since subsurface details are never precisely known in advance of construction, considering them part of the contract can lead to disputes.

6.3 BID PREPARATION

Having read the advertisement for bids and determined that he or she may be interested in bidding, the contractor obtains a set of documents from the designated source. After examining the work to be done, he or she may prepare and submit the proposal (bid) contained therein. In his or her proposal, the contractor lists the prices for which he or she will perform each item of required work.

6.3.1 Examination of Contract Documents

By submitting a proposal, the contractor certifies that he or she understands the documents and will do the work as proposed if selected as the successful bidder. The following excerpt regarding site conditions is from a standard proposal:

> The Bidder further declares that he or she has carefully examined the Contract Documents for the construction of the project; that he or she has personally inspected the site; that he or she has satisfied himself or herself as to the quantities involved, including materials and equipment, and conditions of work involved, including the fact that the description of the quantities of work and materials, as included herein, is brief and is intended only to indicate the general nature of the work and to identify the said quantities with the detailed requirements of the Contract Documents; and that this Proposal is made according to the provisions and under the terms of the Contract Documents, which Documents are hereby made a part of this Proposal.
> The Bidder further agrees that he or she has exercised his or her own judgment regarding the interpretation of subsurface information and has utilized all data that he or she believes pertinent from the Engineer, Owner, and other sources in arriving at his or her conclusions.

By submitting the proposal containing these statements, the contractor has in effect said, among other things, that he or she understands site conditions and their effects on his or her work well enough to submit a valid bid.

6.3.2 Itemization of Costs

The contractor then proceeds to itemize his or her costs. A section of a proposal for a unit-price contract on excavation and embankment is given in Figure 6.5. The proposal of the successful bidder is incorporated in the contract, along with the contract documents. The contract provisions recognize that there may be circumstances under which the proposed costs may change. These circumstances are noted in the contract itself and are fully defined elsewhere in the documents.

ITEM	QUANTITY	UNIT OF MEASURE	UNIT PRICE FIGURES	UNIT PRICE IN WRITING	TOTAL AMOUNT QUANTITY x UNIT PRICE
2A. Earthwork					
Unclassified Excavation					
	79,900	cu.yds.	$ 137	$ One dollar and thirty seven cents	$ 109,463
Removal of Waste Material					
	5,800	cu.yds.	$ 320	$ Three dollars and twenty cents	$ 18,560
Classified Embankment from Unclassified Excavation					
Zone 1	33,000	cu.yds.	$ 168	$ One dollar and sixty eight cents	$ 55,440
Zone 3	16,500	cu.yds.	$ 187	$ One dollar and eighty seven cents	$ 30,855
Classified Embankment Imported					
Zone 1	43,000	cu.yds.	$ 220	$ Two dollars and twenty cents	$ 94,600
Zone 2	37,500	cu.yds.	$ 220	$ Two dollars and twenty cents	$ 82,500
Zone 3	5,500	cu.yds.	$ 220	$ Two dollars and twenty cents	$ 12,100
Zone 4	37,250	cu.yds.	$ 410	$ Four dollars and ten cents	$ 152,725
Zone 5	3,000	cu.yds.	$ 220	$ Two dollars and twenty cents	$ 6,600
				TOTAL	$ 562,843

Figure 6.5 A unit price proposal for excavation and embankment.

Such provisions are a part of all construction contracts, as in the following excerpt, and are not restricted to earthwork alone:

> In consideration of the faithful performance of the work herein embraced, as set forth in these Contract Documents, and in accordance with the direction of the Engineer and to his or her satisfaction to the extent provided in the Contract Documents, the Owner agrees to pay to the Contractor the amount bid as adjusted in accordance with the Proposal as determined by the Contract Documents, or as otherwise herein provided, and based on the said Proposal made by the Contractor, and to make such payments in the manner and at the times provided in the Contract Documents.

6.3.3   Variation of Quantities

Quantities also may vary. This fact is called to the attention of the contractor in the information for bidders section. In some cases, contracts are written so that a quantity change exceeding a specified percentage of the total item is justification for negotiation of new unit-prices. Occasionally, on unit-price contracts, where contractors recognize that the quantity of a work item may have been underestimated, they may unbalance their bids. That is, they will assign a very high unit price to a small quantity and a lower unit price than normal to another quantity, so that the total bid is unaffected. When the work is done, if the underestimated quantity overruns, the high unit price produces a substantial profit. The section on information for bidders usually covers these provisions:

> When the Proposal for the work is to be submitted on a unit price basis, unit price proposals will be accepted on all items of work set forth in the Proposal, except those designated to be paid for as a lump sum. The estimate of quantities of work to be done is tabulated in the Proposal and, although stated with as much accuracy as possible, is approximate only and is assumed solely for the basis of calculation upon which the award of Contract shall be made. Payment to the Contractor will be made on the measurement of the work actually performed by the Contractor as specified in the Contract Documents. The Owner reserves the right to increase or diminish the amount of any class of work as may be deemed necessary, unless otherwise specified in the Special Provisions.

6.4   CONTRACT EXECUTION

When the successful bidder has been determined and the contract executed, the work is undertaken as called for by the documents. The owner will be represented on the project by a resident engineer whose principal function is to ensure that the work complies with specifications.

6.4.1   Contract Specifications

Each general class of work shown on the drawings is covered by a special specification section. Each special specification section is in turn divided into four categories: scope, materials, workmanship, and payment. Their content and purpose in the earthwork section are illustrated by the following paragraphs, excerpted from a contract with reference to the work items in the proposal in Figure 6.5.

2. EARTHWORK

A. Scope.

This section covers the work necessary for the earthwork, complete.

B. Materials.

Unclassified Excavation.   Complete all excavation regardless of the type, character, nature, or condition of the materials encountered. The Contractor shall make his or her own estimate of the kind and extent of the various materials to be excavated in order to accomplish the work.

Classified Embankment.   Material from *Unclassified Excavation* shall be used wherever possible. Additional materials required shall be imported to the project site.

Embankment Zone 1.   Clean granular material uniformly graded from coarse to fine; less than 12 percent finer than No. 200 U.S. standard sieve; maximum size, 3 in. Submit samples to Engineer for approval.

Embankment Zone 2.   Clean granular material uniformly graded from coarse to fine; less than 5 percent finer than No. 200 U.S. standard sieve; less than 10 percent finer than No. 4 U.S. standard sieve; maximum size, 8 in. Submit samples to Engineer for approval.

Embankment Zone 3.   Granular material; less than 30 percent finer than No. 200 U.S. standard sieve. Submit samples to the Engineer for approval.

Embankment Zone 4.   As specified for *Embankment Zone 2*.

Embankment Zone 5.   Sandy loam free from clay, roots, organic matter, and gravel; suitable for topsoil. Submit samples to Engineer for approval.

Waste Material.   Material from *Unclassified Excavation* found unsatisfactory for use in *Classified Embankment* shall be designated *Waste Material*.

C. Workmanship.

General.   The Contractor shall at all times conduct his or her work so as to ensure the least possible obstruction to water-borne traffic. The convenience of water-borne traffic and the protection of persons and property are of prime importance and shall be provided by the Contractor in an adequate and satisfactory manner. If the Contractor blocks the channel, he or she shall remove such obstructions to a width sufficient to allow safe passage, as the water-borne traffic approaches.

The limits of the area to be dredged at the site are shown on the Drawings. The Contractor shall assume full responsibility for the alignment, elevations, and dimensions of the area to be dredged.

Dredging Equipment.   Dredging equipment shall be of adequate capacity to perform the work within the designated time. Dredge may be a suction type, hydraulic type, or clamshell bucket type, or dragline at the Contractor's option.

Tolerance.   Dredge to elevations and cross-sections indicated on Drawings. Do not dredge to depths greater than needed or shown. A tolerance of one foot, plus or minus, from the elevations shown will be allowed. Dredging shall proceed from the top to the bottom of the slope in such a manner that the inclination of the slope does not exceed that indicated. Work laterally on slope to prevent local oversteepening.

Construction of Embankment Area with Dredged Material Conforming to Material Specification.   The dredged material shall be deposited over the embankment area within the limits indicated on the Plans. The embankment shall be constructed to the lines, grades, and cross sections shown on the Plans or established by the Engineer. Failure to conform to the established lines and grades will not be tolerated. Construct all dikes, ditches, culverts, and other related and incidental work necessary to place and drain the embankment as required for a satisfactory job.

If a washout of dredge deposited material should occur, or for any other reason material is deposited outside of the designated area, except those silts and clays that remain in colloidal suspension, this material shall be removed and placed in the designated area at no additional cost to the Owner.

Dredged embankment that is lost prior to completion and acceptance of the work shall be replaced by the Contractor at no additional cost to the Owner. All dredged areas and embankments shall conform to the requirements of the Drawings and Specifications within the tolerances specified.

Grading. Grade the fill area to a smooth surface. Side slopes shall be within 0.5 ft of true grade.

Embankment Zone 1. Place in lifts with maximum compacted thickness of 6 inches. Compact to at least 70 percent relative density according to ASTM D2049.

Embankment Zone 2. Place in uniform lifts. No compaction shall be required. Maximum lift thickness shall be 6 inches.

Embankment Zone 3. Place in uniform lifts. No compaction shall be required. Maximum lift thickness shall be 6 inches.

Embankment Zone 4. Place in accordance with Plans. No compaction shall be required. Place from bottom of slope to water surface.

Embankment Zone 5. Place in uniform lifts. No compaction required. Maximum lift thickness shall be 6 inches.

Waste Material. Remove from project site.

Rate of Excavation and Filling. Excavation from the river bank and placement of embankment materials shall take place at rates that will not impair the stability of the river bank slope. It shall be the responsibility of the Engineer to detect the development of conditions that could lead to instability of the river slope which arise from excessive rates of excavation or filling. The Engineer shall notify the Contractor to stop work or modify procedures as necessary until such conditions as may develop, subside, or are otherwise corrected. This section shall not be interpreted as relieving the Contractor of the responsibility for the consequences of failure to carry out excavation of the slope as specified under Tolerances above.

Compaction Control. The Contractor shall be primarily responsible to assure that the compaction specifications for *Embankment Zone 1* material have been met. Tests and measurements shall be made by the Engineer to verify the compaction obtained.

D. **Payment.**

Unclassified Excavation. Payment for excavation will be based on the unit price per cubic yard stated in the Contractor's Proposal and the number of cubic yards satisfactorily excavated within the authorized limits. No payment will be made for material dredged below the established grade line. Measurement shall be by cross section taken prior to excavation by the Engineer.

Waste Material. Additional payment shall be allowed for removal from the site of *Unclassified Excavation* unsuitable for use in the embankment sections. The additional payment shall be at the unit price stated in the Contractor's Proposal. Measurement shall be made in the haul units at the time the material is removed from the site.

Classified Embankment.   Payment for this item shall be at the unit prices stated in the Contractor's Proposal. Payment shall be based on the number of cubic yards satisfactorily placed within the authorized limits. Measurement shall be by cross section taken of the completed work by the Engineer.

Standby Time.   If a work stoppage or curtailment is necessary under the provisions of Instrumentation or Rate of Excavation and Filling above, payment shall be made to compensate the Contractor for equipment standby time. Payment shall be at the rate of 20 percent of the rate for individual equipment items stated in the current issue of Rental Rates for Equipment without Operators. Payment shall be made only for those items of equipment in use at the time of stoppage or curtailment and which remain on the project site and are unused during the period of delay.

In this specification, the requirements for dredge excavation and subsequent site filling are set forth. When reading this section, the contractor learns first that this section covers all specifications for the earthwork noted in his or her proposal. The material in the excavation is then identified. The embankment material requirements are given. The contractor is given certain restrictions on how he or she may proceed (or not proceed) with the work. In this regard, it should be noted that well-written specifications would not be prepared so as to assume for the owner the responsibility for the proper conduct of the contractor's work. Finally, a method of payment is specified. Thus, by studying the special specifications, the contractor may determine the technical requirements and payment provisions for each item in the proposal.

6.4.2 Disputes

A contract completed without some dispute is rare. In contracts involving soil materials, disputes frequently arise over the effect of site conditions on the progress of the work. In preparing a bid, the contractor has had to consider site conditions. What consideration he or she has actually given may range from almost none, to very detailed, careful study. In any case, when bad site conditions substantially affect the contractor's profit picture, a dispute over what site conditions were represented to be will almost certainly arise.

For whatever reason they arise, disputes are most frequently settled by mutual agreement. Long arguments, voluminous paperwork, and a certain amount of bluffing usually lead one of the parties to the conclusion that his or her position is untenable or that a compromise may be worked out. Settling a dispute in this manner frequently produces a change order (formal change in the contract) agreed to by both parties. For instance, when rock excavation is required, but had not been anticipated, and when it can be shown that a reasonable contractor should not have expected it, a method of payment for the work will have to be agreed on, and a change order may be issued. The settling of disputes by agreement in this manner offers the advantage of timeliness, in that it is done while the work is in progress.

Disputes not settled by mutual agreement may be settled by mediation, arbitration, or litigation. In each instance, disinterested third parties help to decide, or actually rule on, the issues. Litigation is especially costly and apt to take years to conclude. It is, therefore, appropriate only when all other methods have failed. Recently, mediation and arbitration have become popular methods of resolving construction disputes. They offer certain advantages over litigation, in addition to progressing more rapidly and being generally less costly.

Arbitration is a procedure in which the parties in the dispute select a third party to whom arguments will be presented, and who will decide the issue. Proceedings are conducted similarly to courtroom litigation, except that there is no jury, and the rules of evidence are less formal. The arbitrator often questions witnesses, along with counsel for both sides. The arbitrator's decision is usually binding.

Mediation is an even less formal procedure in which the disputing parties agree on a third party who will examine opposing views and try to work out a fair solution or compromise. The mediator is essentially a go-between who develops a position that he or she hopes both sides can live with. This procedure may fail, in which case the dispute may end up in arbitration or in court.

6.4.3 Changed Conditions

It is a nearly universal practice to include in contract documents statements that require the contractor to indicate that he or she (a) has carefully examined all subsurface information available for the project site and (b) knows how these conditions may affect the work. The contractor further must acknowledge that he or she can do the work, in light of these conditions, for the bid price. The following is an example:

> **Site Investigation and Representation.**   The Contractor acknowledges that he or she has satisfied himself or herself as to the nature and location of the work, the general and local conditions, particularly those bearing upon availability of transportation, disposal, handling and storage of materials, availability of labor, water, electric power, roads, and uncertainties of weather, river stages, or similar physical conditions at the site, the conformation and conditions of the ground, the character of equipment and facilities needed preliminary to and during the prosecution of the work and all other matters that can in any way affect the work or the cost thereof under this Contract. The Contractor further acknowledges that he or she has satisfied himself or herself as to the character, quality, and quantity of surface and subsurface materials to be encountered from inspecting the site, all exploratory work done by the Owner, as well as from information presented by the Drawings and Specifications made a part of this Contract. Any failure by the Contractor to acquaint himself or herself with all the available information will not relieve him from responsibility for estimating properly the difficulty or cost of successfully performing the work. The Contractor warrants that as a result of his or her examination and investigation of all the aforesaid data he or she can perform the work in a good and workmanlike manner and to the satisfaction of the Owner.

It is usually further noted that the available subsurface information may not be sufficient to fully describe conditions as they actually may be found to exist.

> **Subsurface and Site Information.**   Test holes have been excavated to indicate subsurface materials at particular locations. This information is shown on the Drawings. Investigations conducted by the Engineer of subsurface conditions are for the purpose of study and design, and neither the Owner nor the Engineer assumes any responsibility whatever in respect to the sufficiency or accuracy of the borings thus made, or of the log of test borings, or of other investigations, or of the interpretations made thereof, and there is no warranty or guarantee, either expressed or implied, that the conditions indicated by such investigations are representative of those existing throughout such area, or any part thereof, or that unforeseen developments may not occur.

By virtue of these paragraphs being included in the contract, the contractor is required to certify that he or she (1) has become fully informed by using available information and (2) is aware that this information may be inadequate to describe site subsurface conditions. The contractor is also asked to certify that, in spite of this possibly inadequate knowledge, he or she can do the work for the price bid. Such language has been used in some cases in an attempt to burden the contractor with the risks inherent in the site. Usually, however, its purpose is to protect the owner or the engineer against frivolous claims.

When difficult ground conditions are encountered during a project, they may be fully predictable from information made available to the contractor and may be readily dealt with by the methods planned. On the other hand, they may be a surprise to everyone concerned, and the planned construction methods may be inadequate. The former case is what should transpire if everyone has done the job well; the engineer in representing the site, and the contractor in interpreting its effects on his or her work. The latter case may develop in spite of everyone's best efforts, simply because natural ground conditions are not totally predictable. In this case, the owner is usually obligated to bear the additional costs involved in developing his or her site, in spite of contractual disclaimers of responsibilities like those described earlier. This is a true change in conditions. Competent people did not expect or anticipate the actual situation, but if they had, the owner would have had to bear the costs involved.

There is also the possibility that a contractor losing money because of difficult conditions will claim that these conditions are different from those originally expected and that he or she is therefore entitled to extra compensation. For this claim to stand, the contractor must show that the conditions encountered were substantially different from those which he or she reasonably could have expected, considering the information available and the contractor's knowledge and capabilities.

In the event of a potential changed condition claim, it is imperative that the facts of the matter be fully documented in a timely fashion. To this end, the best available help should be sought at an early stage. Substantial amounts of money are often involved, and the positions of either side of the dispute will be strengthened by immediate and thorough attention to the problem.

6.5 SUMMARY

Items in earthwork specifications may be many and varied. This chapter has illustrated such specifications by example. Earthwork in construction contracts is often the subject of dispute and controversy, partly because the materials involved, soil and rock, have properties that are not always correctly anticipated, and vary from location to location on a site. Effects of rainfall during construction, and variations in groundwater level are not always foreseen. These disputes are most difficult to resolve when the parties involved are unwilling to assume their proper responsibilities for describing site conditions and assessing their effects on construction operations. Having studied the chapter, the reader should

1. Be able to describe relationships among contracting parties and their agents and the responsibilities of each.
2. Know the contents of the usual construction contract.
3. Understand what constitutes a complete special specification.
4. Know how contract disputes are settled.
5. Understand what "changed conditions" means and the circumstances under which a contract change order may result from them.

REFERENCES

Abbett, R.W., *Engineering Contracts and Specifications*. New York: Wiley, 1967.

Bachner, J.P., *Practice Management for Design Professionals*. New York: Wiley, 1991.

Sadler, W.C., *Legal Aspects of Construction*. New York: McGraw-Hill, 1959.

Sowers, G.F., "Changed Soil and Rock Conditions in Construction," *Journal, Construction Division, American Society of Civil Engineers,* 97, No. CO2, 1971.

CHAPTER 7

Interpretation of Soils Reports

Soils explorations are usually made in advance of construction to assist an engineer in preparing a design. In most cases, the reports of these explorations will substantially assist a contractor in preparing a bid. The usual subsurface exploration and soils investigation is done, however, without the specific needs of the contractor in mind. When the private design engineer has completed the exploration and investigation of the soil, he or she prepares a report for the client, who will be the owner in the eventual construction contract. On government projects in which engineer and owner are one, such work is often handled with internal memoranda. These reports and memoranda contain a great deal of information that will be useful to the contractor in assembling information with which to prepare a bid. The reports and memoranda and all other information that relate to them will usually be made available to the contractor in the period between bid advertisements and bid opening. Selected information may be included in the contract documents. Most of it is referenced therein. The wise contractor will obtain this information, study it, interpret it, and consider it in the preparation of his or her bid. This chapter is intended to be a guide to use of subsurface exploration reports.

The objectives of the chapter are to

1. Provide a clear definition of the purposes of the usual subsurface exploration.

2. Illustrate the various methods of subsurface exploration available, the advantages and disadvantages of each, and the form and utility of presentation of results from each.

3. Describe the format and contents of the usual soils engineering report.

4. Suggest sources of available subsurface information.

7.1 FIELD EXPLORATIONS

Field explorations may be considered to have three fundamental objectives:

1. To locate and define the vertical and horizontal boundaries of the various soil and rock strata that underlie the site of the proposed construction;

2. To locate the groundwater table; and

3. To determine the engineering properties of the subsurface materials.

These tasks may be done either by field testing during the exploration or by testing of samples obtained during the exploration and returned to the laboratory. For the latter, the third objective of the exploration then would be to perform field tests or to secure the samples required for laboratory testing so that the engineering properties of the various strata might be determined. The methods used in accomplishing these objectives will vary according to the manner in which the results of the exploration will be used.

7.1.1 Direct Exploration

The most reliable information may be obtained by directly observing the subsurface materials at a specific site. Direct observation requires excavation of test pits, test trenches, or test holes that are large enough to permit access for visual inspection. In the usual case, direct exploration is limited to fairly shallow depths. Where the groundwater table is not present, excavation may be readily accomplished to depths up to 20 ft using commonly available construction equipment, observing the requirements for trenching discussed in Chapter 10. Deeper excavation for direct observation requires mobilization of large-diameter drilling equipment or hand mining. These techniques will be used only rarely. When the groundwater table is encountered near the surface, direct exploration techniques are usually discarded in favor of some other method.

The principal advantage of direct exploration is that it permits observation of subsurface materials. Soil and rock materials may be observed in their natural state. Minor details of subsurface profiles that may have great significance in engineering and building on a site may be readily seen. Undisturbed samples of subsurface materials may be obtained by removing them at the level desired. Large disturbed samples are also easily obtained. The principal disadvantage of direct exploration is that it is usually restricted in depth. If the depth of the exploration is not restricted by the available equipment or groundwater conditions, then the cost becomes an important factor in deciding whether to use this method.

In addition to sending samples from test pits to laboratories for testing, here are several devices available for quick soil testing at the surface. These are especially important for use with cohesive soils. Some of these devices are as follows:

- *Pocket penetrometers* are direct reading instruments for the purpose of determining the unconfined compressive strength (q_u) of saturated cohesive soil. It consists of a spring-loaded plunger that penetrates the soil. Although these devices are very simple and easy to use, they have a wide error range (±20–40%.)

- *Vane shear testing* equipment consists of a shaft onto which two blades are mounted on the end at a $90°$ angle to each other. The shaft is plunged into the soil and the vanes rotated; the resulting torque is measured and q_u is estimated from that torque. Depending upon how the torque is applied, vane shear testing can also be used to measure the sensitivity of clay (tendency of clay to soften or liquefy when remolded.)

- *Thumb penetration* involves simply pressing the thumb into the soil. Although this has been widely used in the past, the difficulty of calibrating thumbs makes this the least accurate of these methods.

Also, even though they are actually laboratory tests, Atterberg limits (see Chapter 3) are fairly easy to obtain from samples taken at the surface. All of these tests for cohesive soils are especially useful for determining the safe design and height of trenches. (See Chapter 10.)

7.1.2 Semidirect Exploration

To overcome the restrictions inherent with direct exploration, semidirect techniques have become highly developed. These methods are in fact the most widely used techniques for engineering subsurface investigations. A semidirect exploration involves drilling and sampling. A boring is made using one of many methods to gain access to desired locations at depth. Samples are obtained at these locations with techniques that are determined by the purposes for which the samples will be used and the conditions under which they are taken.

A typical *auger drilling* setup is shown in Figure 7.1. The power-driven auger may have either a solid or a hollow stem. Solid-stem augers are used typically in dry regions where the

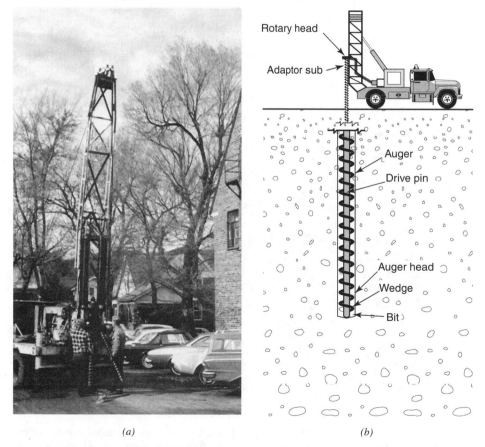

(*a*)   (*b*)

Figure 7.1   Subsurface investigation by drilling and sampling. (*a*) A continuous flight auger drill. (*b*) Schematic drilling setup (Longyear Company, Minneapolis, MN.)

groundwater table will not be encountered. This is because the auger must be removed from the hole prior to attempting to obtain a sample at any elevation. In the event that the groundwater table is high and the subsurface materials are loose, the removal of the auger usually results in the collapse of the hole. Hollow stem augers were developed to permit sampling through the stem of the auger without removing the auger from the boring. The drilling and sampling operation was thus improved since the auger could be left in place. The operation was faster and the stem could function as a casing, always stabilizing the sides of the borehole.

Rotary drilling equipment such as that shown in Figure 7.2 is also commonly used. In this method a clay slurry or driller's mud is circulated down the center of the drill rod and out the end of the roller bit. This slurry picks up the cuttings from the roller bit as it rotates and circulates along the outside of the drill stem returning to the ground surface. The cuttings are deposited in a mud pit and the slurry is returned by way of the pump through the drill stem. A continuous fluid system is thus established, which removes the material cut as the boring advances and stabilizes the sides of the borehole.

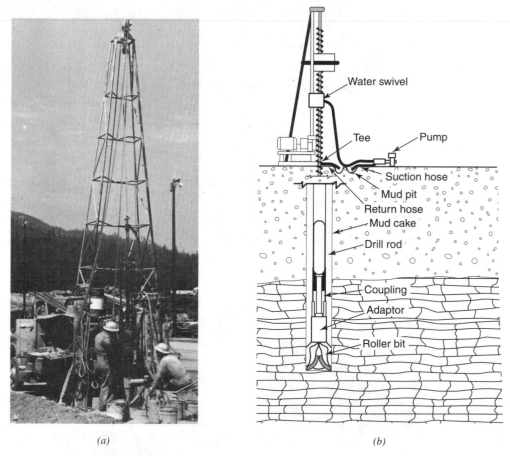

(a) (b)

Figure 7.2 Rotary hydraulic drilling equipment. (*a*) Truck-mounted rotary hydraulic drill. (*b*) Schematic rotary hydraulic drill. (Longyear Company, Minneapolis, MN.)

Figure 7.3   The standard penetration test. The driller raises the hammer and allows it to fall by alternately pulling and releasing the rope around the "cat's head." This process can also be done mechanically.

When the boring reaches the elevation at which a sample is to be taken, one of several sampling methods may be used. *Split barrel sampling* or *standard penetration testing* (ASTM D1586) involves driving a standard sampler with a standard amount of driving energy. A test in progress is shown in Figure 7.3. The resistance to penetration of the sampler is an index of the consistency or relative density of the ground. Standard split samplers have the following specifications and characteristics:

- Outside diameter of sampler: 2 in.

- Length of sample: 30 in.

- Either end of the sampler may be removed and the center section, which is split, opened to reveal any sample that might be retained. Samples removed will be heavily disturbed, and usually suitable only for classification and identification purposes.

- Outside diameter of sample: 1 5/8 in.

- Weight of driving weight: 140 lb

- Drop of driving weight: 30 in. Upon impact, it strikes the collar on the sampling rod.

- For each sample, the sampler is driven for three consecutive 6″ increments. The number of blows required to drive the final two increments is totaled and designated as N, the standard penetration resistance.

Standard penetration resistance has been correlated in a rough way with the properties of cohesionless and cohesive soils, as indicated in Tables 7.1 and 7.2. These values should be considered as approximations only. Other drive sampling methods are used, involving heavier weights, different drops, and larger samplers, and their results have been correlated, in a rough way, with the results of standard penetration tests (SPT). All penetration tests, however, do not produce directly comparable results.

SPT values can vary widely, not only because of variations in the soil, but also because of variations in the execution of the test as well. As a result, SPT results are generally corrected for two important factors: the effect of the overburden and the variations in energy output that depend upon the type of rig actually used to perform the test. Overburden correction considers that fact that, as the sampler is driven into deeper soils with higher effective stresses, the actual N value will increase relative to the same soil at the reference effective stress of 2 ksf. Energy correction considers variations in the impact energy of the driving weight from a reference efficiency of 60% of the safety type of SPT hammer. The latter is especially important because of the increasing popularity of automatic types of SPT hammers that have efficiencies of around 90%. SPT results that have been corrected for at least the latter are referred to as N_{60} values, and most correlations that use SPT data refer to N_{60} values.

In contrast to split-barrel sampling, *thin-walled tube sampling* (ASTM D1587) is accomplished by shoving, rather than driving, a sampling tube into the bottom of a boring. These tubes will typically be 2 to 5 in. in diameter and up to 4 ft in length. The purpose of thin-wall

TABLE 7.1   Relative Density of Cohesionless Soils

Relative Density Designation	Approximate Relative Density, %	N_{60} Standard Penetration Resistance	Approximate Angle of Friction of Soil ϕ, degrees
Very loose	0–5	0–4	25–28
Loose	5–30	4–10	28–30
Medium	30–60	10–30	30–36
Dense	60–85	30–50	36–41
Very dense	>85	Over 50	>41

TABLE 7.2   Consistency of Cohesive Soils

Consistency	Unconfined Compressive Strength, q_u (ksf)	Cohesive Strength c (ksf)	N_{60} Standard Penetration Resistance
Very soft	<0.5	<0.25	<2
Soft	0.5–1.0	0.25–0.5	2–4
Medium (firm)	1.0–2.0	0.5–1.0	4–8
Stiff	2.0–4.0	1.0–2.0	8–16
Very stiff	4.0–8.0	2.0–4.0	16–32
HARD	>8.0	>4.0	>32

Figure 7.4 The Shelby tube thin-walled sampler and standard penetration or "split spoon" sampler (open). The heads of both samplers are provided with a ball check valve to allow water above the sample to escape, as the tube is filled, and to seal the tube on removal of the sampler from the test boring. The latter action results in better sample retention because of the partial vacuum developed.

tube sampling is to obtain specimens for laboratory testing that are undisturbed. No usable information regarding subsurface conditions is obtained during the field operation. Both thin-walled and split barrel samplers are shown in Figure 7.4.

Rock properties, as well as those for soils, must be known for engineering and construction purposes. Different methods are used to drill and sample rock than soils. Softer rock may be drilled with *rotary equipment* (Figure 7.2). Harder rock usually requires coring with diamond or carbide-tipped coring bits. These bits are attached to a core barrel, which in turn is attached to the drilling rod. The typical core barrel is 10 ft long. Different sizes of core barrels are available, but all of them produce a cylindrical sample. A typical core sample is shown in Figure 7.5. The nature of the rock is determined by its mineralogy and degree of weathering, the strength of the intact material, the closeness of any joints or discontinuities, and the type of material that may be present in these discontinuities.

Core recovery (percent of core retained based on length of core barrel run) is one measure of rock quality often provided in an exploration report. Another commonly used method employs the rock quality designation (RQD), which is defined as the sum of the lengths of intact pieces of core at least 4 in. long, divided by the length of the core, expressed as a percentage.

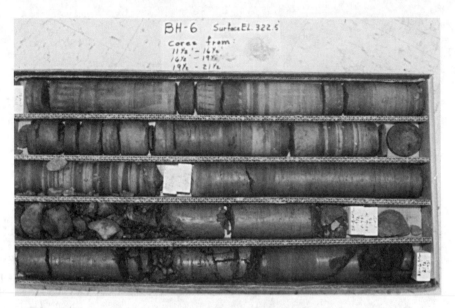

Figure 7.5 A rock core sample. Note the varying lengths of intact pieces, and the fracturing along bedding planes (discontinuities) in the rock.

EXAMPLE Rock Quality Designation

Consider a core same such as is shown in Figure 7.5. The core recovery from a 60″ specimen is as follows:

Core Recovery, In.	Modified Core Recovery, In.
8	8
1	
3	
4	4
2	
4	4
3	
5	5
4	4
3	
2	
5	5
Total 44	30

The core recovery represents the lengths of all of the pieces recovered in a given sample. The modified core recovery represents the lengths of only those pieces greater than or equal to 4″ in length. The total core recovery is the core recovery divided by the total sample length, or $44/60 = 73\%$. The rock quality designation (RQD) is the modified core recovery divided by the total sample length, or $30/60 = 50\%$.

RQD can be evaluated as follows: Very poor, 0–25%; Poor, 25–50%; Fair, 50–75%; Good, 75–90%; and Excellent, 90–100%. So this sample is on the borderline between poor and fair.

RQD is an index of the intensiveness of rock jointing. This intensity of jointing often exerts influence on the properties of the rock mass that is more important in construction than the strength of the intact rock between the joints. Intact strength is often estimated for classification purposes, but also may be determined by running unconfined compression tests on intact specimens.

7.1.3 Indirect Exploration Methods

Indirect exploration methods include those in which subsurface conditions are inferred from in-place measurements, without any direct observation or sampling.

In recent years, *cone penetrometer* soundings have found favor as an exploration method in the United States and Canada. Static cone penetrometer sounding methods originated in Holland, and are often referred to as Dutch cone soundings or probes. They have a number of advantages over the standard penetration testing method described in the foregoing section. For instance, they provide a continuous record of test results that can be used to deduce engineering properties of virtually the entire soil profile. They do not work well in very coarse and hard or dense soils, and do not produce samples for identification or testing.

A standard (ASTM D3441) cone penetrometer and its operation are illustrated in Figure 7.6. A cone-testing apparatus is usually mounted in an enclosed module on a small truck. Within the module, a hydraulically driven ram provides the means to push the cone into the ground, and to measure the forces required to do it. The cone is attached to a rod that has an outer section that slides over an inner core. The cone tip is attached to the inner core. There is a friction sleeve just above the cone tip. The cone assembly is pushed to the test location by advancing the outer rod (position 1 or position 4 in Figure 7.6). From this position the tip is pushed 40 mm by advancing

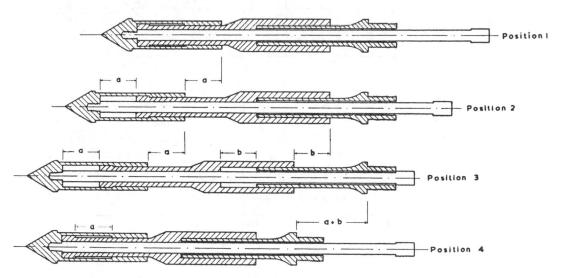

Figure 7.6 Operation sequence for cone penetrometer.

the inner rod, and the cone tip pressure required to do this is measured. This is the tip resistance, q_c. The tip and the friction sleeve are then advanced together another 40 mm, and total tip and friction resistance, $q_c + f_s$, is measured. By subtracting the initial tip resistance from this latter measurement, one obtains the friction resistance, f_s, on the sleeve. With the tip stationary, the outer rod is now advanced, and the penetrometer is in position for a subsequent test.

Cone penetration test results may be used to classify soils according to correlations among soil type determined by actual sampling, cone tip resistance, and friction ratio, f_s/q_c, which is expressed as a percentage. One correlation for doing this is illustrated in Figure 7.7. Figure 7.7 is widely used, but other, similar correlations are available, and all are strictly applicable only to the specific geographic and geologic conditions for which they were developed. In all of the available correlations, coarse soils have lower friction ratios and fine soils have higher friction ratios.

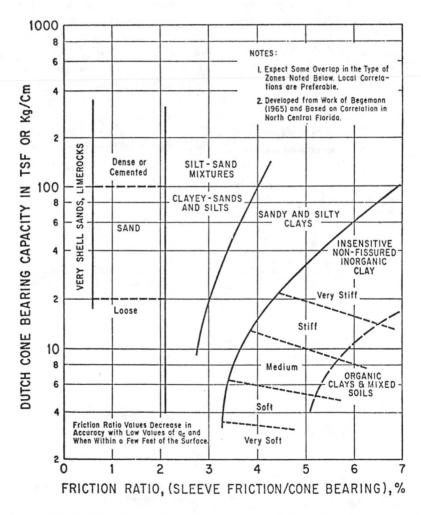

Figure 7.7 Soil classification from cone penetrometer test results.

For clay soils' cohesive or undrained shear strength, c (see Table 5.3) may be determined from cone tip resistance using the relation

$$c = \frac{q_c - p}{N_c},$$ (7.1)

where p is the total overburden pressure at the depth where q_c is measured and N_c is a constant that falls between 12 and 20. An average value of 16 is used for many purposes. Local correlation of strength from laboratory tests and cone resistance is the best way to establish a working value of N_c.

Angle of internal friction for sands may be deduced from cone measurements, but the procedure is somewhat complex and beyond the scope of this text. Elastic modulus for sands, E_s, which is used to make estimates of immediate settlement of footings, has been shown to be about

$$E_s = 2q_c.$$ (7.2)

Higher values of E_s are used for settlement estimates in some instances.

The relationship between cone tip resistance and standard penetration resistance is sometimes useful to know. Figure 7.8 shows the currently available relationship, established by comparing both types of measurement at 16 different sites. To use Figure 7.8, one needs to know the

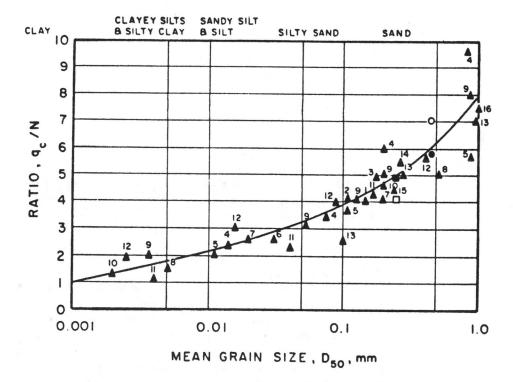

Figure 7.8 Relation between cone penetration resistance, q_c, and standard penetration resistance, N, for various soils. Numbers by data points refer to locations used to develop the correlation, and are not referenced here. N must be in blows/ft and q_c must be in bars, kg/sq cm, or tsf. (Reprinted, by permission, from Robertson, P.K., Campanella, R.G., and Wightman, A., "SPT-CPT Correlations," *Journal of Geotechnical Engineering*, American Society of Civil Engineers, Vol. 109, No. 11, November, 1983.)

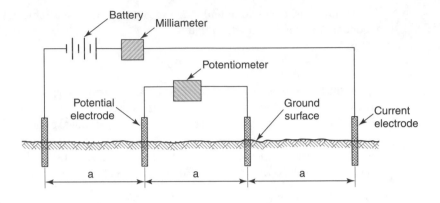

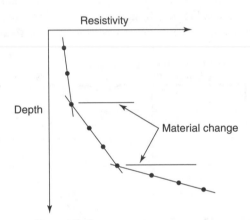

Figure 7.9 Schematic electrical resistivity setup and data reduction. (Reprinted, by permission, from G.B. Sowers and G.F. Sowers, *Introductory Soil Mechanics and Foundations*, Macmillan, New York, 1970.)

mean grain size, D_{50}, of the material for which either test type has been done, or at least its general classification. Given this, and either q_c or N, the other penetration index can be estimated.

Some subsurface explorations are made using geophysical techniques. These techniques involve determination of the properties of substrata indirectly by measuring either their electrical resistivity or the velocity at which a shock wave travels through them.

In the *resistivity method* (see Figure 7.9) two electrodes are inserted in the ground and connected to a current source. Intermediate electrodes are also inserted and connected by a potentiometer. When a reading is to be made, the potentiometer indicates the electrical resistivity of the ground within a depth equal to the electrode spacing a. The electrode spacing is then varied and more measurements are made. The resistivities so obtained are plotted versus depth or electrode spacing. Changes in slope of the resistivity versus depth plot indicate changes in subsurface materials. The resistivity measured is, of course, dependent on the amount of ionized salts and water present in the ground. An indication of the magnitudes of differences to be expected is given in Table 7.3.

TABLE 7.3 Electrical Resistivities of Soils and Rocks

Material	Resistivity (ohm-centimeters)
Saturated organic clay or silt	500–2,000
Saturated inorganic clay or silt	1,000–5,000
Hard partially saturated clays and silts; saturated sands and gravels	5,000–15,000
Shales, dry clays, and silts	10,000–50,000
Sandstones, dry sands, and gravels	20,000–100,000
Sound crystalline rocks	100,000–1,000,000

Source: G. B. Sowers and G. F. Sowers, *Introductory Soil Mechanics and Foundations,* Macmillan, New York, 1970.

Perhaps the most-used geophysical exploration technique with construction applications is the *seismic refraction method.* The essential features of the method are shown in Figure 7.10. A disturbance at the ground surface is produced by striking the ground with a sharp blow using a hammer or setting off an explosive charge. Geophones are used to determine the time taken for the disturbance to propagate several known distances from its source. These times are plotted versus distance, and changes in material are indicated by changes in the slope of this plot. The slope of the time-distance graph is the velocity of propagation of the disturbance. The velocity and thickness of each stratum encountered can be computed. For example, the thickness of the upper stratum is given by the formula

$$H_1 = \frac{A}{2} \sqrt{\frac{V_2 - V_1}{V_2 + V_1}}. \tag{7.3}$$

In the case of seismic exploration, information of considerable utility is obtained, other than stratum boundaries. The wave propagation velocity determined for each stratum may be useful as an index of the resistance to excavation expected. Typical values of propagation velocities are given in Table 7.4. Various equipment companies have correlated seismic velocity

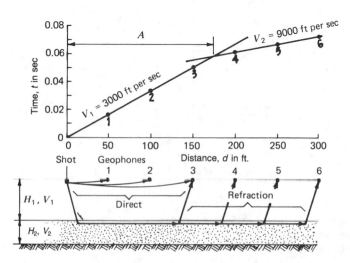

Figure 7.10 Subsurface exploration by seismic refraction. (Rprinted, by permission, from G.B. Sowers and G.F. Sowers, *Introductory Soil Mechanics and Foundations*, Macmillan, New York, 1970.)

TABLE 7.4 Wave Velocities in Earth Materials

Material	Velocity (ft//sec)
Loose dry sand	500–1500
Hard clay, partially saturated	2000–4000
Water, loose saturated soil	5200
Saturated soil, weathered rock	4000–10,000
Sound rock	7000–20,000

Source: G. B. Sowers and G. F. Sowers, *Introductory Soil Mechanics and Foundations,* Macmillan, New York, 1970.

with performance of tractor-ripper combinations. One such rippability chart is given in Figure 7.11. By knowing seismic propagation velocities and the rippability relationship for this particular equipment, a contractor may estimate whether rock on a specific project may be ripped or would require shooting.

One other indirect method that should be mentioned is the *Marchetti dilatometer.* The dilatometer is a flat-bladed device that, through the use of a membrane that slightly penetrates the soil, measures the resistance of the soil to the pressure of the membrane. The blade is pushed into the soil in a manner similar to the cone penetrometer. A wide variety of correlations exists between the various readings of the membrane pressure and different soil properties. Although it is relatively new in the United States, the dilatometer is gaining popularity, as it can be used to estimate properties such as overconsolidation ratio, undrained shear strength and the permeability coefficient and at the same time obtains a more consistent result than the SPT.

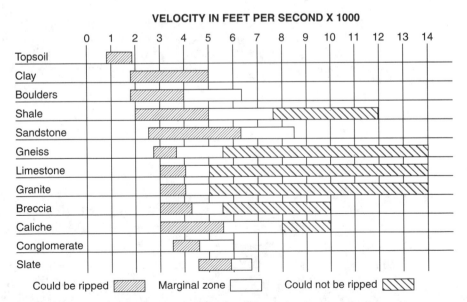

Figure 7.11 Rippability chart for excavation based on seismic refraction exploration. (Reprinted from *Engineering News-Record,* December 10, 1959, copyright © McGraw-Hill, Inc. All rights reserved.)

Indirect exploration methods are not widely used. In some cases, however, the results of these methods have been correlated directly with performance and utility of construction equipment and they may be extremely useful. But, it should be remembered that these methods rely on indirect measurements of properties of subsurface materials and that these indirect measurements are not always reliable. Interpretation of geophysical exploration results is always best done by persons with a thorough background of experience with the method and equipment used. One may confidently use the results of geophysical exploration methods for engineering purposes if these tests are supplemented by tests using direct exploration or semidirect exploration techniques.

7.2 GROUNDWATER CONDITIONS

Explorations made using any of the techniques previously discussed can result in a reliable location of the groundwater table at the time of exploration. In many locations the position of the groundwater table is subject to variation throughout the year. These variations arise from many causes, among which are tides, pumping of wells, seasonal runoff variations, and changes in rainfall patterns. In some instances, *piezometers* are installed in exploratory borings, and the fluctuations in the groundwater table are observed over a period. This information may not always be in the soils engineer's report. In many cases seasonal fluctuations are of sufficient magnitude to permit scheduling of construction to avoid the need for a dewatering system in excavations. When the matter of a few feet difference in groundwater elevation is important in planning construction operations, these seasonal variations and variations arising from other causes should be carefully considered.

In planning subsurface work it is important to consider not only the elevation of the groundwater surface but also the permeability of the soil. The design of a satisfactory dewatering system requires both, in addition to knowledge of location of the various strata. The determination of soil permeabilities may be accomplished in the laboratory or in the field as discussed in Chapter 5, according to the nature and geologic origin of the soils involved. Laboratory determinations are sufficient if soils are very uniform and good quality samples are obtained. Field tests that are properly planned and conducted are more universally applicable and reliable. These tests are done, as previously shown, by pumping from a test well and observing groundwater level changes at other locations, or by borehole permeability tests. Test conditions and observations and appropriate seepage analyses are combined to produce the desired parameters to quantify permeabilities.

Laboratory permeability tests on nonrepresentative samples can produce results that may be completely misleading. Ill-defined or poorly conducted pumping tests are equally as bad. In planning major dewatering operations, careful consideration of the completeness and validity of subsurface information is imperative.

7.3 SOILS ENGINEERING REPORTS

In preliminary planning for any major works, the first step taken is often the site subsurface exploration. Following exploration, alternative designs for the project may be prepared. Results of these explorations usually form the sole basis for the contractor's interpretation of subsurface conditions to be encountered in construction. Soils reports are organized in a format similar to following:

- Scope and purpose
- Introduction
- *Geologic setting*
- *Field studies*
- Laboratory tests
- Analysis
- Conclusions and recommendations
- *Appendix*

The subject headings are self-explanatory. Of these, the contractor will likely find those sections in *italics* to be the most useful. The geologic setting is of most value to the contractor who is familiar with conditions in the general area. Having worked in a similar geologic setting, he or she may fully appreciate the conditions that are to be encountered. The field study section will describe the exploration technique used and important observations made during the actual fieldwork. The most definitive information is usually contained in the appendix, which will include a section showing drilling logs or soil profiles and laboratory test results. These may be separated or they may be included on single summary sheets, one for each boring. A soil profile taken from a typical report is shown in Figure 7.12. Examples of a rock coring log and a cone penetration test log are shown in Figures 7.13 and 7.14.

The soil profiles or logs and associated tables and figures will show the position of the groundwater table, standard penetration resistance test results, boring locations and elevations, and laboratory test results. In practically all cases it will be pointed out in the text of the report that the information shown on the boring logs is representative of conditions at the boring locations only. The assumption that conditions between borings are the same as conditions represented by the borings is one that may not always be warranted. By observing the relative uniformity of conditions at all borings and by knowing something of the geologic history of the particular site, one may better judge the reliability with which one may make such interpolations. It is prudent for the contractor to remember that, when he or she submits his or her bid, he or she usually must sign a statement to the effect that he or she fully understands subsurface conditions and their effects on the progress of his or her work. (See Chapter 6.) If conditions are not clear, he or she is well advised to assume that they are the worst that could be reasonably anticipated and to adjust his or her bid prices accordingly.

7.4 OTHER INFORMATION SOURCES

When for some reason available subsurface information is inadequate, the contractor may find it necessary to obtain his or her own. His or her choices range from seeking a full range of consulting services such as those described in earlier paragraphs, to seeking available information from other sources. This section discusses other possible sources.

Information on subsurface conditions from *private sources* is generally of limited availability. Private studies are the property of those who pay for them; hence, private consultants are obliged to release them only with the owner's permission. There is a wealth of information available from various public agencies, however. This information may be either general or specific in nature.

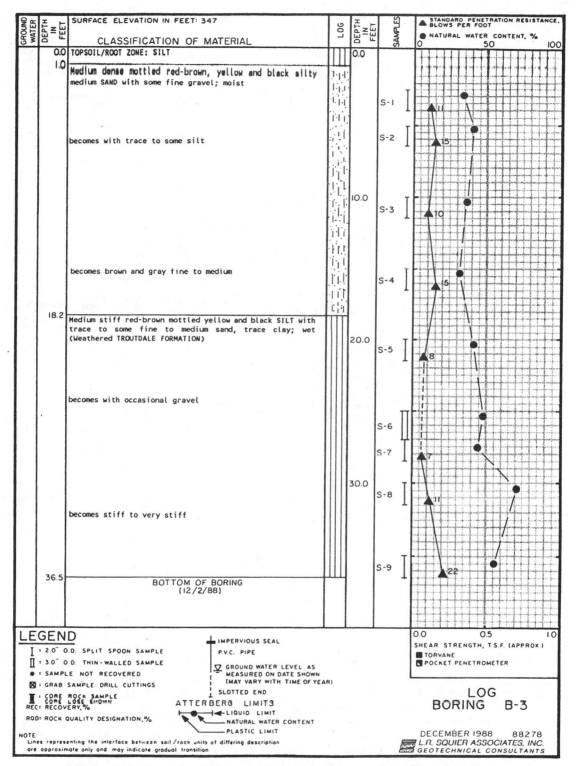

Figure 7.12 Example soil boring log. (Courtesy of Squier Associates.)

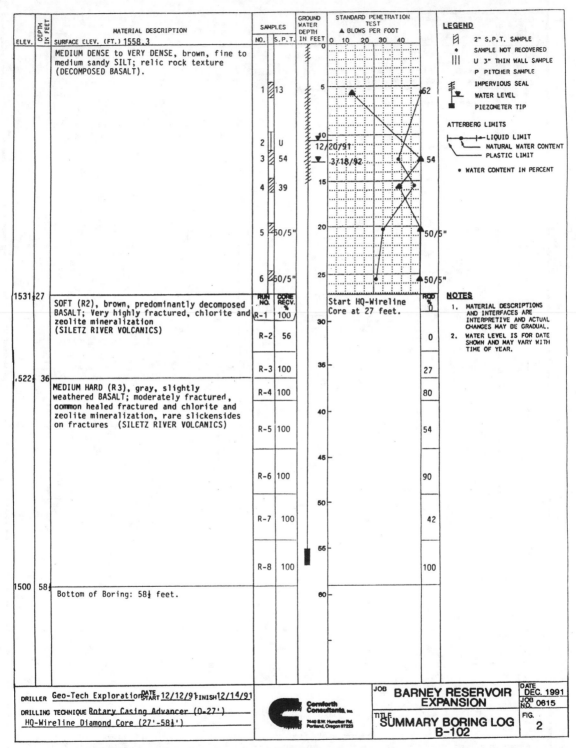

Figure 7.13 Example boring log including rock coring. Note that core recovery and RQD are shown. (Courtesy of Cornforth Consultants, Inc.)

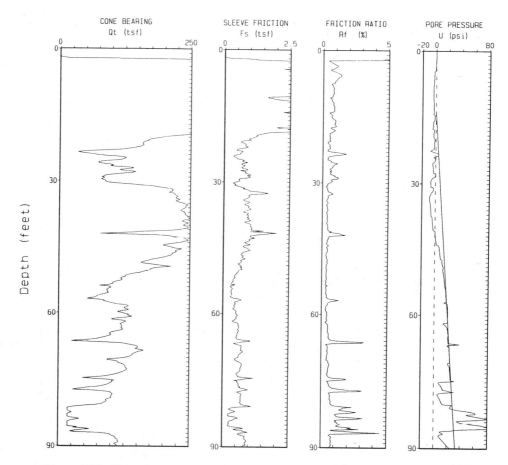

Figure 7.14   Example cone penetrometer log. (Courtesy of Geotechnical Resources, Inc.)

Certain general subsurface information is generated by federal and state agencies in the course of their normal duties. Such information may usually be obtained at a good library or from the agencies themselves. The *U.S. Geological Survey* (U.S.G.S.) publishes various maps, bulletins, and reports defining topography, geologic formations, and groundwater conditions of most sections of the country. This information is often of such large scale and low definition that it is of limited application for planning at a specific site. It is readily available, however, and should not be overlooked as a starting point for obtaining further information from other sources. Many *state geologic agencies* produce information similar to that provided by the U.S.G.S. In some cases, states have produced reports and maps that define in great detail subsurface conditions on a small enough scale to provide reliable information at specific locations. This is particularly true in highly developed or rapidly growing areas. Groundwater records and well logs are always available from some governmental unit. Typically, a state engineer's or geologist's office maintains records on wells drilled in that state. These are usually filed by section, township, and range, or according to the owner's name at the time the well was drilled.

Descriptions of soils, water levels, and bedrock locations contained therein are often useful. One should always be cautious in relying totally on such information, however, and consider its reliability in the light of how, by whom, and for what purpose it was obtained.

For specific site information, one may often obtain very detailed and valuable data from those governmental agencies having engineering or public works sections that have done work in a nearby area. The files of these agencies are public information, and those persons involved are generally cooperative and helpful in providing access. Practically no modern major civil works are executed without first obtaining some subsurface information. If these works are nearby or closely related to proposed work for which there is inadequate subsurface information, the contractor should seek out and make use of that which may be readily available as a result of earlier public work.

7.5 SUMMARY

Field explorations are usually made for design purposes. Only in rare circumstances are they made at the direction of a contractor, who is reliant on the work of others and is usually obliged to bid only with the knowledge they have communicated. He or she is, therefore, obligated to acquire and maintain a reasonable degree of proficiency in interpreting soils engineering reports and to observe the effects of various subsurface conditions on his or her operations. With such a background, the contractor will be able to recognize adequate and inadequate information and thereby, one hopes, avoid the headaches arising from the need to argue for a claim of extra costs that result from changes in anticipated subsurface conditions.

After studying this chapter the reader should solve the accompanying problems. Having completed them, the reader should

1. Know what information to look for in a soils engineering report, and understand what it means.

2. Be able to locate existing subsurface information for a specific site when none has been provided.

REFERENCES

American Society for Testing and Materials, *1988 Annual Book of ASTM Standards*, Vol. 04.08, Soil and Rock, Building Stones; Philadelphia: Geotextiles,

Begemann, H.K., Improved Method of Determining Resistance to Adhesion by Sounding through a Loose Sleeve Placed Behind the Cone, Proceedings, Third International Conference on Soil Mechanics and Foundation Engineering, Vol. 1, pp.213–217, 1953.

Cheney, R.S., and Chassie, R.G., *Soils and Foundations Workshop Manual*. Washington, DC: U.S. Department of Transportation, FHWA, 1982.

"How to Determine Rippability," *Engineering News Record,* December 10, 1959.

Occupational Safety and Health Administration. *OSHA Technical Manual*. TED 1-0.15A. Washington, DC: Occupational Safety and Health Administration, 1999.

Peck, R.B., Hanson, W.E., and Thornburn, T.H., *Foundation Engineering* (2d ed.). New York: Wiley, 1974.

Pile Buck Steel Sheet Piling Design Manual. Palm City, FL: Pile Buck, 1987.

Robertson, P.K., Campanella, R.G. and Wightman, A., "SPT-CPT Correlations," *Journal of Geotechnical Engineering*, American Society of Civil Engineers, Vol. 109, No. 11, November, 1983.

Schmertmann, J.H., *Guidelines for Cone Penetration Test, Performance and Design*, U.S. Department of Transportation, Federal Highway Administration, FHWA-TS-78-209, Washington, DC, 1978.

Sowers, G.B., and Sowers, G.F., *Introductory Soil Mechanics and Foundations.* New York: Macmillan, 1970.

Terzaghi, K., and Peck, R.B., *Soil Mechanics in Engineering Practice.* New York: Wiley, 1948.

PROBLEMS

1. Select a site in your locale, assuming that it is to be used for a borrow source. Further assume that the borrow site must produce 30,000 yd^3 of compacted fill. Prepare a map showing the site location and a plan view of the borrow area. From such public records as may be available and from your own observations and analysis, describe subsurface conditions at the site and any limitations that may be imposed on the borrow operations by those conditions. Summarize your findings in a brief letter report or record memorandum.

2. For a local site of your own choosing, prepare a profile sheet indicating the sequence of subsurface materials you would expect to find to a depth of at least 50 ft. Describe the materials as completely as possible. Document the information on which your findings are based. Supplement your findings with personal observations during a site visit. A letter or memorandum should be prepared to summarize and transmit your findings.

CHAPTER 8

Embankment Construction and Control

Embankments are nearly always compacted when incorporated in engineered works. Some deep embankments are placed to full depth and compacted later. Compacting soil is a simple process. Energy in some form is applied to make the soil more dense. This process results in the expulsion of air from the soil–water–air system. In the special case of saturated sand, only a two-phase system of high permeability is involved, and water is expelled from the soil voids. While the obvious effect of soil compaction is densification, or an increase in unit weight, the purpose of compaction is to produce a soil mass with controlled engineering properties. Certain engineering properties of compacted soils are determined not only by density, but also by the means and conditions under which the density was obtained.

Soil improvement through the application of additives during placement is usually intended to enhance engineering properties, but may also result in a change in the workability of the soil that will ease handling during construction.

Knowledge of the fundamentals of soil compaction and stabilization is essential to the understanding of specifications for these items of work in any contract, and particularly on large embankment projects like the dam shown in Figure 8.1. It will also greatly aid in the selection of equipment for actual construction.

This chapter deals with laboratory and field compaction techniques, methods of compaction control, and chemical stabilization of soils.

The objectives of the presentation are to

1. Illustrate typical moisture–density relationships for coarse and fine soils, considering different levels of compactive effort.

2. Show the effects of various compaction conditions on engineering properties of fine and coarse soils.

Figure 8.1   Construction of a major zoned embankment.

3. Present standard laboratory compaction procedures.

4. Familiarize the reader with the intent, form, and content of the usual compaction specification.

5. Show how shrink and swell calculations should be made.

6. Indicate the appropriate use of specific types of compaction equipment and methods applicable to given soil conditions.

7. Explain the details of various methods for making and interpreting compaction control tests.

8. Introduce the reader to the fundamentals of chemical soil stabilization.

9. Discuss deep subsoil improvement by compaction and other means.

8.1 MOISTURE–DENSITY RELATIONSHIPS

The amount of densification obtained by any compaction process is dependent on the amount of energy used, the manner in which it is applied, the type of soil involved, and the soil water content. This section discusses the relationships among these various factors.

8.1.1 Cohesionless Soils

The single-grained nature of the structure of cohesionless soils depends on grain-to-grain contact forces arising from gravity for its stability. If these contacts are disturbed, particle weight causes the structure to assume a new, more stable, and denser configuration. It is important to note that the initial destruction of grain-to-grain contact will result in densification. On the

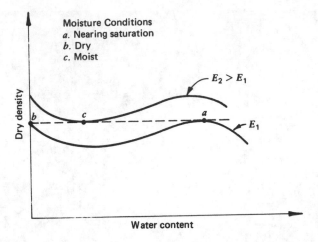

Figure 8.2 Idealized moisture–density relationships for clean sand.

other hand, application of a steady force to such a soil structure results primarily in an increase in the grain-to-grain forces with little permanent densification. For these reasons, it has been found that cohesionless soils are most efficiently compacted by vibrations, which disturb grain-to-grain contacts in the loose state.

Idealized moisture–density curves for clean sand are shown in Figure 8.2. The curves are representative of the results of compacting the same soil using two different energy levels, E_1 and E_2. Several points on the figure are worthy of special note. Most obviously, it is evident that greater energy input results in greater densification. However, logically, as higher energy levels are used, greater compaction results and further densification becomes increasingly difficult. Densification effects are, therefore, by no means proportional to the level of energy input. In addition, the curves are S-shaped. That is, highest densities are obtained for a given compactive effort or energy level when the water content is either very low (approaching zero) or very high (approaching saturation). Thus, if a given density is specified for construction, that represented by the horizontal dashed line, it may be obtained in one of several ways. The possibilities are indicated by points a, b, and c. For instance, compactive effort will be minimized if moisture conditions are adjusted to be either very dry or very wet (a or b) More compactive energy will be required for moist conditions (c). The choice in practice is dictated by the economics of providing moisture control or compaction energy. The characteristic S-shape of the curves is the result of surface tension in the soil moisture that results in apparent cohesive strength, as explained in Chapter 5 and, therefore, resistance to compaction. When the soil is dry or saturated, these effects do not exist, the strength of the soil structure is at a minimum, and greatest density for a given compaction energy level is achieved. When the soil is moist, capillary forces impart strength (apparent cohesion) to the soil, which results in increased interparticle forces and reduced densification.

8.1.2 Cohesive Soils

Moisture–density curves characteristic of a cohesive soil are shown in Figure 8.3. As in the case of cohesionless soils, increased compactive energy is seen to result in increased densification. Curves representing the different compactive efforts are of similar shape. Maximum densification for a given effort is obtained at some intermediate moisture content. Points a and b are referred to as having the coordinates *maximum dry density* and *optimum water content* for a given compactive effort. It is believed that this optimum condition results from resistance to densifica-

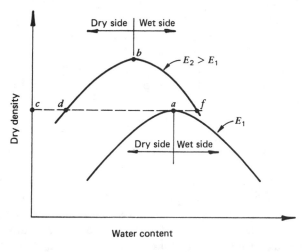

Figure 8.3 Idealized moisture–density relationships for silty clay.

tion, on the dry side by forces in tightly held soil water (see Figure 3.5), and on the wet side by the high degree of saturation and low permeability of the mass. It is apparent that a specified density such as that at c may be obtained with widely different compactive efforts and under widely different moisture conditions. The possibilities in this case are indicated by points a, d, and f for the two compactive efforts shown. If we want to achieve density c, we can do it with energy E_2 by compacting with any water content between d and f. Or, we may use E_1 at water content a. The proper choice of moisture content is dictated by the economics of moisture control and available compaction energy. If, as is often the case, a range of moisture contents is specified along with minimum density, the range of available choices of compactive efforts is narrowed.

8.2 PROPERTIES OF COMPACTED SOILS

8.2.1 Relative Density

Engineering properties of cohesionless soils are primarily a function of relative density. Relative density may be defined in terms of void ratio as

$$D_r = \frac{e_{(\text{max})} - e}{e_{(\text{max})} - e_{(\text{min})}} \times 100\% \tag{8.1}$$

or in terms of unit weights as

$$D_r = \frac{\gamma_d(\text{max})}{\gamma_d} \frac{\gamma_d - \gamma_d(\text{min})}{\gamma_d(\text{max}) - \gamma_d(\text{min})} \times 100\% \tag{8.2}$$

The following definitions apply:

$$e = \text{void ratio measured,}$$
$$e_{(\text{max})} = \text{maximum void ratio determined by standard test method,}$$
$$e_{(\text{min})} = \text{minimum void ratio determined by standard test method,}$$
$$\gamma_d = \text{dry density measured,}$$
$$\gamma_d(\text{max}) = \text{maximum dry density determined by standard test method,}$$
$$\gamma_d(\text{min}) = \text{minimum dry density determined by standard test method.}$$

Typical values of relative density relating to other soil properties are given in Table 7.1.

Tests exist to determine the relative density of a soil. These tests employ loose pouring into an open container to determine minimum density (ASTM D4254), and vibration of the container and contents to determine maximum density (ASTM D4253; see Appendix A.6). Because of the specialized nature and cost of the equipment necessary for the maximum density test, some firms and agencies use the modified test (ASTM D1557) as a substitute. Data from three test methods for maximum density on six granular soils are shown in Figure 8.4. Note that the maximum density by ASTM D1557 is very nearly the same as by ASTM D4253. It should be noted that relative density values are very sensitive to slight changes in the laboratory values of maximum and minimum density (see Equation (8.2)). These values must always be determined by standard methods if relative density is to be used as a construction control index.

As relative density increases, soil strength increases and compressibility decreases. Permeability is reduced. It will be recalled that cohesionless soil structure is single grained. With this type of structure there is usually no important difference between a natural deposit or a fill, provided both are at the same relative density. Since the objective of compaction is to produce a fill with controlled properties, usually only density is specified for these types of soils. Seldom will control of water content be specified in compaction specifications. In those cases where water control is specified, it is usually done to ensure maximum compaction for the energy available.

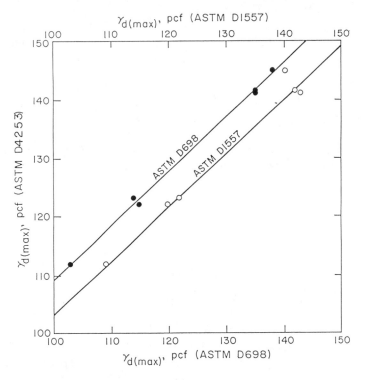

Figure 8.4 Comparison of results from maximum density tests. Note that maximum densities by ASTM D4253 and ASTM D1557 are approximately equal, and that both exceed that by ASTM D698 of about 7 pcf.

8.2.2 Water Content

Engineering properties of plastic soils are very much affected by compaction water content. Water content during placement influences the structure of clay soils sufficiently to have a significant effect on strength, permeability, and compressibility. Studies of compacted soils at the microlevel have shown that soils compacted on the dry side of optimum moisture content have a flocculent structural arrangement of particles. On the wet side of optimum moisture content, the structure is dispersed. A schematic representation of these findings is shown in Figure 8.5. Points *a* (flocculent) and *b* (dispersed) are at the same density. Soil structure influences engineering properties; therefore, compaction water content must be controlled along with density if the influence of structure is of importance in the design of an embankment.

Wet side as opposed to dry side compaction, for instance, may reduce the coefficient of permeability of a soil dramatically, as shown in Figure 8.6, even though density is constant. The larger pore sizes in the flocculent structure, even though few in number, permit more flow than a larger number of smaller openings in the dispersed structure. Where permeability control, notably in the core section of earth dams, is important, a specification for wet side compaction may be made.

Deviator stress–strain curves (see Figure 5.14) from triaxial shear tests on soils compacted at various water contents are shown in Figure 8.7. Compaction on the wet side of optimum water content (Sample Nos. 4, 5, 6) yields a flexible material that has low strength. High flexibility is indicated by the flat slope (modulus) of the stress–strain curves. Wet-side compaction of an earth dam core produces an impervious barrier to water flow, which can yield without cracking as the dam deforms. Its low strength must be compensated for by stronger construction in the outer zones or shells of the structure. Dry-side compaction (Sample Nos. 1 and 2) yields a structure that is brittle (i.e., has a steeply sloping stress–strain curve) but of high strength.

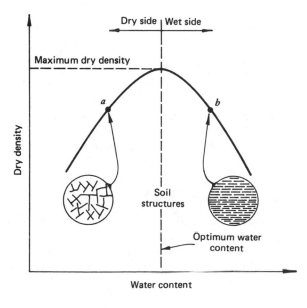

Figure 8.5 Effects of water content during compaction on soil structure.

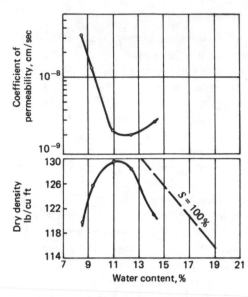

Figure 8.6   Effects of water content during compaction on soil permeability.

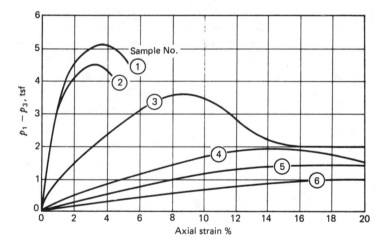

Figure 8.7   Stress–strain relationships for compacted clay soil. (Reprinted, by permission, from Seed, H.B. and Chan, C.K., "Structure and Strength Characteristics of Compacted Clays," *Journal, Soil Mechanics and Foundations Division*, American Society of Civil Engineers, Vol. 85, SM5, 1959.)

8.3   LABORATORY COMPACTION PROCEDURES

Standardized laboratory procedures are needed to clearly specify construction requirements. They also are useful in studies related to the effects of compaction techniques on soil properties. Different agencies use slightly different test methods, but all methods are similar. They specify the soil to be used and compactive effort to be exerted in developing the moisture–density relationship.

Figure 8.8 The standard laboratory compaction test.

For plastic soils, the most widely used test method is the standard compaction (ASTM D698) test procedure, shown in Figure 8.8. It is sometimes referred to as the Standard Proctor test, after the pioneer of these methods of control. The Standard Proctor test, however, is slightly different from ASTM D698. The modified compaction test (ASTM D1557) was developed to permit compaction control where very high densities are to be specified. The standard and modified test methods are summarized in Table 8.1 and (in more detail) in Appendix A.5.

TABLE 8.1 Comparison of Compaction Test Procedures

Designation	Standard ASTM D698	Modified ASTM D1557
Mold		
Diameter (in.)	4	4
Height (in.)	4 5/8	4 5/8
Volume (ft^3)	1/30	1/30
Tamper		
Weight (lb)	5.5	10.0
Free drop (in.)	12	18
Face diameter (in.)	2	2
Face area (in.^{2})	3.1	3.1
Layers		
Number, total	3	5
Surface area, each (in.^{2})	12.6	12.6
Compacted thickness, each (in.)	1 5/8	1
Effort		
Tamper blows per layer	25	25
(ft-lb/ft^3)	12.375	56.250

Note therein that the modified procedure incorporates a significantly higher energy level than the standard test. Its use will result in significantly higher density than the standard test at the same water content. For most soils, the maximum dry density from the standard test is 90 to 95 percent of the corresponding value from the modified test. The results of a standard compaction test are presented in Figure 8.9. The zero air voids line (coordinates of water content and dry density for the saturated condition) Equation (8.3) is shown for reference:

$$\gamma_{d(zav)} = \frac{G_s \gamma_w}{1 + wG_s}. \tag{8.3}$$

For cohesionless soils, special laboratory compaction techniques are necessary. Tests using the basic procedures outlined in Table 8.1 are sometimes specified, although often inappropriately.

8.4   COMPACTION SPECIFICATIONS

The special specifications for embankments in the construction contract may be one of two general types, prescriptive and performance. For either type of specification, compaction control is ultimately based on a comparison of field results with laboratory tests.

8.4.1   Relative Compaction

For cohesionless soils, relative density (Equation (8.1) or Equation (8.2)) is usually specified. For cohesive soils, relative compaction

$$RC = \frac{\gamma_d}{\gamma_{d(max)}} \times 100\% \tag{8.4}$$

and water content are specified. Relative compaction specifications are sometimes used for cohesionless soils as well.

The distinction between relative density and relative compaction is not always clear. The two concepts are compared graphically in Figure 8.10. It should be noted that it is possible for field relative density to be either less than zero percent or greater than 100 percent, and that relative compaction may exceed 100 percent. Tests have shown that, in a large number of cases involving sands and gravel, minimum density corresponds to about 80 percent relative compaction based on the ASTM D1557 method.

8.4.2   Prescriptive Specifications

A prescriptive specification tells the contractor what to do and how to do it. If he complies, the owner is obligated to accept the result. This type of specification should produce the lowest of bids, since the contractor knows exactly what must be done to accomplish the work. To develop a prescriptive specification, the contractor builds a test fill and the adequacy of his compaction method is established by tests on that fill. Alternatively, through his experience with a particular soil, the engineer may know what is required to achieve adequate compaction and write a performance-type specification. The specification in either case is written after direct or

MOISTURE — DENSITY RELATIONSHIP

TEST FOR _Lone Oak Road Embankment_

TEST BY _FJB_   DATE _June 26, 1992_

SAMPLE IDENTIFICATION _Red-brown Silty Clay_

COMPACTIVE EFFORT _ASTM D698_

TRIAL NUMBER	1	2	3	4	5	6
Wt. of Wet Soil & Cylinder	12.92	13.12	13.32	13.35	13.28	
Wt. of Cylinder	9.66	9.66	9.66	9.66	9.66	
Wt. of Wet Soil	3.26	3.46	3.66	3.69	3.62	
Vol. of Cylinder	1/30	1/30	1/30	1/30	1/30	
Unit Wet Weight	97.9	103.9	109.9	110.8	108.7	
Unit Dry Weight	81.5	84.1	86.1	83.4	79.4	
Pan Number	161	181	189	185	170	
Wt. of Wet Soil & Pan	80.58	92.92	88.92	101.65	93.45	
Wt. of Dry Soil & Pan	72.57	81.43	76.68	84.49	76.99	
Wt. of Water	8.01	11.49	12.24	17.16	16.46	
Wt. of Dish	32.73	32.46	32.44	32.37	32.40	
Wt. of Dry Soil	39.84	48.97	44.24	52.12	44.59	
Water Content, %	20.1	23.5	27.7	32.9	36.9	

Figure 8.9 Standard laboratory compaction test results.

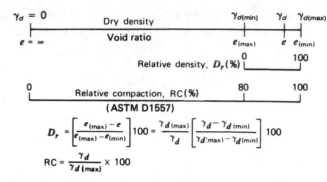

Figure 8.10 Relative compaction–relative density relationships. (Reprinted, by permission, from Lee, K.L., and Singh, A., "Compaction of Granular Soils," *Proceedings, 9th Annual Engineering Geology and Soils Engineering Symposium*, Boise, Idaho, 1971.)

previous field experience has demonstrated the sufficiency of the compaction method to be specified. The following is an example of a prescriptive specification:

2-03.3(14)A Rock Embankment Construction. Rock embankments shall be constructed in horizontal layers not exceeding eighteen (18) in. in depth, except that when the average size of the rocks exceeds eighteen (18) inches, the layers may be as deep as required to allow their placement. Occasional rocks exceeding the average size may be disposed of as approved by the Engineer instead of being incorporated in the embankment.

Each layer shall be compacted with at least one full coverage with a 50-ton compression type roller or four full coverages with a 10-ton compression type roller for each 6-in. depth of layer or fraction thereof. The number of coverages for compression type rollers, including grid rollers, weighing more than 10 tons and less than 50 tons shall be as directed by the Engineer. Rollers shall be so constructed that they will exert a reasonably uniform pressure over the area covered. In lieu of the foregoing, each layer shall be compacted with four full coverages with an approved vibratory roller for each 6-in. depth of layer or fraction thereof. Rolling may be omitted on any layer or portion thereof only when in the judgment of the Engineer it is physically impractical of accomplishment. In addition to the above rolling, each layer shall be further compacted by routing the loaded and unloaded hauling equipment uniformly over the entire width of the embankment.

The material shall be placed carefully so that the larger pieces of rock or boulders are well distributed. The intervening spaces and interstices shall be filled with the smaller stone and earth as may be available so as to form a dense, well-compacted embankment. On projects where earth embankments are to be constructed under Method A, rock embankments shall be compacted by uniformly routing loaded and unloaded hauling equipment over the entire width of the layer being compacted.

In making rock embankments, the Contractor will be required to bring up such fills to within 2 ft below subgrade as designated by the Engineer. The top 2 ft of the subgrade shall be constructed in accordance with the method of compaction specified for the project from rock not to exceed 4 in. in size and/or from granular material to be obtained from the roadway excavations or from borrow pits as approved by the Engineer. The finer materials from rock excavations shall be saved as far as practicable for use in topping out rock fills and backfilling over the subgrade excavation in rock cuts.

When it is specified that the subgrade shall be trimmed with an automatically controlled machine, no rocks or stones larger than 2 in. shall be left within 4 in. of the subgrade. (From *Standard Specifications for Road and Bridge Construction,* State of Washington Department of Highways, 1963.)

fications

on the other hand, requires that the contractor achieve a specific
tor is told what he or she must accomplish. The contractor's
re some experimentation, and the bid might be correspondingly
compaction, the contractor is told to achieve a specified relative
n and that tests will be conducted during construction to ensure
specification is the more common of the two types:

ankment Construction.   Earth embankments shall be constructed in
rm thickness by one of the three methods, A, B, and C, described in
er all methods the layers shall be carried up full width from the bottom
slopes of all embankments shall be compacted to required density as part
action work.

tated in the Special Provisions, earth embankments shall be constructed

g operations, the surface of embankments and excavations shall be
s section. All ruts and depressions capable of ponding water shall be

enter of embankment layers shall be constructed higher than the sides. Side
e constructed with the intersection with the original ground as the high
all uniformly slope to the outer side with a slope not to exceed 1 ft in 20 ft.

ng Earth Embankments.   Except when Method A is specified, earth
compacted with modern, efficient, compacting units satisfactory to the En-
units may be of any type provided they are capable of compacting each
e specified density. The right is reserved for the Engineer to order the use
of any particular compacting unit discontinued if it is not capable of compacting the material to
the required density in a reasonable time.

The determination of field in-place density shall be made in accordance with methods and
procedures approved by the Engineer.

Method A.   Under Method A, earth embankments shall be compacted in successive horizontal
layers not exceeding two (2) feet in thickness, and each layer shall be compacted by routing the
loaded haul equipment over the entire width of the layer. When permitted by the Engineer, side hill
fills too narrow to accommodate the hauling equipment may be placed by end dumping until the
embankment material can be spread to sufficient width to permit the use of hauling equipment
upon it. Thereafter the remainder of the embankment shall be placed in layers and compacted as
above specified. Suitable small mechanical or vibratory compactor units shall be used to compact
the layers adjacent to structures that are inaccessible to the loaded haul equipment.

Method B.   Under this method each layer of the top 2 ft of embankments shall be compacted to
95 percent and each layer of embankments below the top 2 ft shall be compacted to 90 percent of
the maximum density as determined by compaction control tests specified in Section
2-03.3(14)D. Horizontal layers shall not exceed 8 in. in loose thickness, except that the layers of
the top 2 ft shall not exceed 4 in. in loose thickness. Moisture content of the embankment material
at the time of compaction shall be as specified in Section 2-03.3(14).

At all locations that are inaccessible to compaction rollers the embankments shall be compacted in layers as required herein and shall be compacted to the required density by the use of
small mechanical or vibratory compactor units.

Method C. Under this method each layer of the entire embankment shall be compacted to 95 percent of the maximum density as determined by compaction control tests specified in Section 2-03.3(14)D. The horizontal layers shall not exceed 8 in. in loose thickness except that the layers of the top 2 ft shall not exceed 4 in. in loose thickness. The moisture content of the embankment material at the time of compaction shall be as specified in Section 2-03.3(14).

At all locations that are inaccessible to compaction rollers the embankment shall be constructed in layers as required herein and shall be compacted to the required density by use of small mechanical or vibratory compactor units. (From *Standard Specifications for Road and Bridge Construction*, State of Washington, Department of Highways, 1963.)

Having established water contents and densities corresponding to the engineering properties desired in the fill, the engineer prepares a compaction specification. The essence of the specification is that a certain percentage of relative density or relative compaction will be required for specific soils. Water content may also be specified if compaction water content influences the fill properties. For example, the U.S. Bureau of Reclamation requires compaction along the lines indicated in Table 8.2 for dams less than 50 ft high.

The specification in Table 8.2 for cohesive soils recognizes the need to control both density and water content. The specification for sands recognizes the importance of density and the need for nearly full saturation during construction to achieve it. Both specifications become more restrictive as dam height increases and the need for controlling engineering properties becomes more important.

TABLE 8.2 Example of Earthwork Compaction Requirements for Earth Dams

Compaction Requirements	*Water Control*
Cohesive Soils	Optimum $\pm$ 2 percent
0–25 percent retained on No. 4; RC = 95 percent	
26–50 percent retained on No. 4; RC = 92.5 percent	
More than 50 percent retained on No. 4; RC = 90 percent	
Cohesionless Soils	Very wet
Fine sand	
0–25 percent retained on No. 4; $D_r = 75$	
Medium sand	
0–25 percent retained on No. 4; $D_R = 70$	
Coarse sand and gravel	
0–100 percent retained on No. 4; $D_r = 65$	

8.5 SHRINK AND SWELL FROM BORROW TO FILL

Up to this point, we have simply considered the compactive characteristics of the soil itself. We now need to consider the effects of transport of soil from one place to another, as most soil compaction efforts involve some kind of fill and grade work. This is a critical issue in construction management, because miscalculations in this type of operation can lead to contract disputes and expense on the part of owner or contractor.

The process of earthwork generally involves three steps:

- Removal of the soil from the borrow area. The soil removed from this point is referred to as the cut soil.
- Transport of the soil from the borrow area to the area where the soil is to be deposited and compacted.
- Placement and compaction of the soil at the desired location. The soil here is the fill soil.

Generally speaking, the unit weight of the soil decreases (the soil expands) as it is removed from the borrow area and placed on the trucks. When the soil reaches the fill location, the soil is compacted and the unit weight increases. Usually, soil materials are placed in a compacted fill at greater density than that in their natural state in the borrow area. Intact rock usually is denser in its natural state than in a fill constructed with the rock after it has been ripped or shot. The amount of shrink or swell from borrow to fill is important, because payment for fill is usually based on the fill volume quantity, yet cost of doing the work is affected by the excavation and haul quantities, which are different. The quantities during these operations can be computed rationally, using information from subsurface exploration reports.

EXAMPLE   Shrink and Swell from Borrow to Fill

Consider a case where the subsurface exploration has shown that the soil dry density and water content in a borrow area for a job are 83 pcf and 13 percent, respectively, and that a 100,000 yd^3 fill must be built under a specification that calls for minimum relative compaction of 95 percent. Test results in the soils engineer's report show that the maximum dry density of the soil is 112 pcf; moreover, the saturated water content for this maximum dry density is 15.2%. Compute the following:

- Total volume and weight of the borrow soil.
- Shrinkage factor.
- The amount of water required (if any) for proper compaction in the fill.
- The number of truckloads for transport from borrow to fill, assuming each off-road truck can haul 100 kips of soil.

In looking at the problem statement, the first thing we notice is that we are transporting two substances: soil and water. Both of these substances have weight and volume. From Equation (3.5),

$$W_{t_{cut}} = W_{s_{cut}} + W_{w_{cut}}, \tag{8.5}$$

$$W_{t_{fill}} = W_{s_{fill}} + W_{w_{fill}}, \tag{8.6}$$

and, by definition,

$$V_{t_{cut}} = \frac{W_{t_{cut}}}{\gamma_{t_{cut}}} = \frac{W_{s_{out}}}{\gamma_{d_{cut}}}, \tag{8.7}$$

$$V_{t_{fill}} = \frac{W_{t_{fill}}}{\gamma_{t_{fill}}} = \frac{W_{s_{fill}}}{\gamma_{d_{fill}}}, \tag{8.8}$$

where

- W_t = total weight of cut or fill
- W_s = weight of soil of cut or fill
- W_w = weight of water of cut or fill
- V_t = total volume of cut or fill
- γ_t = total unit weight of cut or fill
- γ_d = dry unit weight of cut or fill

If we apply the definition of water content (see Table 3.1) to Equations (8.5) and (8.6), they reduce to

$$W_{t_{cut}} = W_{s_{cut}}(1 + w_{cut}) \tag{8.9}$$

$$W_{t_{fill}} = W_{s_{fill}}(1 + w_{fill}) \tag{8.10}$$

Substituting the definition of dry unit weight in Equations (8.7) and (8.8) into both of these equations yields

$$W_{t_{cut}} = V_{t_{cut}}\gamma_{d_{cut}}(1 + w_{cut}), \tag{8.11}$$

$$W_{t_{fill}} = V_{t_{fill}}\gamma_{d_{fill}}(1 + w_{fill}). \tag{8.12}$$

This substitution is important because all of the specifications for cut and fill are given in dry unit weight.

Since the specifications call for a minimum relative compaction of 95%, the specifications therefore require that the minimum density of the fill be $(112)(0.95) = 106.4$ pcf. Solving Equation (8.8) for the dry weight of the fill, we get

$$W_{d_{fill}} = (106.4)(100,000)(27) = 287,280,000 \text{ lbs.}$$

This is also the weight of the borrow soil as well. Since we know the dry unit weight of the borrow soil, from Equation (8.7), the total volume of the borrow soil is

$$V_{t_{cut}} = \frac{287,280,000}{83} = 3,461,205 \text{ ft}^3 = 128,193 \text{ yd}^3.$$

The shrinkage factor can be defined as

$$SF = \left(\frac{\gamma_{d_{fill}}}{\gamma_{d_{cut}}} - 1\right) \times 100\% \tag{8.13}$$

Substituting, we compute

$$SF = \left(\frac{106.4}{83} - 1\right) = 28.2\%.$$

The definition of the shrinkage factor given in Equation (8.13) is based on shrinkage from borrow to fill. In some cases, it can be reversed (i.e., from fill to borrow). In other cases, the shrinkage factor is expressed as a decimal, or simply the ratio of the two dry unit weights. The contractor needs to be clear on which definition is being used in the specifications that he or she is looking at.

The haul estimate might be based on yardage or tonnage. If it is to be based on yardage, then a swell factor from borrow to the haul unit will have to be derived. If the swell factor from borrow to haul is reliably quantified, the problem can be solved in this way. A more easily quantifiable way of solving the problem is to do so based on the weight rating of the haul units. To arrive at this result, we must first know the total weight of the borrow, which from Equation (8.9) is

$$W_{t_{cut}} = W_{s_{cut}}(1 + w_{cut}) = 287,280,000(1 + 0.13) = 324,626,400 \text{ lbs.}$$

Since each truck can haul 100 kips, the number of trips is

$$n_{trips} = \frac{324,626,400}{1000} = 3,246 \text{ trips.}$$

Finally, we need to consider the amount of water needed for compaction at the fill. Since the weight of the soil is the same, the additional weight of the water needed (if any) will be the difference of the total weights. From Equation (8.10), the weight of the fill is 330,946,560 lbs. The difference between the two is 6,320,160 lbs., or 757,661 U.S. gallons.

We need to note the following concerning borrow and fill calculations:

- Shrinkage calculations are very sensitive to errors, and these errors can magnify themselves rapidly because of the large scope of the work. For example, the weight difference between the borrow and fill is only 1.95% of the borrow weight, yet this translates rapidly into a large amount of water. Failure to properly estimate this expense can lead to serious economic problems during a job. All factors need to be properly quantified during the estimating process. The contractor even needs to minimize the round-off error of the calculations.

- This example assumes that the contractor is under a performance specification and has some flexibility in coming up with the borrow volume. Other, more prescriptive proposals might have an excavation plan in them according to which the contractor is given a borrow area with the proper amount of soil. It is especially important in this case for the contractor to thoroughly review the cut and fill specifications to make sure the excavation plan is realistic, and that the contractor will "make the cut" and have sufficient soil for the fill requirements.

8.6   COMPACTION EQUIPMENT FOR SHALLOW LIFTS

To meet compaction specifications, the contractor must select the proper equipment to do the job, and use it correctly. It is the engineer's responsibility to see that the specifications are met.

Selection of equipment is keyed to the nature of soil to be compacted, the degree of compaction required, and the space available in which to do the job. It is well known that sands are most efficiently compacted by vibration, while cohesive soils are better compacted by pressure that is maintained for some time. This is logical, considering earlier discussions of soil structure and plasticity. Some indication of the acceptability of different types of large compaction equipment is given by Figure 8.11. Among the various types of compactors available, there are many variations of size and shape of the compacting element. Figures 8.12, 8.13, and 8.14 show several of the types of rollers indicated in Figure 8.11.

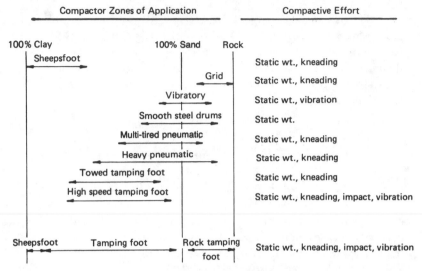

Compactor Zones of Application	Compactive Effort
Sheepsfoot	Static wt., kneading
Grid	Static wt., kneading
Vibratory	Static wt., vibration
Smooth steel drums	Static wt.
Multi-tired pneumatic	Static wt., kneading
Heavy pneumatic	Static wt., kneading
Towed tamping foot	Static wt., kneading
High speed tamping foot	Static wt., kneading, impact, vibration
Sheepsfoot Tamping foot Rock tamping foot	Static wt., kneading, impact, vibration

Figure 8.11 Compaction equipment selection guide. This chart contains a range of material mixtures from 100 percent clay to 100 percent sand, plus a rock zone. Each roller type has been positioned in what is considered to be its most effective and economical zone of application. However, it is not uncommon to find them working out of their zones. Exact positioning of the zones can vary with differing material conditions.

(a)   *(b)*

Figure 8.12 Compaction equipment for coarse materials. (*a*) Smooth drum vibratory roller for compacting granular materials. (*b*) A grid roller for compacting rock.

Figure 8.13   Sheepsfoot Roller.

Figure 8.14   A self-propelled tamping roller.

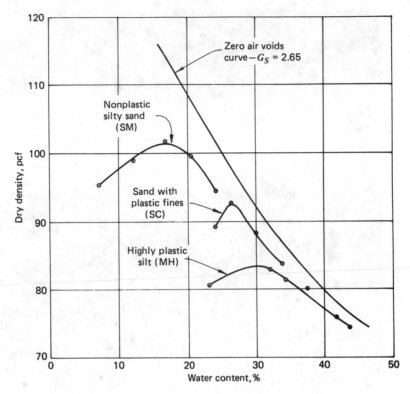

Figure 8.15   Moisture–density curves indicating the effects of soil plasticity differences.

It would be desirable, in advance of construction, to know how the properties of the soil to be placed relate to the optimum conditions for compaction. This information provides a realistic basis for selecting equipment and scheduling to allow for additional watering or drying of embankment materials. In general, it is found that the shape and position of the moisture–density curve for cohesive soils depend on soil plasticity characteristics. These effects are illustrated by Figure 8.15. The more plastic materials do not compact to high densities. The densities obtained for them are insensitive to moisture content changes as illustrated by the relative flatness of the moisture–density curve. The soils of lower plasticity must be compacted within narrow ranges of water content to attain optimum conditions. Figure 8.16 relates optimum water content for a large number of soils to their plasticity characteristics. Figure 8.17 empirically relates density at optimum water content to optimum water content. Using Figure 8.16, we find that the soil's optimum water content may be estimated by means of classification indices. This may be compared with field water content in a borrow area to assess the workability of the material. Figures 8.16 and 8.17 may be used with field test data to make preliminary estimates of material quantities.

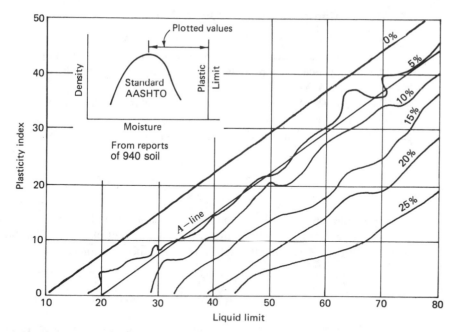

Figure 8.16 Optimum moisture content related to soil plasticity. (Reprinted, by permission, from McDonald, J.K., "Soil Classification for Compaction," *Proceedings, 10th Annual Engineering Geology and Soils Engineering Symposium*, Moscow, Idaho, 1972.)

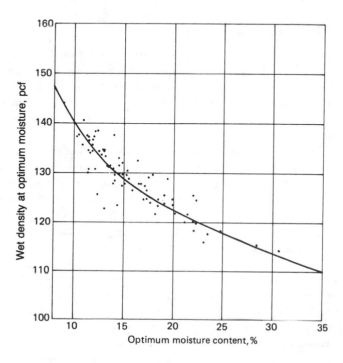

Figure 8.17 Empirical relationship of optimum moisture content and wet density at optimum moisture content. (Reprinted, by permission, from Hilf, J.W., *A Rapid Method for Construction Control for Embankments of Cohesive Soil*, American Society for Testing and Materials, Special Technical Publication, No. 232, 1957.)

EXAMPLE   Compaction of Cohesive Soil

Suppose that laboratory tests on soil to be used in a fill show that its liquid limit and plastic limit are 46 and 30, respectively. Its natural water content is about 28 percent and its dry density, in the borrow area, is about 86 pcf. Figure 8.16 indicates that the optimum water content for this soil (an ML in the Unified Classification System) is about 10 percent below its plastic limit, or about 20 percent. We learn, in addition, from Figure 8.17 that the wet density of the soil at its optimum water content is about 122 pcf. Solving Equation 8.10 for a unit volume, we find that the dry density of the fill at optimum water content would thus be about

$$\frac{W_{s_{\text{fill}}}}{(1 + w_{\text{fill}})} = W_{s_{\text{fill}}} = \frac{122}{(1 + 0.20)} = 101.7 \text{ pcf.}$$

From the foregoing illustration, we have learned that the soil will be very wet and of optimum water content and will require drying for efficient placement. Furthermore, excavation might be difficult. Assuming that the compaction specification calls for 95 percent of standard maximum dry density, or

$$0.95 \times 101.7 = 96.6 \text{ pcf,}$$

it follows from Equation (8.13), that the shrinkage factor from borrow to fill should be about

$$\text{SF} = \left(\frac{\gamma_{d_{\text{fill}}}}{\gamma_{d_{\text{cut}}}} - 1\right) \times 100\% = \left(\frac{96.6}{86} - 1\right) \times 100\% = 12.3\% \tag{8.13}$$

8.7   COMPACTION CONTROL TESTING

The ultimate purpose of compaction control efforts is to ensure that the owner in the contractual relationship receives a fill that will serve his or her purpose. The engineer will sample the fill periodically to ensure compliance with the specifications. The fill may be checked after a certain number of cubic yards has been placed, after a new lift has been placed, or sometimes after each shift. On large jobs, the contractor's operation may sometimes be made more efficient by careful interpretation of these test results.

Compaction control tests involve determination of the in-place density and water content of the fill. A number of methods are available. Field density is determined either directly or indirectly. Conventional tests such as the sand-cone (ASTM D1556, see Appendix A.1), Washington Densometer (ASTM D2167), or oil-replacement method involve excavation of a small hole, the material from which is weighed. The volume of the hole is found by filling it with sand of a known density, a balloon full of water, or oil. The weight of soil to volume of soil ratio is the wet density. Water content of the soil removed is also determined and the dry density is calculated using Equation (3.4). The rapid determination of moisture content is essential so that the contractor will have his fill approved or disapproved before placing more material. Figure 8.18 illustrates the results for the sand cone test. A test setup is shown in Figure 8.19.

It should be noted that the usual field density test procedure involves compacting one cylinder (a check mold) of the soil removed for the test to insure that it is the same soil that the laboratory standard moisture-density curve being used as a reference for relative compaction

FIELD DENSITY — SAND CONE METHOD

TEST FOR __Site Fill - Test 3__

TEST BY __TKN__   DATE __6 JUNE 1994__

SOIL DESCRIPTION __gravelly sand__

TEST LOCATION __see sketch__

FIELD DENSITY		CYLINDER	
1. Wt. Sand + Jar	11.90	1. ASTM Designation	D 1557
2. Wt. Residue + Jar	5.78	2. Wt. Soil + Mold	9.63
3. Wt. Sand Used (1) - (2)	6.12	3. Wt. Mold	9.36
4. Wt. Sand in Cone & Plate	3.41	4. Wt. Soil	4.74
5. Wt. Sand in Hole (3) - (4)	2.71	5. Wet Density, pcf (4) ÷ Vol. Mold	142.2
6. Density of Sand	89.5	6. Moisture Content	6.76
7. Wt. Container + Soil	4.57	7. Dry Density, pcf (5) ÷ [1 + (6)]	133.2
8. Tare Wt. Container	0.16		
9. Wt. Soil (7) - (8)	4.41	MOISTURE CONTENT PAN NO. __110__	
10. Vol. of Hole (5) ÷ (6)	0.0303	WW __127.25__ DW __121.30__	
11. Wet Density, pcf (9) ÷ (10)	145.54	DW __121.30__ TW __33.25__	
12. Moisture Content, %	6.56	WATER __5.95__ SOIL __88.05__	
		PER CENT MOISTURE __6.76__	
13. Dry Density, pcf (11) ÷ [1 + (12)]	136.58	NOTES:	
Representative Curve No.	2		
14. Optimum Moisture, %	7.0		
15. Maximum Dry Density, pcf	133.3		
16. Relative Compaction, % (13) ÷ (15)	102.0		

FIELD MOISTURE DETERMINATION

Method __oven__

Percent Moisture __6.56__

NOTES:

172'

↑ NORTH

prop. line

156'

⊗

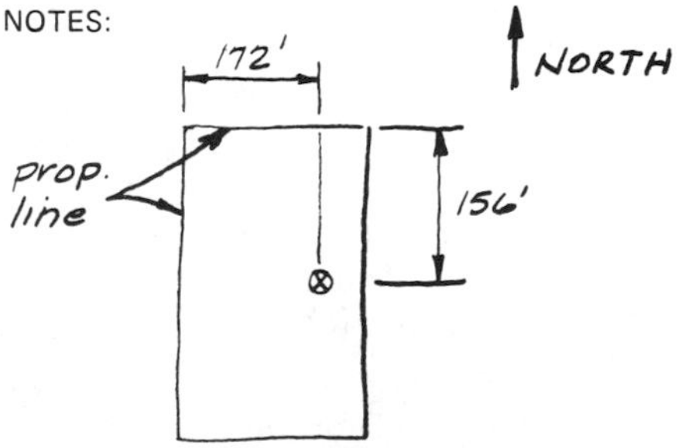

Test @ EL. 26.6

depth in fill = 18"

Figure 8.18 Field density test results using the sand cone method.

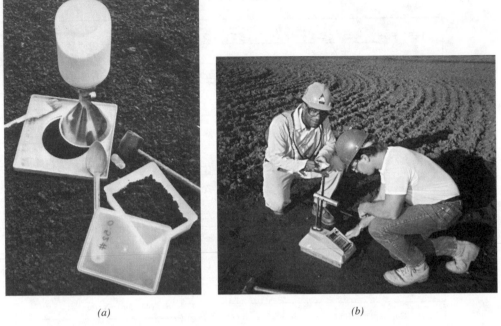

(a)   *(b)*

Figure 8.19 Common methods of field density testing. (*a*) The sand cone test.
(*b*) Nuclear density testing.

calculations represents. Figure 8.20 shows laboratory curves for Soil No. 1 and Soil No. 2. These soils had a similar appearance, yet their maximum densities differed significantly. When check molds were run during field density testing on the project, it was clear that the soils tested in the field did not have the same moisture–density curves as those tested in the laboratory, except in two cases. The check mold points did not fall on the previously established curves. Since we know, however, that similar soils will have similarly shaped moisture–density curves, approximate moisture density curves for each soil tested in the field may be sketched as shown, using the check mold point for each test and the two laboratory curves as a guide. The maximum dry densities and optimum water contents established by this procedure are often good enough for determining whether specifications have been met, without the need to establish a laboratory curve for each field density test sample.

In the right kinds of soils, undisturbed samples of regular shape may be taken by sampling tubes or hand trimming. This permits a volume determination to be made without replacement by some other substance. It is only necessary to calculate volume directly. An indirect method that employs radioactive isotopes (ASTM D2922), however, is more convenient and rapid, especially on larger jobs, than this or the other methods discussed above. Because of its rapidity, the nuclear density test has become very widely used.

Nuclear density gauges (see Figure 8.19*b*) consist of two essential parts, a radiation source and a radiation counter. The source is placed either on the fill surface or at some depth. To make the density determination, a gamma radiation source is used. The emissions sensed by

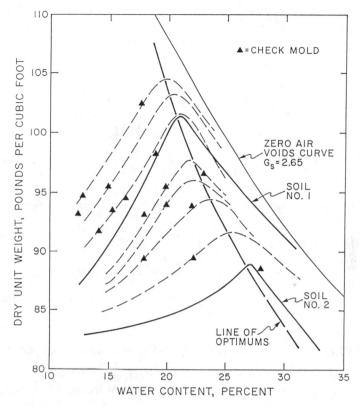

Figure 8.20 Use of check mold points to estimate maximum dry density and optimum water content.

the counter are inversely proportional to the density of the material (soil) through which they pass. The count rate is calibrated to reflect this property. For a water content determination, a high-energy neutron source is used. The emissions are reflected and counted at a rate proportional to the soil water content. With proper calibration, the method gives accurate results. Its principal benefits, however, are that the results are available immediately and that large numbers of tests may be conveniently made. Nuclear density gauges give the best results in soils that are fairly uniform and without large rocks. The results are generally unsatisfactory in soils that have a high content of hydrocarbons; this can include organic soils or soils that have been contaminated by hydrocarbons, such as gasoline.

The standard compaction test (ASTM D698) is run on material passing either the No. 4 sieve or the three-fourths-in. sieve. Any larger material is removed. Yet field density tests involve samples that contain this fraction, which is not included in establishing the laboratory standards for the project. What effect does this have on the compliance with specifications? The next example illustrates the point. The procedure follows ASTM D4718, a method commonly used for correction for the oversize fraction.

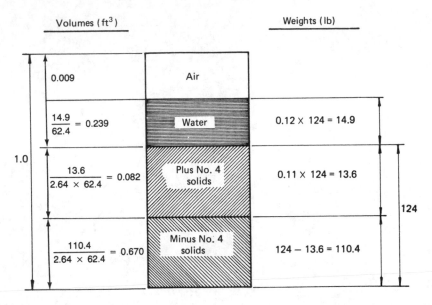

Figure 8.21   Phase diagram illustrating an example of the oversize correction.

EXAMPLE   Correction for Oversize Fraction in Compaction Tests

Assume that a field density test has been run on a soil with 11 percent of its weight retained on the No. 4 sieve. The test showed that the dry density was 124 pcf and that the water content was 12 percent for the total sample. The specifications, based on ASTM D698, require a minimum dry density of 122 pcf and maximum water content of 12 percent. Were they satisfied? Figure 8.21 shows a breakdown of the constituents, using a specific gravity of 2.64 for the solids. The assumption must be made in correcting for oversize material that the water is associated with a certain size range of the solids in some regular fashion. Commonly, it is assumed that all of the moisture is held by the minus No. 4 material. Alternatively, the moisture content of the coarser fraction may be measured. For purposes of this illustration, if the water content of the plus No. 4 soil is taken as zero, for the minus No. 4 material,

$$w = \frac{14.9}{110.4} \times 100\% = 13.5\%,$$

and the dry density is

$$\gamma_d = \frac{110.4}{1 - 0.082} = 120.3 \text{ pcf.}$$

Therefore, the soil fraction for which the specification was developed is wetter than allowed, and has not been sufficiently compacted. The specification is not met; yet, for the total sample, it appeared that it had been.

The following example illustrates how field and laboratory test results may be used to modify an unsatisfactory compaction operation. It should be remembered that to change compaction results we may change compaction energy or water content or both.

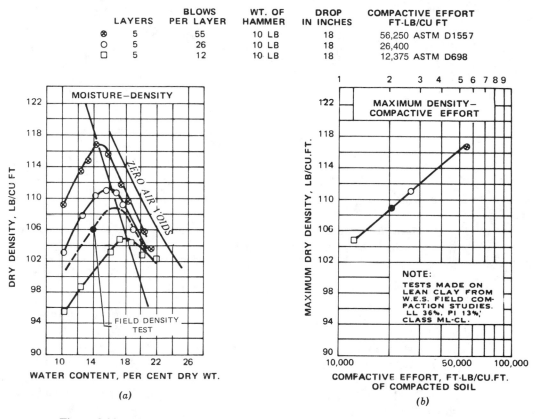

	LAYERS	BLOWS PER LAYER	WT. OF HAMMER	DROP IN INCHES	COMPACTIVE EFFORT FT-LB/CU FT
⊗	5	55	10 LB	18	56,250 ASTM D1557
O	5	26	10 LB	18	26,400
□	5	12	10 LB	18	12,375 ASTM D698

Figure 8.22 Illustrative examples of modification of compaction procedure based on laboratory tests and field density tests. Note tests were run using 6-in. mold. Actual compaction energy varies slightly from that for 4-in. mold. (From *Soil Manual* (MS-10), published by the Asphalt Institute.)

EXAMPLE Modification of an Unsatisfactory Compaction

Consider Figure 8.22. It has been experimentally demonstrated that a semilogarithmic plot of maximum dry density versus compactive effort, such as that shown in Figure 8.22*b*, is linear. Assume that the soil for which these results were obtained is to be compacted to 100 percent of maximum density according to ASTM D1557 and at optimum water content. Further assume that the project has begun and that a field density test shows that the fill has a dry density of 106 pcf and a water content of 14 percent. How should the compaction procedure be changed to produce the desired result? In Figure 8.22*a*, the coordinates representing the fill conditions have been plotted, and the compaction curve corresponding to the compactive effort being used has been sketched. For this curve, the maximum dry density for the field compactive effort is about 109 pcf. From Figure 8.22*b* the field compactive effort may be estimated by intersecting the laboratory curve at this density. Since the field effort is equivalent to about 20,000 ft-lb/ft³ and the effort in the ASTM D1557 test

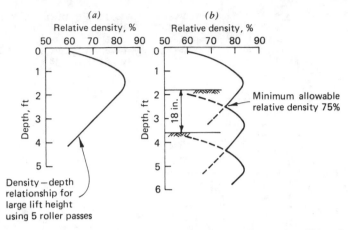

Figure 8.23 Depth–density relationships for compaction of a granular soil by vibratory roller. (Reprinted, by permission, from D'Appolonia, D.J., Whitman, R.V., and D'Appolonia, E., "Sand Compaction with Vibratiory Rollers," *Journal, Soil Mechanics and Foundations Division*, American Society of Civil Engineers, Vol. 95, SM1, 1969.) (*a*) Single pass. (*b*) Multiple lifts.

on which the specification is based is about 56,000 ft-lb/ft^3, the field effort will have to be increased by about 2.5 times to achieve the desired results. Thus, if the roller has been making two passes over the fill, the procedure should be changed so that it makes at least five passes before compaction is checked again. Note that in this case water content should also be increased about 1 percent, to comply with specifications. It is, of course, possible that the roller being used will not be capable of obtaining the required density with the lift thickness being used.

The foregoing paragraphs have considered compaction in shallow lifts on the order of 2 ft thick or less. Surface rolling is not usually effective below this depth. Figure 8.23 illustrates the depth–density relationship resulting from vibratory compaction of a cohesionless material. Maximum compaction usually occurs a foot or two below the ground surface. To achieve some specified density then, the lift thickness should not exceed the depth to the minimum acceptable density on the plot. Figure 8.23*b* shows the compaction achieved in a number of lifts, illustrating the need for lift thickness control. In a cohesive soil, compaction equipment such as the sheepsfoot roller compacts from the bottom of the lift up, rather than from the top down. Large-tired pneumatic equipment works from the top down. In the former case, the lift thickness is limited by the length of the feet on the compactor. In the latter case, the width of the tire and the weight of the roller determine how deeply the soil will be compacted.

8.8 CHEMICAL STABILIZATION OF FILLS

There are a host of chemicals and other products that can be used to improve soil properties for a specific purpose. None is a universal cure-all for improving engineering properties or construction workability. With certain soils, certain additives have a long and widely known history of success. The principal additives employed are Portland cement and lime. The benefits of chemical stabilization are described generally in this section. Results that are more specific are described in Section 12.3.

Cement is added to soil to improve strength and durability. The usual application is to pavement construction where *soil–cement* is substituted for aggregate base. Other successful uses have included slope protection for dams and levees and impermeable linings for canals and reservoirs. Cement usually proves effective as a stabilizing agent for coarse soils with a low percentage of fines or for fine soils having little plasticity. To produce true soil–cement (having high strength and adequate durability), the percentage of cement required is usually between 5 and 10 for sands and gravels and between 10 and 20 for silts and clays. Strengths approaching about one-half that of concrete are not uncommon, with values of seven-day, unconfined compressive strength in the 500 to 700 psi range generally required to meet durability requirements. Lower cement contents are sometimes used to produce *cement-modified soil*, which is similar to soil-cement but does not meet durability requirements.

As for concrete, strength for soil–cement increases with time. Because development of strength begins as soon as water and cement are mixed, construction with soil-cement must be done according to a carefully controlled schedule. Long time delays between addition of water and compaction are unacceptable. Construction with soil–cement may be done by a batch process or a mix-in-place process, according to the requirements and materials for a particular job. In the batch process, constituents are obtained from stockpiled materials, mixed, transported, and placed. A uniform product can result if close control on all phases of the work is maintained. The essential elements of a mix-in-place operation include loosening of the material to be stabilized, addition of cement, addition of water, blending, and compaction. The opportunity for nonuniformity is greater on a mix-in-place job, and sometimes greater cement content than the actual percentage required is specified to compensate.

Lime is an effective stabilizing agent for fine soils with high plasticity. The addition of quicklime (CaO) or hydrated lime [$Ca(OH)^2$] results in both a cementing action and an alteration of the physical chemistry of the soil–water system, which both have beneficial effects. The calcium in the lime reacts with available silicas in the soil to produce cement, which increases strength. In some cases, additional silica (pozzolan) is added which enhances this effect. In highly plastic clays, the calcium from the lime is attracted to the negatively charged soil particle surfaces (see Figure 3.5) which makes their repulsive potential less strong. The result is a collapse of the double layer accompanied by decreased plasticity and improved workability of the soil. The amount of lime added does not usually exceed about 10 percent. Laboratory studies are used, as is also the case for soil–cement, to determine the optimum percentage. Results of one such series of tests are shown in Figure 8.24.

Hydration of quicklime by the water in a soil according to the reaction indicated by Equation (8.14)

$$CaO + H_2O \rightarrow Ca(OH)_2 + Heat \tag{8.14}$$

can result in improved soil workability on a very wet construction site. Equation (8.5) also indicates that, pound for pound, quicklime has a greater potential for beneficiating soil than hydrated lime because of the proportionately greater amount of calcium (71 percent of total weight versus 54 percent) it contains. Because of this, on a large job the disadvantages of the added precautions that must be taken in handling quicklime may be offset by savings in haul costs.

Lime stabilization may be accomplished with less stringent schedule requirements than for cement stabilization, especially when the primary benefits are derived from alteration of soil–water chemistry. To this end, the lime–soil mixture is sometimes lightly compacted after blending, then remixed after a curing period, and compacted to the specified density.

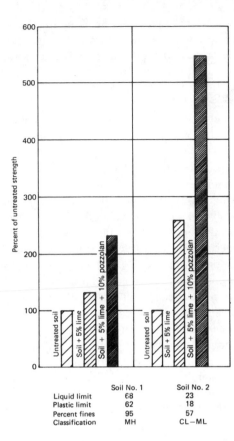

Figure 8.24 Summary of results of tests on lime-pozzolan stabilization. Soils compacted near plastic limit. Samples stored seven days before testing. Strength determined by unconfined compression test.

The curing period allows for a more thorough dispersion of calcium ions through the soil. In most instances, lime stabilization is accomplished in a mix-in-place operation.

Figure 8.25 shows a large stabilization project where cohesionless sand dredge spoils were stabilized with lime, cement, and pozzolan to form the base for a heavy industrial pavement in the backup area at a container ship terminal.

8.9 STABILIZATION OF FILL FOUNDATIONS

When fills are placed on compressible foundation soils, the weight of the fill induces settlement as discussed in Section 5.5. The purpose of the fill is usually to provide a surface for construction of structures or pavements. The amount of settlement is determined by the load resulting from the fill and the compressibility and thickness of the foundation soil. The rate at which the settlement progresses is determined by the permeability (which determines the coefficient of consolidation) of the foundation soil and the efficiency of the subsurface drainage. In the event that the settlement is large and takes a long time to occur, structures built on the fill surface may be damaged. Various methods are employed to preclude this from happening.

Figure 8.25   Stabilization of sands using lime, cement, and pozzolanic additives. Construction was the batch process. (Courtesy of M.L. Byington.)

8.9.1 Surcharge Loading

On one project, laboratory testing and analysis showed that a planned site fill would induce some 6.5 in. of settlement at the site of a major metropolitan sewage-treatment plant. Furthermore, it was estimated that these settlements would develop over about a two-year period. Since this time exceeded the planned time for construction, and since there could be no assurances that the proposed structures would not suffer damage from these deep-seated settlements, it was decided to *preload*, or *surcharge*, the site. In this process, a fill of greater depth than required for site development is imported with the result that greater settlements are induced at a more rapid rate than would be the case for the site fill only. The predicted relationships for this case are shown in Figure 8.26. The plan was to observe actual settlements, remove the surcharge, and proceed with construction when the settlement reached that resulting from the site fill only. The result would be a site with fill-induced settlements complete and foundation soils precompressed above the load that results from the site fill and structures. One year was allowed for the process. Actual measurements showed that the soil was much more compressible than had been anticipated, and the settlements induced were very large. Fortunately, as is usually the case, the settlements took place more rapidly than predicted. During the year allowed before removal of the surcharge and initial work on the structures, nearly 3 ft of settlement occurred, approximately half of which resulted from the site fill. The surcharge was removed and construction proceeded as planned, with no subsequent difficulty with the structures.

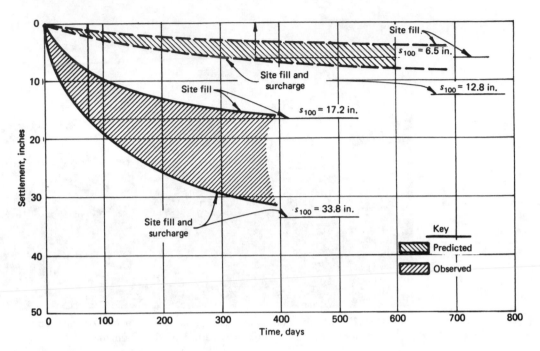

Figure 8.26 Example of time–settlement relationships for site fill effects.

8.9.2 Vertical Drains

Surcharging accelerates settlements because of the weight of extra fill added. Settlement may also be accelerated by improving subsurface drainage using vertical sand drains. In this process, holes are drilled in compressible, relatively impervious soils before site fill is added. The holes are relatively closely spaced, and they are filled with pervious soils. The drains so constructed greatly decrease the drainage path length in the subsoil and, as a result, speed consolidation and settlement. Figure 8.27 shows equipment used to construct over 7000 such drains to depths exceeding 150 ft on one job. This equipment consists essentially of a hollow-stem auger attached to a sand hopper that can be sealed and pressurized. The auger is drilled to the desired depth and then retracted without rotation as the sand in the hopper is forced out the auger stem to fill the resulting void. Figure 8.28 shows a nomograph which allows estimation of the performance of a vertical drain system.

An alternative to sand drains that has been widely adopted involves various fabric or plastic drains, usually referred to as *wick drains*. These drains are punched into the soil with about the same spacing as sand drains and function in a similar manner. They offer a much-simplified method of construction. Required equipment is shown in Figure 8.29.

8.10 DEEP COMPACTION OF SANDS

If soils are to be compacted to great depths, say in lifts of up to 30 or 40 ft, special procedures are required. Coarse soils have been successfully densified by controlled explosive charges set in patterns at depth. More commonly, densification can be obtained by driving displacement

Figure 8.27   Vertical sand-drain construction equipment. Sand in the hopper on the upper end of the auger is discharged under compressed air as the auger is withdrawn.

piling, by vibroflotation, or by use of a vibrating mandrel that is driven and withdrawn. Deep dynamic compaction, involving the dropping of large weights from great heights, may also be used. With this technique, a crane is used and the weight is lifted by the crane and dropped onto the ground. The weight is dropped in a grid pattern for compaction over a large area. Displacement piling are simply timber piling that serve no purpose other than to occupy void space. If the density of a deposit is known and the desired density is established, the spacing of piling

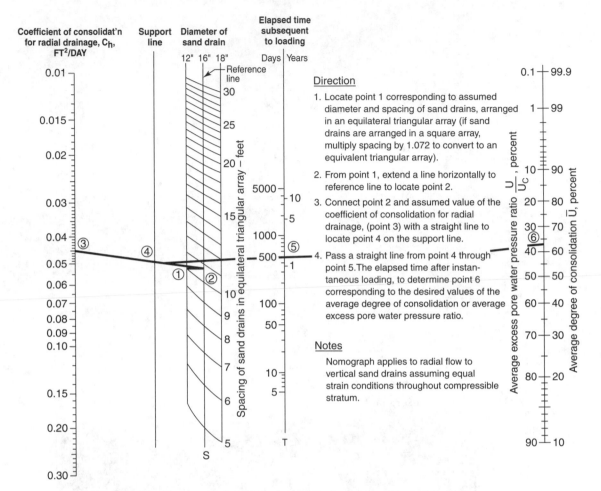

Figure 8.28 Nomograph for Estimating the Performance of Vertical Drain Systems. (after NAVFAC DM 7.01)

required to achieve it may be computed. A compaction-piling job is illustrated in Figure 8.30. Vibroflotation is a patented process in which a vibrating probe is sunk with the aid of water jets in the deposit to be densified; the equipment is shown in Figure 8.31. As the equipment is withdrawn, the vibration continues and sand is added at the surface to occupy the volume of the depression resulting from densification. A vibrating probe such as that shown in Figure 8.32, used without the aid of water jetting, produces similar results but does not appear to be as efficient. It is, however, much less expensive. The deep compaction methods all work best in submerged clean coarse soils. Deep compaction is often specified using prescriptive requirements. The contractor is told what to do in terms of probe spacing, for instance. The engineer will then determine the acceptability of the work, usually with a subsurface exploration operation using standard penetration tests or a cone penetrometer to estimate relative density. If more work is required, provision should be made for additional compensation. In some instances, a performance specification is written and the contractor must know what can be accomplished with a

(a)   *(b)*

Figure 8.29   Equipment for inserting vertical wick drains. (*a*) Crane and boom. (*b*) Rolled wick fabric supply.

Figure 8.30   Densification of loose sand behind a bulkhead using compaction piling. (Courtesy of E.G. Worth.)

(a)   *(b)*

Figure 8.31 Vibroflotation equipment for densification of deep sand deposits.
(*a*) Probe mounted on crane. (*b*) Probe during insertion. Note material on front loader
for replacement. (Courtesy of Tom Huntsinger.)

given procedure to bid the work properly. Figure 8.33 illustrates average results obtained on
several jobs of this nature. It should be noted that maximum density achieved by deep probing
occurs at a depth of about 15 ft, and that densification is more difficult to achieve on slopes. Although Figure 8.32 shows a pipe pile used as a probe, probes used can actually be a number of
shapes, such as H-beams and customized configurations. Probe systems can also be equipped
with instrumentation that measures the soil response to the vibration and adjusts the frequency
and amplitude of the vibrations to optimize the compaction performance. This is referred to as
resonance compaction. Figure 8.33 may be used to estimate work requirements on a job with
performance-type specifications. It is important to note, however, that the efficiency of the vibroflotation and probing operations are time dependent and that probing time requirements
may differ at different sites, even though probe spacing may be constant.

Figure 8.32   Using a 30-in. pipe probe and vibratory pile hammer to densify loose sand.

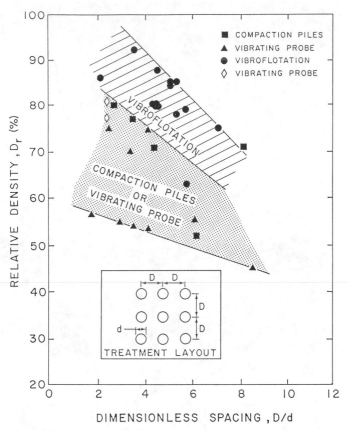

Figure 8.33 Results of various deep compaction methods for loose submerged sands. (Reproduced, with permission, from American Society of Civil Engineers.)

8.11 SUMMARY

Soils are compacted to produce a material with known properties, which, in turn, determines the suitability of an embankment. Test methods are available to serve as a standard for construction control. These same tests can be used to produce samples for laboratory studies to correlate soil engineering properties and more easily measured index properties such as density and water content.

Following study of this chapter, the reader should

1. Be able to explain why moisture–density curves for various soils exhibit the shape they do.
2. Know why soils are compacted.
3. Know in a general way how compaction influences engineering properties of soils.
4. Be familiar with common laboratory compaction testing methods.
5. Be able to perform calculations associated with compaction testing and presentation of compaction test results.

In the field, soil may be compacted in a number of ways. It should always be remembered that the purpose of compaction is to produce a manufactured fill with controlled engineering properties. The conditions under which the work is undertaken, the equipment used, and the procedures followed, all have a bearing on the final product. In many cases, the contractor is obligated by the contract language to assume the responsibility for the consequences of an unsatisfactory fill. It is, therefore, to the contractor's benefit to build it properly as well as efficiently.

The reader should now

1. Know the meanings of the terms *relative density* and *relative compaction.*
2. Be able to list the provisions of a well-written compaction specification.
3. Understand the ambiguities in a poorly written specification.
4. Know what types of compaction equipment and methods are applicable to given soil conditions.
5. Know how compaction control tests are run.
6. Understand how compaction procedures might be modified to produce a desired result.

In some contracts, provision is made for soil improvement by means other than compaction. The reader should

1. Understand the reasons for stabilization of soils by addition of lime or cement.
2. Know whether lime or cement may be the most appropriate stabilizing agent in a given situation.
3. Understand the principles involved in preloading and the use of vertical drains.

REFERENCES

Aldrich, H.P., "Precompression for Support of Shallow Foundations," *Journal, Soil Mechanics and Foundations Division, American Society of Civil Engineers*, 91, No. SM2, 1965.

Anderson, R.D., "New Method for Deep Sand Vibratory Compaction," *Journal, Construction Division, American Society of Civil Engineers*, 100, No. C01, 1974.

Asphalt Institute, *Soils Manual MS-10*, College Park, Md., 1969.

Caterpillar Tractor Company, *Caterpillar Performance Handbook* (3d ed.), 1973.

Coduto, Donald P. *Geotechnical Engineering: Principles and Practices*. Upper Saddle River, NJ: Prentice Hall, 1999.

D'Appolonia, E., *Loose Sands—Their Compaction by Vibroflotation*, American Society for Testing and Materials, Special Technical Publication No. 156, 1953.

Hall, C.E., "Compacting a Dam Foundation by Blasting," *Journal, Soil Mechanics and Foundations Division, American Society of Civil Engineers*, 88, No. SM3, 1962.

Highway Research Board, "Cement-Treated Soil Mixtures," *Highway Research Record* 36, 1963.

Hilf, J.W., *A Rapid Method for Construction Control for Embankments of Cohesive Soil*, American Society for Testing and Materials, Special Technical Publication No. 232, 1957.

Johnson, A.W. and Sallberg, J.R., "Factors That Influence Field Compaction or Soils," *Highway Research Bulletin* 319, Highway Research Board, National Research Council, 1962.

Lambe, T.W., "Soil Stabilization," in *Foundation Engineering*, G.A. Leonards, Editor, New York: McGraw-Hill, 1962.

Lee, K.L., and Singh, A., "Compaction of Granular Soils," *Proceedings, 9th Annual Engineering Geology and Soils Engineering Symposium*, Boise, Idaho, 1971.

Leycure, P. and Schroeder, W.L., *Slope Effects on Probe Densification of Sands, in Soil Improvement, a Ten Year Update*, Geotechnical Special Publication No. 12, American Society of Civil Engineers, 1987.

McDonald, J.K., "Soil Classification for Compaction," *Proceedings, 10th Annual Engineering Geology and Soils Engineering Symposium*, Moscow, Idaho, 1972.

McDowell, Chester, "Stabilization of Soils with Lime, Lime-Flyash, and Other Lime Reactive Materials," *Highway Research Board Bulletin* 231, 1959.

Portland Cement Association, *Soil-Cement Laboratory Handbook.* Chicago, 1959.

Richart, F.E., "A Review of Theories for Sand Drains," *Transactions, American Society of Civil Engineers*, 124, 1959.

Schroeder, W.L., and Worth, E.G., "A Preload on Fine Silty Sand," *Proceedings of Specialty Conference on Performance of Earth and Earth-Supported Structures*, Vol. 1, American Society of Civil Engineers, 1972.

Seed, H.B., and Chan, C.K., "Structure and Strength Characteristics of Compacted Clays," *Journal, Soil Mechanics and Foundations Division, American Society of Civil Engineers*, Vol. 85, No. SM5, 1959.

United States Bureau of Reclamation, *Earth Manual*, 1974.

Washington Department of Highways, *Standard Specifications for Road and Bridge Construction*, 1963.

Welsh, J.P. (Ed.), *Soil Improvement, A Ten Year Update*, Geotechnical Special Publication No. 12, American Society of Civil Engineers, 1987.

PROBLEMS

1. Data points from a standard moisture–density test on a fine-grained soil are given in the following table:

Water Content (%)	Total Density (pcf)
20.0	99.6
23.1	105.9
25.6	111.2
28.5	115.1
31.3	113.4
34.0	111.5

Reduce the data and plot the water content versus dry density curve. Clearly indicate values for optimum water content and maximum dry density from this test.

2. A new employee compacted one cylinder of the soil described in Problem 1 to compare new results with those shown. The data obtained included a water content of 32.5 percent and a total density of 121.9 pcf. Comment on these new findings.

3. Given the definition of relative density in terms of void ratio (Equation (8.1)), derive Equation (8.2) in terms of unit weights.

4. Derive the general relationship

$$w = S\left(\frac{\gamma_w}{\gamma_d} - \frac{1}{G_s}\right)$$

between the water content and the dry unit weight for any point in Figure 8.8. Now derive Equation (8.3) from this relationship, and then plot the curves for the zero air voids curves for specific gravity of solids values of 2.6, 2.7 and 2.8. Discuss the possible significance of these curves.

5. For the soil described in Problem 1, assume that a field density test on a fill lift indicates a water content and dry density of 23.2 percent and 88.6 pcf, respectively. Would specifications calling for a minimum relative compaction of 96 percent of standard maximum dry density and a water content 62 percent of optimum be satisfied?

6. Maximum and minimum dry densities for a sandy soil according to ASTM D4253 and D4254 are 96.3 pcf and 82.6 pcf, respectively. Compute the relative density of this soil if its in-place dry density were
   a. 90.2 pcf
   b. 81.6 pcf
   c. 99.1 pcf

7. Laboratory compaction test results for a plastic soil are given in the following table:

Test Method ASTM D698		Test Method ASTM D1557	
Water Content (%)	Dry Density (pcf)	Water Content (%)	Dry Density (pcf)
11.8	100.2	12.0	116.2
13.6	102.8	13.6	117.6
15.7	105.3	14.8	117.9
17.7	106.4	16.1	116.5
19.6	103.7	17.2	113.6
20.8	100.8		

Specifications call for a fill dry density of 114 pcf and further require that the fill be compacted at optimum water content for the field compactive effort. For the following field density test results on a fill, explain how you would alter the field placement operation:
   a. $\gamma_d = 115$ pcf; $w = 12.5$ percent
   b. $\gamma_d = 112.5$ pcf; $w = 13.5$ percent
   c. $\gamma_d = 110$ pcf; $w = 14.5$ percent
   d. $\gamma_d = 111$ pcf; $w = 17.3$ percent

8. A subsurface exploration report shows that the average water content of a fine-grained soil in a proposed borrow area is 22 percent and that its average natural dry density is 82 pcf. This soil is to be placed in a compacted fill with a dry density specified at 96 pcf. How many cubic yards of borrow will be required to produce 50,000 yd³ of fill? What is the shrinkage factor for the borrow?

9. A field density test has been run on a soil containing 18 percent of its weight in particles retained on the No. 4 sieve. The dry density of the total sample was 128.7 pcf and the water content was 7 percent. The water content of the gravel fraction was 3 percent. The specific gravity of solids was 2.69. Compute the water content and dry density of the soil fraction of the fill that passes the No. 4 sieve.

10. Atterberg limits and natural water contents for five fine-grained soils are listed in the following table:

Soil No.	Liquid Limit (%)	Plastic Limit (%)	Natural Water Content (%)
1	10	10	4
2	33	18	20
3	46	34	12
4	74	31	32
5	74	49	16

Classify the soils in the Unified Soil Classification System, estimate their optimum water contents and maximum dry densities, and indicate approximately how much each soil must be wetted or dried if it is to be efficiently compacted.

CHAPTER 9

Dewatering

Many excavations are carried below groundwater level. Techniques for dealing with the problems that result are dependent on the excavation dimensions, the soil type, and the groundwater control requirements, among other factors. The simplest dewatering operations are carried out with little planning. Major operations in difficult conditions require advanced engineering and construction methods.

This chapter is intended to provide the reader with

1. An awareness of the principal dewatering methods available and their applicability to various soil conditions.

2. Knowledge of some common problems associated with various dewatering methods.

3. The fundamental requirements for a dewatering system that is reliably designed.

9.1 SPECIFICATIONS

Dewatering is usually considered the contractor's responsibility. The contractor chooses the method to be used, to be responsible for design of any appropriate system, and to ensure that it is operated correctly. In many circumstances, the safety of the project, and the partially completed work, rests with the contractor's ability to conduct a successful dewatering operation. Dewatering specifications are, therefore, usually brief. The emphasis is on the assignment of the responsibility to the contractor for dewatering and protecting the work thus far completed. The following sample specification illustrates this point:

1B-23.1 General.   The contractor shall be responsible for the design, construction, operation, maintenance, and removal of pumping facilities, cofferdams, bulkheads, piping, etc., necessary to unwater various required work areas or divert existing water flows. Contractor shall submit plans of cofferdams and pumping facilities for care of water and for commencement of such construction. The Contracting Officer's approval of such plans shall not be construed in any way as relieving the contractor of his responsibilities for care of water, diversion, and cofferdamming, but will be primarily to ascertain compliance with the requirements of the contract (maintenance of anadromous fish run, reasonable precaution, the exercise of sound engineering judgment, and prudent construction practices), overloading or misuse of existing or new structures, the adequacy and safety of any such works, and potential damage or undermining of existing or completed works. (From *Spring Creek National Fish Hatchery Construction Contract, Skamania County, Washington*, Department of the Army, 1971.)

9.2   UNDERWATER EXCAVATIONS

In highly pervious soils or in those situations where it is not desirable to lower the groundwater table, excavations may be completed underwater. In these situations, it is necessary to first enclose the work area with some type of impervious structure. Any excavation required is completed within this structure, and the structure bottom is then sealed against the intrusion of water. When the seal has been completed, the water remaining within the structure is removed and construction is completed.

9.2.1   Caissons

The caisson method of construction involves excavation from within the permanent structure. In this method, the structure is either built in place if the site is on land or floated into position if the site is in water. When the structure or partial structure is in position, excavation from within begins. As excavation proceeds, the structure sinks because of its own weight, or weight that may be added, and the process is continued until the final foundation elevation is reached. The process is shown schematically in Figure 9.1. This method has been used successfully for small structures only a few feet in diameter and for large structures over 100 ft in width. In some instances, the top of the caisson is sealed and the work within is done under elevated air pressure. The use of compressed air permits unwatering of the caisson as excavation proceeds. As long as the air pressure within the caisson is greater than the water pressure at its base, intrusion of water is prevented.

In very soft ground, it is sometimes difficult to maintain the alignment of the caisson as the excavation proceeds. The sinking is accompanied by lurching a few inches from side to side as the structure moves through the ground. Large movements can overstress the structure and damage it. In other cases, the frictional resistance between the caisson and the surrounding soil must be overcome by special measures to cause the caisson to sink. Adding weight is one possibility. In other cases, a lubricant, usually a bentonite clay slurry, is injected through the caisson walls at the soil-structure interface. Jetting (see Section 11.4.3) may be used in cohesionless soils. The frictional resistance between the structure and the soil can be estimated in advance of construction, but such estimates are approximate. The bid price should, therefore, include the costs of measures needed to overcome these uncertainties.

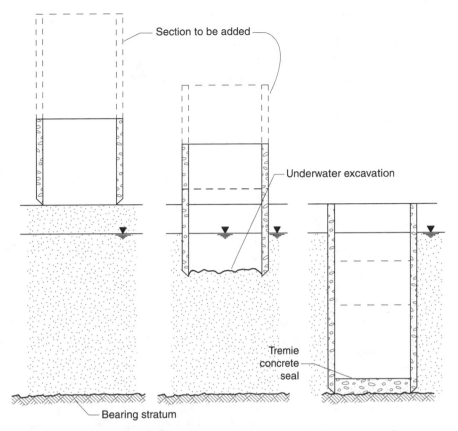

Figure 9.1 Schematic sequence of caisson installation. Part of the structure is completed before excavation begins. As excavation proceeds underwater, additional structure or weight may be added. When excavation is complete and the bearing stratum is reached, a tremie concrete seal is placed. The caisson is then unwatered and construction within completed.

When a caisson is unwatered before the seal is in place, seepage enters the bottom by upward flow from the surrounding groundwater as shown in Figure 9.2. In cohesionless soils, this upward flow can induce a *quick* condition or loss of strength at the bottom of the excavation.

The physical cause of the quick condition may be demonstrated by the analogy illustrated in Figure 9.3, in which the surface of the sand corresponds to the surface of the underwater excavation in Figure 9.2. The total vertical stress at depth L in the sand is

$$p_v = a\gamma_w + L\gamma, \tag{9.1}$$

and the water pressure is

$$u = (H + a + L)\gamma_w. \tag{9.2}$$

By subtracting Equation (9.2) from Equation (9.1), we obtain the vertical effective stress:

$$p_v' = L\gamma - (H + L)\gamma_w. \tag{9.3}$$

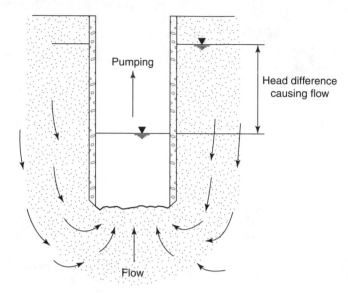

Figure 9.2 Partially unwatered caisson with no seal, showing seepage from surrounding ground.

Since for cohesionless soils, strength is proportional to effective stress, a condition of zero strength exists if p'_v is zero. Thus, from Equation (9.3), circumstances that may bring the zero strength condition about are those when

$$\frac{H}{L} = \frac{\gamma - \gamma_w}{\gamma_w}. \tag{9.4a}$$

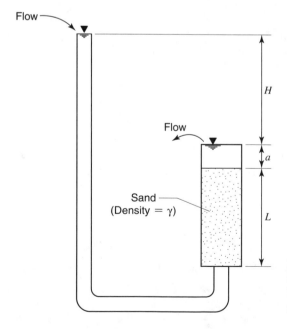

Figure 9.3 Illustration for demonstration of quick condition brought about by upward seepage in cohesionless soils.

That is, when the hydraulic gradient resulting from upward seepage is sufficiently large to reduce effective stress to zero. Because the total density γ of sands is approximately twice that of water, the critical value of the hydraulic gradient is about 1. To prevent development of a quick condition, the head difference causing flow should be kept low. One way to accomplish this is to include a factor of safety for the allowable hydraulic gradient in a given situation; thus, Equation (9.4a) should be modified for design purposes to

$$\left(\frac{H}{L}\right)_{\text{allowable}} < \frac{\gamma - \gamma_w}{\gamma_w \text{FS}}. \tag{9.4b}$$

A factor of safety of 1.5 to 2 is used in this case. In some cases, to ensure against upward flow, the water level inside is actually maintained above the surrounding groundwater level by pumping water into the caisson.

Downward seepage along the outside of the caisson produces an increased lateral pressure on the structure. The combination of loss of support at the bottom and this increased pressure can be damaging to the structure if not accounted for in design. A further detrimental effect of this approach to dewatering a caisson is the loss of ground associated with seepage from the outside and excavation from within. As a general rule, caissons should not be used where adjacent structures exist that are susceptible to damage by loss of ground from beneath their foundations.

9.2.2 Cofferdams

Cofferdams are structures built in place to exclude water and earth from an excavation. In those instances where the distance across the excavation is sufficiently small to permit internal bracing, single-walled cofferdam construction is used. The typical application of single-wall cofferdam construction is for placement of small bridge piers. This type of operation is illustrated in Figure 9.4. In the beginning, a steel sheet pile enclosure is constructed in water. A tremie concrete seal is then poured unless pilings are to be driven as shown in Figure 9.4. Driving of piling would precede the placing of the seal in that instance. When the seal is complete, internal bracing is installed and the cofferdam is unwatered. Construction of the pier is completed in the dry.

Cellular cofferdams are used only in those circumstances where the excavation size precludes the use of cross-excavation bracing. In that circumstance, the cofferdam must be stable by virtue of its own resistance to lateral forces. A large cellular cofferdam is shown in Figure 9.5. Individual cells within a cellular cofferdam may be of various shapes. The usual shape is the circular cell, which offers the advantage of being able to be filled as soon as it is completed. Adjacent noncircular cofferdam cells must be filled simultaneously to prevent collapse of their common walls. Unlike the single-wall cofferdam, the cellular cofferdam may be unwatered as soon as the enclosure is complete. For cofferdams on pervious foundations, this unwatering often requires the use of a supplementary dewatering system of the type discussed in later sections of this chapter. When the cofferdam has been unwatered, excavation work may then proceed in the dry.

9.3 SEEPAGE BARRIERS

For large areas where pervious soils extend to great depths, it may be impractical to construct a cofferdam only above the ground surface. In some circumstances, conditions dictate that the cofferdam extend below the ground surface to provide a barrier against seepage. No seal is constructed on the bottom of the excavation, but supplementary dewatering may be required inside

Figure 9.4 A typical single-walled cofferdam for construction of a bridge pier foundation. Access to this cofferdam is over a temporary timber-decked trestle. Pilings are being driven prior to placement of the tremie seal. Note internal struts and walers.

the enclosure. The usual applications of the methods to be discussed include dewatering large excavations for structures and dams. The same methods are also often used for the construction of permanent seepage-control facilities.

9.3.1 Slurry Trench Methods

In recent years, the slurry trench method has been successfully developed to deal with particularly troublesome dewatering and excavation support problems. The method involves constructing an impervious barrier beneath the ground surface. The process is illustrated schematically in Figure 9.6. As excavation for the wall is progressing, the material removed is

Figure 9.5   A large cellular cofferdam for construction of Phase I, Lock and Dam 26 (Replacement), on the Mississippi River near Alton, Illinois. (Photo courtesy of the Corps of Engineers, U.S. Army.)

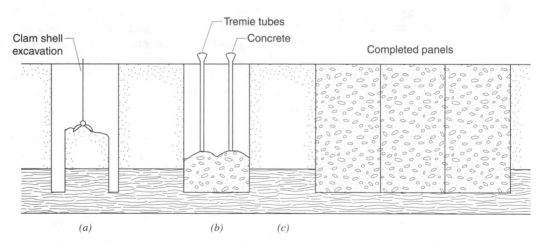

Figure 9.6   Illustration of seepage barrier construction by the slurry wall panel method. Excavation of a panel by clamshell proceeds at (*a*) with the excavation walls supported by slurry. At (*b*) a completed panel excavation is backfilled with concrete by the tremie method. The intermediate panel at (*c*) will then be completed by the same procedures as construction progresses from left to right.

replaced with heavy clay slurry. The lateral pressure from the slurry is sufficient to support the excavation walls. When the excavation has been completed, concrete placement proceeds by the tremie method from the bottom to the top of the excavation. The slurry displaced by this operation is collected for reuse. When the concrete has cured, the construction site is enclosed within a rigid, impervious barrier. Alternative procedures employ excavation by dragline and backfill using clay–soil mixtures or cement–bentonite mixtures. Each procedure has advantages and disadvantages that may make it more applicable to a specific site. The work is generally performed by specialty subcontracting firms. Barriers constructed by these methods have been proven effective in nearly all foundation materials and have been constructed to depths exceeding 200 ft.

In Figure 9.7, excavation for a large project was preceded by the slurry trench-installation shown.

9.3.2 Freezing

Freezing of an impervious groundwater barrier is a means of eliminating flow to an excavation and has been used in both shallow and deep excavations. On one project, an ice barrier 8 ft thick, 130 ft deep, and about 230 ft long was formed to aid the excavation of the cutoff trench for a dam. This method was employed only after a previous attempt using upstream interceptor wells had failed.

9.3.3 Grouting

Grouting is another highly specialized, usually expensive method for retarding the flow of water. Grout curtains are used in permanent works to construct cutoffs for groundwater and sometimes have been employed as construction aids in dewatering. The process involves injection of chemical or cement grouts into the voids of pervious soils. When these grouts solidify, they form an impervious barrier. The success of the operation will depend on the distribution of the grout injected. Of course, as it is injected, the grout will follow the more pervious passages within the soil. Grouts will have varying capabilities to penetrate soils, depending on their composition. The final grout curtain position will be affected by the presence of groundwater flow. These effects are illustrated in Figure 9.8. Grouting is very much an art and does not lend itself readily to analytical treatment. The work is planned but then controlled in the field as it progresses. Surprises and increased costs are generally the rule rather than the exception. A massive grout curtain some 835 ft deep was included in the construction of the Aswan High Dam in Egypt.

9.4   OPEN SUMPS

Open sumps are probably the most extensively used dewatering method. Sumping is simple and cheap and can be carried out with very little planning, provided that it will work at all. It is well suited to some situations and totally inapplicable to others.

(a)

(b)

Figure 9.7 Seepage barrier constructed by the slurry wall panel method for the excavation of the second powerhouse for the Bonneville Dam on the Columbia River in Washington. The wall is approximately 1 mile long and extends through ancient landslide debris containing cobbles and boulders to a depth of about 180 ft.
(*a*) Exposed section with excavation beginning in background. (*b*) Close-up of exposed section, showing quality of tremie concrete and wall surface resulting from placement of concrete against trench walls.

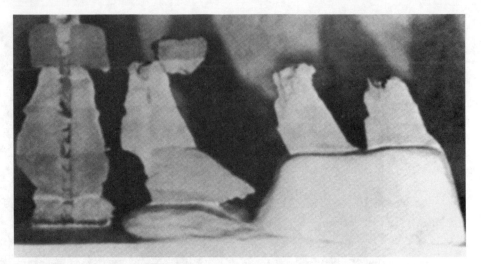

Figure 9.8 Chemical grout pattern affected by soil permeability to grout and groundwater flow. (Reprinted, by permission, from Karol, R.H., "Chemical Grouting Technology," *Journal, Soil Mechanics and Foundations Division, American Society of Civil Engineers*, Vol. 94, No. SM1, 1968.)

9.4.1 Cohesive Soils

Sumping methods work best in soils that are nearly impervious and when an excavation must be made only a short distance below the groundwater level. The procedure is illustrated in Figure 9.9. The bottom of the work area is graded to drain to a central location where pumps are installed. In homogeneous clay soils, the excavation can usually proceed rapidly enough so that the inflow of water does not present a problem. Therefore, it is usually necessary only to install a moderate amount of pumping capacity, and the installation may be made after the excavation is complete.

Even though seepage may be readily controlled by sumps, uplift pressures may cause the base of an excavation in fine soils to heave. Heave can occur as the uplift pressure resulting from lowering the groundwater level (Equation (9.2)) approaches the total pressure (Equation

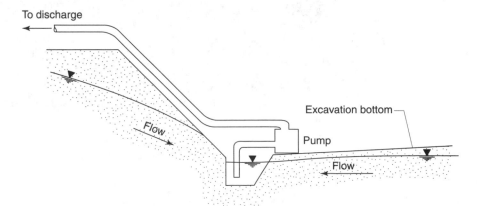

Figure 9.9 Schematic arrangement for dewatering by sumping.

(9.1)) in the overlying soil. Generally, well dewatering systems (Section 9.5) are required to control heave where groundwater lowering is large.

9.4.2 Cohesionless Soils

Cohesionless soils are usually of sufficiently high permeability that the success of a sumping operation will depend on the acceptability of comparatively large pumping capacities and certain problems that may arise from movement of soil particles to the sumps. In gravels and coarse sands, the principal problem to be dealt with is the need for large pumping capacity. In finer sands, the pumping capacity required is reduced, but the slopes of the excavation generally tend to ravel and run toward the sumps as the result of seepage pressures due to the groundwater lowering. This process is referred to as backward erosion, internal erosion, or piping depending on whether the particles are moving from the slope surface or from within the soil being dewatered. The effects of seepage along the slopes are discussed in Chapter 10. One unsatisfactory consequence of the movement of soil to the sump is loss of ground from the surrounding area. Cases have been recorded where the ground was actually pumped from beneath adjacent structures by this mechanism (internal erosion or piping). One means of controlling loss of ground resulting from internal erosion is to provide sumps with protective filters. A filter is a barrier to the movement of these soil particles. It may consist of a properly graded soil blanket, a metal screen, or a filter fabric. When sumped excavations are provided with filters, it is usually necessary to enclose the sump with a perforated conduit.

The quantity of seepage toward a sump may be estimated from the expression

$$q = \frac{k(H^2 - h^2)}{2(D - d)},$$ (9.5)

where q is the flow rate per unit length of sump trench and other terminology are as defined by the sketch in Figure 9.10. The usual application of Equation (9.5) would be where the amount of groundwater lowering, $H - h_t$, was known and the influence distance D was estimated. In that case, $h = h_t$ at $d = 0$ and

$$q = \frac{k(H^2 - h_t^2)}{2D}.$$ (9.6)

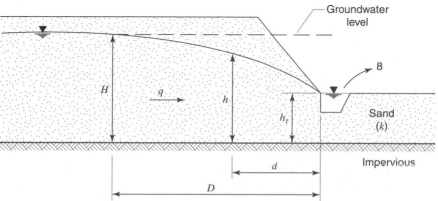

Figure 9.10 Terminology for analysis of seepage toward a sump.

The resulting discharge should be multiplied by trench length to provide the total pumping requirement.

9.5   WELLS AND WELL-POINT SYSTEMS

When a well is pumped, the groundwater surface in the surrounding area is lowered. The amount of lowering is dependent on the pumping rate, the size of the well, the permeability of the ground, and the distance from the well. Provided that the permeability of the ground can be established with sufficient accuracy, the draw-down curve may be reliably computed. These calculations may be used to plan dewatering operations within the zone of influence of the well.

9.5.1   Single Wells

The draw-down curve for a single well penetrating an open aquifer was shown in Figure 5.6. Usually, the draw-down from a single well is not of sufficient magnitude to dewater an excavation. However, single wells are often pumped in tests done prior to bidding to determine the coefficient of permeability of the ground. In these tests, all of the parameters in the governing equation for discharge are measured except the coefficient of permeability which may then be calculated. This information will be furnished to the contractor for his prebid planning. With the information so obtained, the contractor may design one of the several types of dewatering systems described in this chapter.

9.5.2   Multiple Wells

When deep wells are spaced sufficiently close so that their zones of influence overlap, the draw-down at any point is greater than the draw-down resulting from a single well. Wells of this nature may be located so that the shape of the draw-down surface fits the needs of a dewatering operation. These calculations are complex. For an open aquifer such as that indicated in Figure 9.11, the expression

$$h = \sqrt{H^2 - \sum_{i=1}^{n} \frac{q_i}{\pi k} \ln\left(\frac{R}{r_i}\right)} \qquad\qquad 9.7$$

applies. The integer i indicates the number of any well, so that q_i is the rate at which it is pumped and r_i is the distance from the well to the point where h, the water surface height due to pumping all the wells, is to be computed. To apply Equation (9.7) for more than two or three wells involves lengthy calculations, for which a computer is appropriately employed. The draw-down surface computed for two interfering wells is shown in Figures 9.11 and 9.12, relative to the dimensions of a planned excavation.

When deep wells are used to dewater an excavation and the draw-down of the surrounding ground surface is large, several problems may arise. Loss of ground by piping through the well screens is a possibility that can be overcome by proper design of the well screens to act as filters. Potentially more serious is the possibility that the zone of influence of the wells may extend to such a great distance that it affects the underground water supply in the surrounding area. Of further practical importance, one must consider the effects of groundwater lowering on

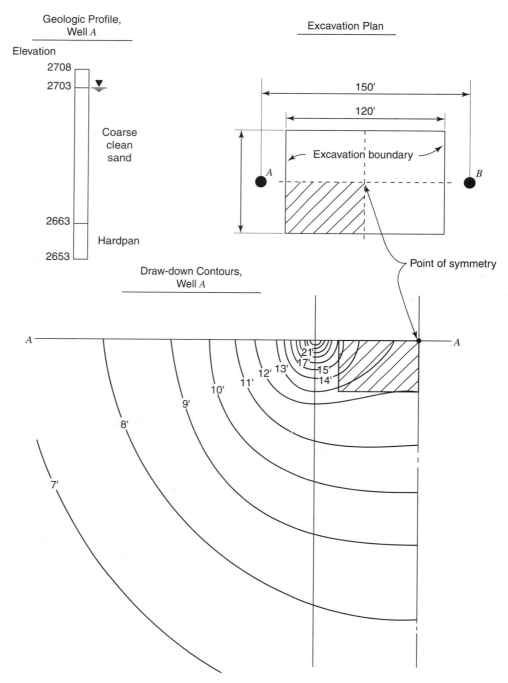

Figure 9.11 Plan view showing excavation boundary and well layout for dewatering. Draw-down water surface contours are shown for one-quarter of the excavation area.

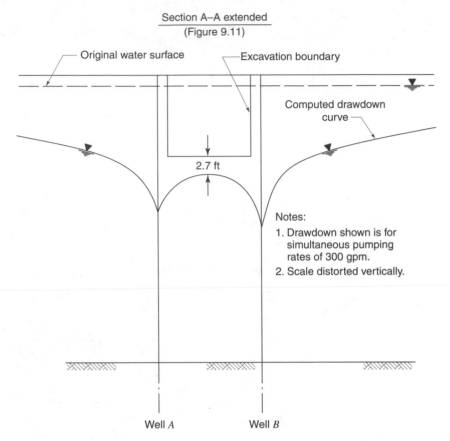

Figure 9.12 Section through excavation and wells, showing groundwater drawdown curve computed.

producing compression settlements of the surrounding ground surface. Lowering of the groundwater table by 10 ft is equivalent to placing a surface load on the soil of 624 psf. The calculation of pressure changes is illustrated in Chapter 5. When the surrounding area is underlain by compressible soils, this increase in pressure produces settlements in a pattern related to the amount of groundwater lowering; an extreme example of such settlements is shown in Figure 9.13, in which the groundwater lowering is shown to result from water supply wells and not construction dewatering. The magnitude of the settlement induced by groundwater lowering will be dependent on the compressibility of the soil and the amount of groundwater lowering. Settlement calculations were discussed in Chapter 5.

9.5.3 Well Points

When the required groundwater lowering at a site is not large, and in some other circumstances, the use of a well-point dewatering system is advantageous. A well-point dewatering system consists of a series of closely spaced small-diameter wells installed to shallow depths. These wells are connected to a pipe or header that surrounds the excavation and that is attached to a

Figure 9.13   Compression settlement resulting from groundwater lowering in Mexico City. Top of well casing shown was originally at the ground surface. (Photo courtesy of J.R. Bell.)

vacuum pump. A typical well-point system is illustrated in Figure 9.14. The amount of groundwater lowering produced by well points can be readily estimated using Equation 9.6 and taking the pumping rate for each well point as the well-point spacing times q, the rate per unit length of sump. More detailed calculations are required, however, to evaluate the shape of the drawdown water surface in the vicinity of the line of well points because each well point produces a localized water surface lowering. The method involves procedures for multiple wells discussed

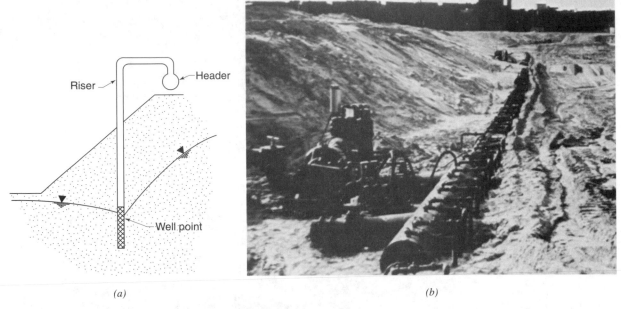

Figure 9.14 A single-stage well-point system, showing the header and riser pipes. (*a*) Schematic. (*b*) Installation. (Courtesy of E.G. Worth.)

in the previous section. A quick estimate of the feasibility of using well points may be obtained by application of the nomograph in Figure 9.15. A distinct advantage of the well-point method is its adaptability in those situations where the calculations have not proven reliable. In these circumstances, the spacing of individual well points around the header can be decreased until the desired amount of lowering is obtained. There is, of course, a maximum amount of lowering that may be obtained in soils of a given permeability. Because it is a vacuum system, the amount of lowering for a single-stage well-point system is theoretically limited to a head of water corresponding to one atmosphere of pressure. Because of losses in the lines and well points, and other considerations, the practical limit of groundwater lowering for single-stage well-point systems is 15 to 20 ft. Where greater depths of dewatering are required, multiple-stage well-point systems may be employed or deep well systems with submersible pumps can be used.

Because well points are installed behind the toe of an excavation slope, they have the effect of creating a stable condition in the slope, since the seepage will be to the well points or into the slope. Well points may, thus, be used to stabilize a slope where slope instability has been induced by a sumping system.

Well-point equipment is specialty equipment and not usually owned by the general contractor. It may be rented or, alternatively, a specialty subcontractor may be employed for this particular phase of the construction contract. Several such contractors with national and world-wide experience are available. Their services include a complete range of planning, construction, and operation activities.

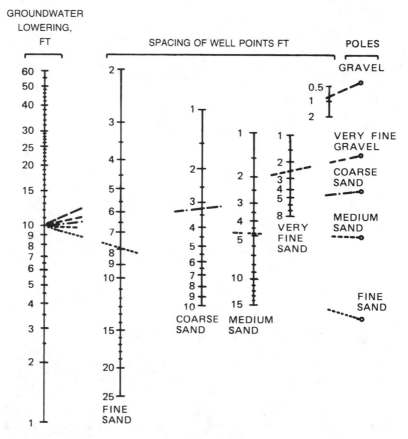

Figure 9.15 Nomograph for estimating feasibility of well-point dewatering in uniform clean sands and gravels. This nomograph should be considered as approximate only, and more detailed analyses should be completed for specific conditions at each actual project site. (Courtesy of Moretrench American Corporation.)

9.6 ELECTROOSMOSIS

Electroosmotic dewatering systems have been employed to dewater fine-grained soils. They are usually applicable to silts within a certain plasticity range, although the specific range of applicability is not clearly defined. An electroosmotic system is basically a well-point system in which the gradients causing flow are supplemented by the use of direct current electricity. A system such as this is illustrated schematically in Figure 9.16. Electrodes are inserted midway between adjacent well points. Direct current is applied so that the polarities of the well point and the electrode are as shown. The resulting electrical gradient presumably causes the positive ions in solution in the groundwater to migrate toward the negatively charged well point. The mobility of the water to the well point is thereby improved. By properly locating the well-point electrode system, the stability of an excavation slope may also be improved.

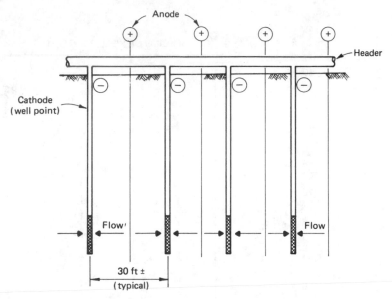

Figure 9.16 Electroosmotic dewatering system (schematic).

9.7 PLANNING DEWATERING OPERATIONS

The dewatering methods previously discussed are applicable to certain specific soil conditions and excavation sizes. Their suitability for various soil types is illustrated in Figure 9.17. Methods involving wells and well-point systems are employed where the soils to be pumped are predominantly sand and gravel. Freezing and grouting and seepage barriers may be used in these same soils. Open sumps by themselves are most applicable to clay soils; however, in the design of dewatering systems, it is the usual practice to sump excavations dewatered by other methods. Such supplementary sumps are used to control rainfall runoff within the excavation area and to mop up that seepage which may penetrate through the principal dewatering system.

The analysis of a dewatering system requires knowledge of the permeability of the soil to be dewatered. The permeability of all but the simplest soil deposits is difficult to determine reliably. Therefore, dewatering systems are typically conservatively planned. Any analysis made for design of a dewatering system should include consideration of the effects of either underestimating or overestimating this variable. The design should then be formulated so that if the on-site conditions encountered prove to be different from those expected, the basic method of dewatering selected will still apply with some modification. In the event that deep wells are used, for instance, the pattern should permit the addition of more wells if the number originally planned is insufficient.

In practically all cases, a sudden rise in the groundwater level within a dewatered excavation cannot be tolerated. For this reason, extra pumps should be installed within the pumping system to provide standby capacity in the event of pump failure. Similarly, where electric power is used to drive pumps, generators powered by on-site sources should be available for rapid employment.

The planning of a successful dewatering operation requires the following information:

1. Ground conditions at the site to be dewatered. The essential information includes the sequence of soil strata to be encountered, their permeabilities, and the position of the natural groundwater table during the construction period.

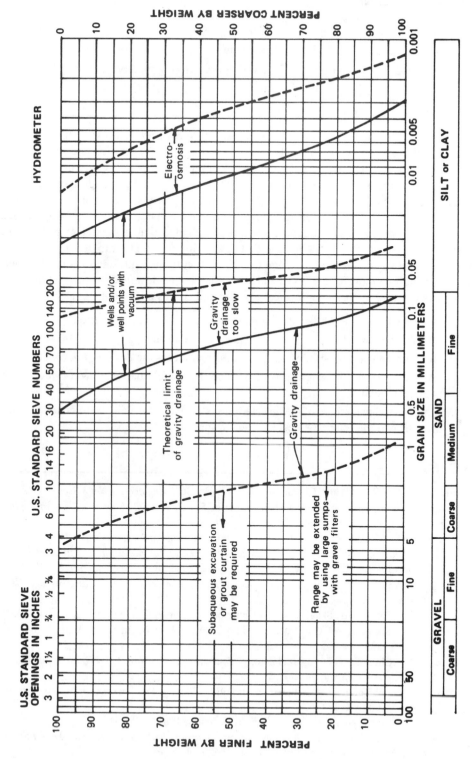

Figure 9.17 Applicability of dewatering methods (Courtesy of Moretrench American Corporation.)

2. The amount of groundwater lowering required.

3. The duration of the period of dewatering.

4. The applicability and costs of the various dewatering methods that may be employed.

9.8 SUMMARY

Failure in a major dewatering effort can result in great financial loss to the contractor in that partially completed work can be destroyed. Similarly, poorly planned and executed dewatering operations can affect the performance of the completed project by altering subsurface conditions. Dewatering conducted without regard to its effects on surrounding properties can result in damages and subsequent claims against the contractor and other parties to the construction contract. Planning for dewatering should, therefore, best be undertaken by knowledgeable people within the contractor's organization. If they are not available, outside help should be sought. In either event, it must be recognized that dewatering is not a task to be undertaken lightly, nor is the planning for such an operation to be based on limited information and the hope for success. A method must be employed and a system must be designed that can reliably accomplish the work within the budget allocated. Having studied this chapter, the reader should

1. Know which types of dewatering methods are appropriate for various soil conditions.

2. Be able to anticipate problems associated with dewatering.

3. Recognize what constitutes adequate information for the design of dewatering systems.

4. Appreciate that a dewatering system, to be successful, may require considerable modification as it is built.

5. Understand that failure of the dewatering system cannot usually be tolerated and that the system should, therefore, be designed accordingly.

REFERENCES

Bowman, W.G., "Record Grout Curtain Seals Nile's Leaky Bed," *Engineering News Record*, February 29, 1968.

Casagrande, Leo, "Electro-osmotic Stabilization of Soils," *Journal, Boston Society of Civil Engineers*, January, 1952.

D'Appolonia, D.J., D'Appolonia, E., and Namy, D., "Precast and Cast in Place Diaphragm Walls Constructed Using Slurry Trench Techniques," *Proceedings of Workshop on Cut and Cover Tunneling*, Report FHWA-RD-74-57, Federal Highway Administration, Washington, D.C., 1974.

"Electro-osmosis Stabilizes Earth Dam's Tricky Foundation Clay," *Engineering News Record*, June 23, 1966.

"High Gorge Dam, Skagit River, Washington," *Engineering News Record,* January 22, 1952.

Kapp, M.S., "Slurry Trench Construction for Basement Wall of World Trade Center," *Civil Engineering*, April, 1969.

Karol, R.H., "Chemical Grouting Technology," *Journal, Soil Mechanics and Foundations Division, American Society of Civil Engineers*, 94, No. SM1, 1968.

Mansur, C.I., and Kaufman, R. I., "Dewatering," in *Foundation Engineering*, G. A. Leonards, Editor. New York: McGraw-Hill, 1962.

Powers, J.P., *Construction Dewatering.* New York: Wiley, 1981.

Sanger, F.J., "Ground Freezing in Construction," *Journal, Soil Mechanics and Foundations Division, American Society of Civil Engineers,* 94, No. SM1, 1968.

Terzaghi, K., and Lacroix, Y., "Mission Dam, An Earth and Rockfill Dam on a Highly Compressible Foundation," *Geotechnique,* 14, No. 1, 1964.

Todd, D.K., *Groundwater Hydrology*. New York: Wiley, 1959.

White, R.E., "Caissons and Cofferdams," in *Foundation Engineering*, G.A. Leonards, Editor. New York: McGraw-Hill, 1962.

PROBLEMS

1. Figure 9.18 is representative of soil conditions at a construction site. If you are the contractor, select a dewatering method you would consider appropriate for each of the projects described below. Consider the information given in this chapter, earlier chapters and any other information that may be appropriate. Describe in detail how your system would be built and operated. Justify clearly any assumptions you make. Discuss what difficulties you may encounter and how you should plan for them.

   a. A circular reservoir 100 ft in diameter is to be built with a bottom slab 25 ft below present ground surface. The groundwater level must be held at least 2 ft below the slab until the reservoir is filled.

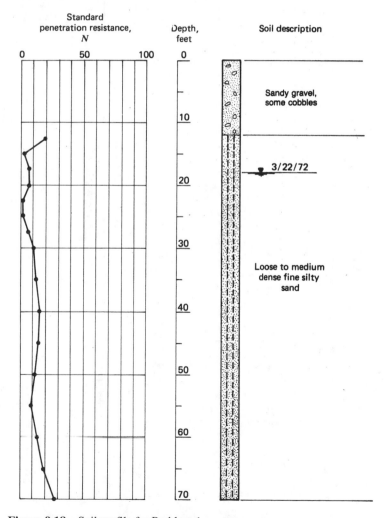

Figure 9.18   Soil profile for Problem 1.

    b. A bridge pier with a driven pile foundation is to be built with the bottom of the pile cap 40 ft below present grade. The cap is 20 ft by 30 ft in plan view.

2. Repeat Problem 1 for the following projects with soil conditions as indicated in Figure 9.19.
    a. 1200 ft of an 18-in. sewer line is to be placed in a trench about 16 ft below ground surface.
    b. A 30-ft-diameter concrete pump station is to be founded on the basalt rock at 65 ft in depth.
    c. Conditions as in *a*, except that the pipeline is 28 ft below the ground surface.

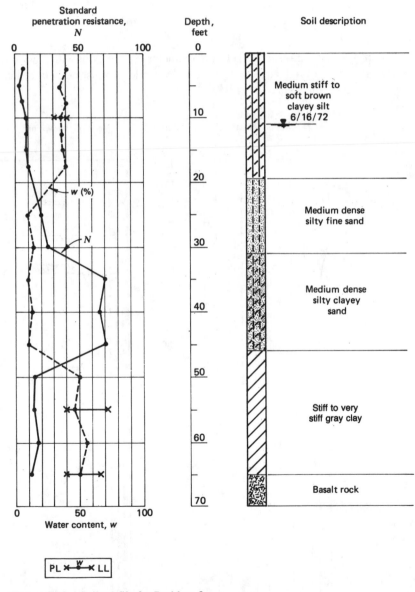

Figure 9.19   Soil profile for Problem 2.

CHAPTER 10

Excavations and Excavation Supports

Excavation slopes would sometimes be unstable at inclinations permitted by the space available for construction. In other instances, there is a need to support structures adjacent to excavations. The state of the art in excavation support technology has developed according to the needs and capabilities of the construction industry. Statutes are stringent regarding the personal safety of construction workers. For these and many other reasons, the construction contract involving significant excavation usually contains some provision covering excavation supports.

This chapter introduces various aspects of the excavation support problem. It is intended to give the reader a background bearing on the principal considerations in planning excavation support work.

The topics to be covered are

1. Specification requirements in contracts.

2. Stability analysis for unsupported slopes.

3. Methods for support of shallow and deep cuts.

4. Lateral earth pressure analysis.

5. Planning for excavation supports.

6. Ground movements adjacent to braced cuts.

10.1   SPECIFICATIONS

The detail contained in specifications varies according to the complexity of the support system required for execution of the contract. When an excavation must be supported, care in the design and erection of the support system is of utmost importance. This is, of course, true because failure of excavation supports often leads to loss of life. However, for shallow excavation supports, the engineering required for successful completion of the work is not usually as comprehensive as is the case with deeper excavations. Generally, excavation support problems become more complex as the depth of excavation exceeds 10 to 20 ft.

Because excavation supports are not usually part of the permanent structure, the details of such systems will not be specified in the normal construction contract. The details of design and the responsibility for planning and construction are, therefore, left to the contractor. This is especially true on smaller jobs, where the specifications can give the contractor considerable latitude in the design and construction of the temporary works. While providing workable excavation supports, the knowledgeable contractor can take advantage of experience and do so in an economical fashion, thereby gaining a bidding advantage over competitors. On jobs that are more complex and when the excavation support system is to become part of the permanent structure, the specifications will be more detailed. The design and execution of temporary works will call for more oversight from both the owner and the engineer; however, the contractor still bears the responsibility for the successful design and execution of temporary works on the jobsite.

10.2   DESIGN OF EXCAVATION SLOPES

When the construction area is large enough, excavation walls may be sloped in lieu of providing structural support. Selection of a suitable slope angle involves knowledge of the properties of the soil at the site and application of the principles of soil mechanics. The use of sloped, rather than supported, excavation walls appears to be attractive from a cost point of view. It should be remembered, however, that the savings resulting from not building a support structure are at least partially offset by the increased quantity of excavation required.

One ongoing issue that anyone involved in construction must deal with is the relationship between regulatory requirements and sound engineering design judgment. This section, for example, deals with slope stability from an engineering and design standpoint, which, in this case, is generally the contractor's responsibility; the following section deals with trench safety from a regulatory one. Although the contractor is obligated to meet all regulatory requirements, meeting those does not relieve him or her of the consequences of slope or other failure. Moreover, the methods described in this section are generally applicable to slopes that are not necessarily trenches, while those in the next section deal with trenches where workers may have to be located. Ultimately, the contractor must insure that both requirements are met, and the problems at the end of the chapter should be solved with this in mind.

10.2.1   Slope Stability

Slopes may fail because of a number of mechanisms, depending on the nature of the soil involved and the arrangement of natural earth materials at the site in question. A number of these mechanisms are shown in Figure 10.1. Slope failures occur, of course, because forces tending to cause instability exceed those tending to resist it. The driving forces are represented by a component of soil weight downslope, and the resisting forces are represented by the soil

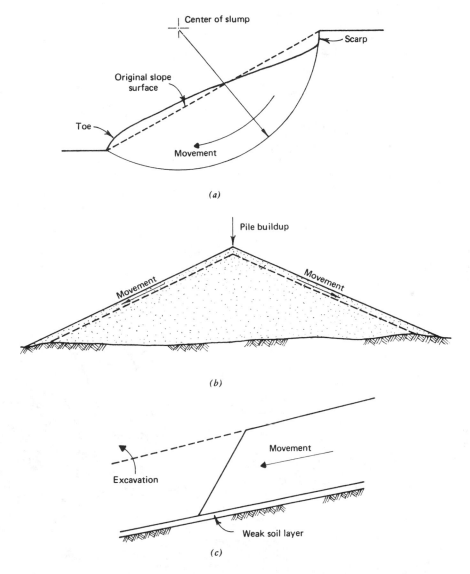

Figure 10.1 Some slope failure mechanisms. (*a*) Rotational slump in homogeneous clay. (*b*) Translational slide in cohesionless sand or gravel. (*c*) Slip along plane of weakness.

strength acting in the opposite direction. For the case of the rotational slump, driving and resisting moments about the center of slumping rather than forces are considered. In any case, the factor of safety (FS) for the slope is expressed as the ratio of the resisting forces or moments to the driving forces or moments. When the factor of safety is 1 or less, the slope must fail. When the factor of safety exceeds 1, the slope is theoretically stable. In designing cut slopes, the usual factor of safety required is between 1.3 and 1.5. In order to estimate the factor of safety for a slope, there are four required types of information:

1. the soil and water profile involved,
2. the kinematics of the potential slope failure,
3. the strengths and weights of the soils, and
4. the proposed slope geometry.

With this information, the critical potential failure surface can be located by engineering analysis. This is a simple matter in homogeneous materials. The difficulty of the problem increases with more complex subsurface conditions. The chart shown in Figure 10.2 can be used to estimate the factor of safety for cut slopes in homogeneous soils with cohesion and friction and a slope angle from the horizontal, β. The variable used to express the effects of both the height of the slope and the properties of the soil is referred to as the stability number and is given by

$$N_s = \frac{\gamma H_c}{c}. \tag{10.1}$$

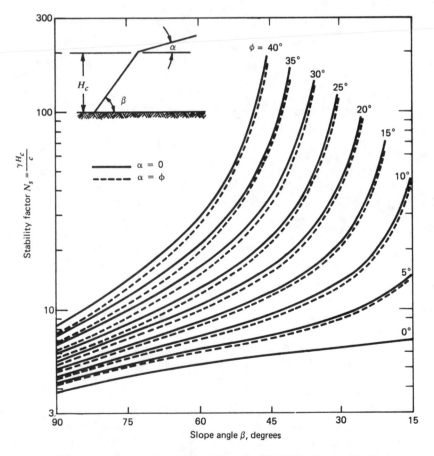

Figure 10.2 Chart for stability numbers for soils with friction and cohesion. (Reprinted, by permission, from *Foundation Engineering Handbook*, edited by Hans F. Winterkorn and Ysai-Yang Fang. Copyright © 1975 by Litton Educational Publishing, Inc. Reprinted by permission of Van Nostrand Reinhold Company.)

For a clay soil that behaves as if $\phi = 0$, the stability number is 3.85 for a vertical-sided trench. The values of the stability number shown in Figure 10.2 are critical values, corresponding to a factor of safety of 1. For the $\phi = 0$ case, the factor of safety of a stable slope can be calculated by finding the ratio of the critical height to the actual height or by similarly comparing the cohesion available to the critical cohesion.

EXAMPLE Slope Stability of a Slope with a Purely Cohesive Soil

The stability number of a slope cut at $\beta = 65°$ in a clay for which $\phi = 0$ is 5, if the upslope angle α is zero. If the soil has $\gamma = 110$ pcf and $c = 650$ psf, and if the slope height is 20 ft, then

$$H_c = \frac{cN_s}{\gamma} = \frac{650 \times 5}{110} = 29.5'.$$

Therefore,

$$FS = \frac{29.5}{20} = 1.48.$$

Alternatively, the critical cohesion for the slope would be

$$c_c = \frac{\gamma N_c}{N_s} = \frac{110 \times 20}{5} = 440 \text{ psf},$$

and again,

$$FS = \frac{650}{440} = 1.48.$$

If we now assume a required factor of safety, say, 1.5, and a range of consistencies as described in Table 5.3, we can tabulate theoretically safe depths for vertical cuts using the method just described. The results of such calculations are shown in Table 10.1.

Study of Table 10.1 makes it apparent that slope failures are probable in shallow excavations only for very soft to medium homogeneous clays. It can further be shown, using Figure 10.2, that significant improvement in the factor of safety for a slope of given height can be made by flattening the slope angle from 90° to about 45°.

Stability calculations for slopes in soil having both friction and cohesion can also be made using Figure 10.2, but the procedure requires several trials and is slightly more complex.

TABLE 10.1 Theoretically Safe Heights for Homogeneous Clay Cut Slopes with Vertical Sides, Estimated from Figure 10.2

Soil Consistency	Unconfined Compressive Strength, q_u (psf)	Cohesion, c (psf)	Safe Height, H (ft)
Very soft	<500	<250	<5
Soft	500–1000	250–500	5–10
Medium	1000–2000	500–1000	10–20
Stiff	2000–4000	1000–2000	20–40
Very stiff	4000–8000	2000–4000	40–80
Hard	>8000	>4000	>80

EXAMPLE Slope Stability of a Slope with a Soil with Both Cohesion and Internal Friction

We demonstrate this procedure by computing the factor of safety for a 30-ft-high slope with $\phi' = 30°$, $c = 300$ psf, and $\gamma = 115$ pcf cut in level ground at $\beta = 60°$. First, we define the factor of safety as the ratio of actual soil strength on the slope to that strength required to maintain the slope at a safety factor of 1. Furthermore, we require that the factor of safety with respect to soil frictional strength be the same as that due to cohesion. Therefore,

$$FS = FS_c = FS_\phi = \frac{c}{c_{req'd}} = \frac{\tan \phi'}{\tan \phi'_{req'd}} \tag{10.2}$$

To begin the calculation, we choose an approximate value of FS_c, say, 1.5. Therefore, the strength value $c_{req'd}$ to be used in stability calculations is

$$c_{req'd} = \frac{c}{FS_c} = \frac{300}{1.5} = 200 \text{ psf.}$$

From Equation (10.2),

$$N_s = \frac{\gamma H}{c_{req'd}} = \frac{115 \times 30}{200} = 17.2.$$

From Figure 10.2, for $\beta = 60°$ and $N_s = 17.2$, $\phi = 32°$. This value of f is that required to maintain $FS_\phi = 1$ in the slope if the actual cohesion c is 200 pcf. The real cohesion, of course, is $c = 300$ psf, but the lower value is used in the calculations to arrive at a factor of safety. The factor of safety with respect to friction is

$$FS_\phi = \frac{\tan \phi'}{\tan \phi_{req'd}} = \frac{\tan 30°}{\tan 32°} = \frac{0.577}{0.625} = 0.92.$$

Since $FS_\phi 50.92 \neq FS_c = 1.5$, we must make additional trials. Assuming that

$$FS_c = 1.20,$$

$$c_{req'd} = \frac{300}{1.2} = 250 \text{ psf,}$$

and

$$N_s = \frac{115 \times 30}{250} = 13.8.$$

From Figure 10.2, $\phi_{req'd} = 28°$, and

$$FS_\phi = \frac{0.577}{0.532} = 1.08.$$

Still, FS_ϕ is not equal to FS_c and an additional trial is needed. If that trial is made using $FS_c = 1.13$, FS_ϕ will be shown to also be 1.13. The theoretical safety factor for the slope is, therefore, $FS = FS_c = FS_\phi = 1.13$. This magnitude of the safety factor would indicate that, while the slope is theoretically stable, there is not sufficient soil strength to assert that it is stable enough to be as safe as normal practice would require.

By this procedure, we have shown that the slope would be at critical height if its actual strength parameters were $c_{req'd}$ and $\phi'_{req'd}$, but since they are higher, according to Equation (10.2), the slope is at less than critical height and is stable at the factor of safety defined by the calculations. It should be noted that these calculations can be done more rapidly using computer software that models both the driving weight of the soil and the resisting cohesion and internal friction.

If there is seepage along the slope, the factor of safety may be estimated as shown above, except that in the calculations the internal friction angle ϕ' is replaced by ϕ'_s, where

$$\phi'_s = \frac{\gamma - \gamma_w}{\gamma}\phi'. \tag{10.3}$$

It can be shown that slopes in dry or submerged cohesionless soil are stable when the slope angle is less than the soil's angle of internal friction. The factor of safety is expressed as

$$FS = \frac{\tan \phi'}{\tan \beta}. \tag{10.4}$$

The presence of seepage emerging from a slope reduces the factor of safety to half the value without seepage (Equation (10.4)). A slope in partially saturated or moist sand behaves essentially as a cohesive material with friction. It may be cut, therefore, at a comparatively steep slope angle. Such slopes owe their stability to the existence of apparent cohesion. It has been previously noted (see Section 5.6.1) that apparent cohesion may be destroyed, either by drying or by saturation. Apparent cohesion, therefore, should not be relied on to maintain the overall stability of unprotected excavation slopes. A range of stable slope angles calculated for a theoretical factor of safety of 1.3 is shown in Figure 10.3. In most cases, sand slopes disturbed by excavation are at a low relative density. Low relative density results in a low angle of internal friction. Such slopes should, therefore, be designed using low assumed values of internal friction angles.

The discussion in this section has been presented to illustrate simple slope mechanics and methods of slope stability analysis. The undertaking of such analyses is in most cases considerably more complex than for the hypothetical circumstances illustrated here. Accordingly, these analyses should not be undertaken by those who do not have a thorough grounding in the principles of soil materials and slope behavior.

10.2.2 Slope Protection

In some materials, slopes that exhibit overall stability may deteriorate from the surface inward when they are exposed by construction. The result is a sloughing of soil materials into the excavation, which is troublesome and inconvenient. In these circumstances, it is sometimes

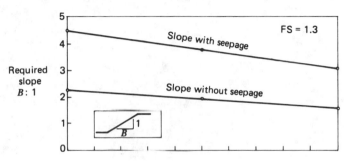

Figure 10.3   Stable slope angles in cohesionless sand.

advantageous to provide temporary slope protection. Generally, these types of protection are in the form of a coating or other impervious material applied to the slope.

Silty soil materials are subject to erosion during rainstorms. A number of spray-on products are available that bind the soil particles on the surface together and, at the same time, permit a permanent vegetative cover to develop on the slope. These materials may often be specified as part of the completed contract, particularly on flatter slopes. For temporary excavations, it would be unlikely that slope protection would be considered as a separate item in the contract. Where appropriate, however, the contractor should consider its use and adjust his bid items to account for the additional expense. An illustration of the use of slope protection is shown in Figure 10.4. In this excavation, the slopes consist of an almost cohesionless silty material. Direct rainfall on such slopes causes them to erode rapidly. Drying results in destruction of apparent cohesion and a similar erosion or sloughing. The plastic covering shown prevents changes in moisture content on the surface of the slope and, therefore, its stability is maintained.

In special situations where slope sloughing and raveling result in displacement of large blocks of material from the slope surface, a chain link fence has been used to provide safety in the excavation. The fencing material is simply draped over the slope surface. This application has been particularly successful where the soil tends to fail in large chunks or when the slope contains significant amounts of loose, large rock.

10.3   SHALLOW TRENCHES AND EXCAVATION SAFETY

Collapse of shallow, unsupported excavations results in many construction deaths each year. Proper design and construction of these trenches can avoid many of these fatalities. The Occupational Safety and Health Administration (OSHA) has extensive regulations on the configuration of trenches; failure to adhere to these can result in both civil and criminal penalties for the contractor. In addition, various states have adopted statutes requiring that the slopes of excava-

Figure 10.4   Use of plastic sheets for slope protection in a shallow excavation.

tions beyond some minimum depth either be supported or be inclined at an angle that will preclude their failing. The design, construction, and regulation of personnel activities around these trenches are thus governed more by regulatory requirements than the design methods of the previous section. What follows is a summary of OSHA regulations; as with any safety, health or environmental regulatory issue, the contractor should directly consult the current regulations themselves when implementing them on the jobsite.

OSHA categorizes soils and rocks used for trenches into four broad categories; these are detailed in Table 10.2. Where a layered geologic structure exists, the soil must be classified on the basis of the soil of the weakest layer. Determination of the soil type can be done by the methods described earlier in this book.

TABLE 10.2 OSHA Classification of Soil Types (After OSHA Technical Manual, Section V, Chapter 2)

Soil or Rock Classification	Unconfined Compressive Strength, q_u	Examples	Other Characteristics
Stable Rock	N/A	Granite, Sandstone	Natural solid mineral matter that can be excavated with vertical sides and remain intact while exposed. Determination may be difficult unless the existence of cracks is known and if these cracks run away from the excavation.
Type A Soils	>3 ksf (144 kPa)	Clay, silty clay, sandy clay, clay loam, silty clay loam, sandy clay loam	Cannot be: 1. Fissured 2. Subject to any kind of vibration 3. Previously disturbed 4. Part of a sloped, layered system where the layers dip in to the excavation on a slope of 4 : 1 (horizontal : vertical) 5. Has seeping water
Type B Soils	1 ksf (48 kPa) $< q_u <$ 3 ksf (144 kPa)	Angular gravel, silt, silt loam	Can include: 1. Previously disturbed soils unless otherwise classified as Type C 2. Soils that meet the Type A classification but are fissured or subject to vibration 3. Dry unstable rock 4. Layered systems sloping into the trench at a slope less than 4H : 1V, but only if the material can be classified otherwise as a Type B Soil
Type C Soils	<1 ksf (48 kPa)	Gravel, sand, soil from which water is freely seeping, and unstable submerged rock	Includes material in a sloped, layered system where the layers dip into an excavation or have 4H : 1V or greater.

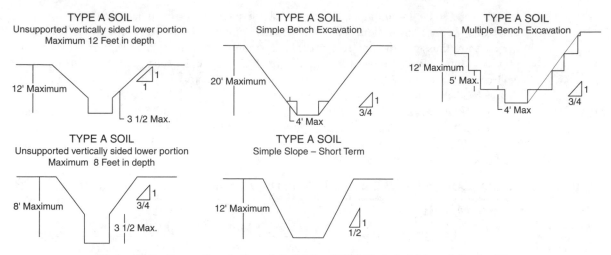

Figure 10.5 Excavations in Type A Soil (after OSHA Technical Manual, Section V, Chapter 2)

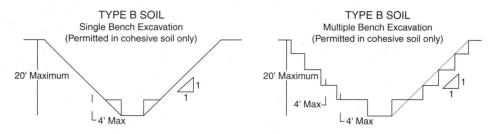

Figure 10.6 Excavations in Type B Soil (after OSHA Technical Manual, Section V, Chapter 2)

The soil or rock type determines the configuration of the trench. Figure 10.5 shows excavation configurations for Type A soil, Figure 10.6 for Type B soil, and Figure 10.7 for layered soils. Table 10.3 shows the allowable slopes for each soil type. Inspection of these figures will show that, in many cases, benching is permitted with excavations. For the first bench, the bottom vertical height of the trench must not exceed 4 feet (1.2 m). Subsequent benches may be up

TABLE 10.3 Allowable Slopes for Various Soil Types (After OSHA Technical Manual, Section V, Chapter 2)

Soil Type	Height to Depth Ratio (H : V)	Slope Angle, Degrees
Stable Rock	Vertical	90°
Type A	3/4 : 1	53°
Type B	1 : 1	45°
Type C	1 1/2 : 1	34°
Type A (Short Term, maximum excavation depth = 12 feet)	1/2 : 1	63°

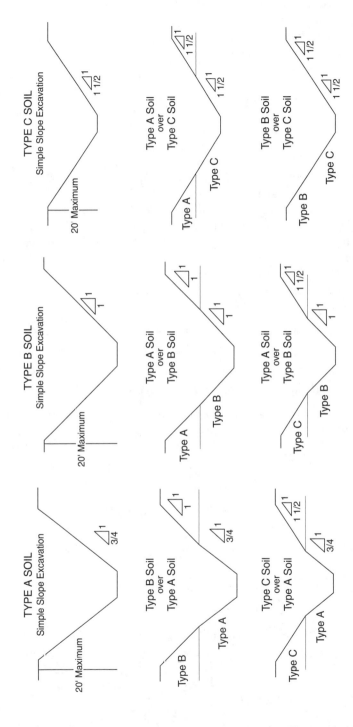

Figure 10.7 Excavations in Layered Soils (after OSHA Technical Manual, Section V, Chapter 2)

to a maximum vertical dimension of 5 feet (1.5 m) in Type A soils and 4 feet (1.2 m) in Type B soils. The maximum total trench depth is 20 feet (6 m). All of the benching must stay below the maximum allowable slope for the particular soil type. Trench excavation for Type B soils is permitted in cohesive soil only.

10.4 SUPPORT FOR SHALLOW TRENCHES

Sloping and benching for trenches requires jobsite space that may or may not be readily available. These space requirements may be mitigated by the use of shoring for the vertical portion of trenches. Shoring enables the more extensive use of vertical cuts, although these vertical cuts must comply with OSHA requirements. Two types of shoring are in common use today, timber shoring and hydraulic shoring.

Timber shoring is shown in Figure 10.8. It consists of vertical planks which are both aligned and supported by the wales and spaced across the trench by the struts or crossbraces. Timber shoring is fairly easy to construct but is labor intensive and more subject to damage than metal shoring.

Hydraulic shoring (generally made of aluminum, sometimes steel) is becoming more popular. A key advantage of hydraulic shoring is that personnel do not have to enter the trench to install or remove it. Other advantages include:

- Light weight; can be installed by one worker;
- Insure even distribution of pressure along trench line through gage regulation;
- Able to have their trench faces preloaded to use the soil's cohesion and prevent movement; and
- Can be adapted to various trench lengths and widths.

Typical configurations of hydraulic shoring are shown in Figure 10.9. All shoring should be installed from the top down and removed from the bottom up. As with any other piece of hydraulic equipment, hydraulic shoring requires the same standard of care for cleanliness of the hydraulic system and maintenance of the cylinders, hoses, and connections. Another variation

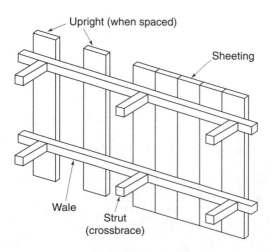

Figure 10.8 Timber shoring (after OSHA Technical Manual, Section V, Chapter 2)

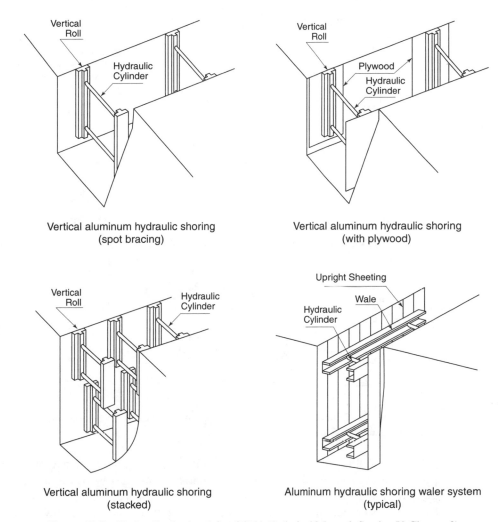

Vertical aluminum hydraulic shoring
(spot bracing)

Vertical aluminum hydraulic shoring
(with plywood)

Vertical aluminum hydraulic shoring
(stacked)

Aluminum hydraulic shoring waler system
(typical)

Figure 10.9   Hydraulic shoring (after OSHA Technical Manual, Section V, Chapter 2)

of hydraulic shoring is *pneumatic shoring*, which operates on the same principle except that compressed air instead of hydraulic fluid is used to actuate the cylinders. Since most jobs have at least a small (approximately 175 cfm) air compressor on the job, supplying air is not a significant problem.

In many cases *trench boxes* (see Figure 10.10) can be used, especially if the primary purpose is to protect workers from cave-ins. These can in some cases be stacked to allow for higher vertical cuts. It is important that the space between the trench and trench shield be as small as possible to prevent destabilization of the shielding system due to soil movement. It is also important not to use trench shields beyond their rated design capacity. Trench shields can be used in conjunction with sloped excavations as shown in Figure 10.11. Trench shields should be seated at the bottom of the trench.

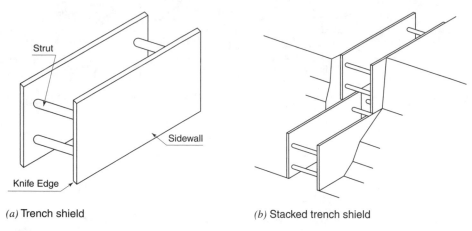

(a) Trench shield   *(b)* Stacked trench shield

Figure 10.10   Trench Shields (after OSHA Technical Manual, Section V, Chapter 2)

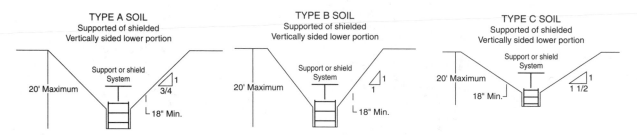

Figure 10.11   Combined slope and shield configurations (after OSHA Technical Manual, Section V, Chapter 2)

One additional consideration in trenching is the placement of spoil, both temporary and permanent. As shown in Figure 10.12, temporary spoil must be placed at least 2 feet (0.6 m) away from the surface edge of the excavation; this distance is measured from the edge of the spoil, not its crown. Spoil should also be placed so that it channels rainwater and other runoff away from the excavation and also so that it does not fall into the excavation itself. Permanent spoil should be placed at a considerable distance from any excavation, as it can act as a surcharge and change the effective stress of the soil under it, thus degrading the ability of an excavation to repose without a sheeting or shoring system.

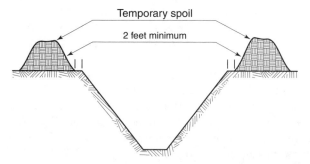

Figure 10.12   Placement of temporary soil near an excavation (after OSHA Technical Manual, Section V, Chapter 2)

10.5 SUPPORT FOR DEEP CUTS

When excavation depths exceed 10 to 20 ft, generally some specialized planning for the support of such cuts is needed. Engineering for braced cuts has not progressed to the point where it is a matter of a systematic, mathematical analysis. The details of the behavior of cut bracing systems are still not yet completely understood, though much is known about them. Engineering of a deep-cut support system is very much a matter for experienced engineers and construction personnel. The design must account for the forces in the support system and the movements of the adjacent ground that will result from the excavation. Selection of a dewatering method is often a major consideration. This section discusses some of these problems.

10.5.1 Lateral Earth Pressure Calculations

Virtually any engineering material will exercise lateral contraction when placed in tension or contraction when placed in compression. The level of this movement—and any resulting stress—is a function of the nature of the materials and the confinement of the material. Soils are no exception to this, although the method of computation and application of the theory for soils is different from that for conventional elastic materials.

Lateral pressure in an undisturbed soil deposit may be considered as proportional to the vertical pressure, and is given by the equation

$$K = \frac{p_h}{p_v}, \tag{10.5}$$

where

- $K =$ lateral earth pressure coefficient
- $p_h =$ horizontal soil stress
- $p_v =$ vertical soil stress (generally equal to the vertical effective stress p')

The relative magnitude is dependent on the origin of the soil deposit and its stress history. To estimate the lateral pressure distribution, calculations of vertical pressure distribution are first to be made in the manner described in Section 5.2. Since water has no strength, lateral water pressure and the vertical water pressure are the same.

The lateral earth pressure coefficient—and thus the relationship between vertical and horizontal stresses—varies, depending upon the condition of the loading, and generally is classified in one of three ways:

1. *At-rest earth pressures.* These are pressures that occur either in the mass of the soil with no retaining walls or where the walls do not move. The at-rest lateral earth pressure coefficient K_o for normally consolidated soils (soils which have never been preloaded or overconsolidated[1]) generally varies from 0.3 to 0.5 and can be estimated by the equation

$$K_o = 1 - \sin \phi' \tag{10.6}$$

---

[1]Preloading or overconsolidation, discussed in Chapter 5, refers to those situations where an existing soil deposit has been subjected to greater vertical load than presently exists. This situation generally arises where the surface of the deposit has undergone major erosion or where it has carried the added weight of ice during glacial periods. It is also possible to create the effects of overconsolidation by drying in a severely arid climate. In these situations, the lateral pressure exerted within a soil deposit may exceed the vertical pressures.

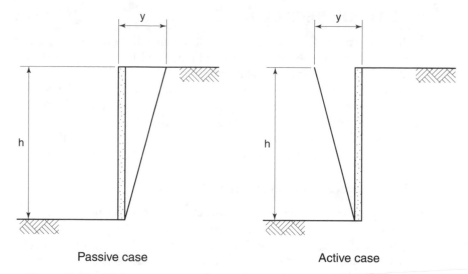

Figure 10.13   Wall movements for active and passive earth pressures (after EM 1110-2-2502)

2. *Active earth pressures.* These take place when the retaining wall moves away from the soil, as shown in Figure 10.13. These are generally less than at-rest pressures, as the wall movement relieves some of the lateral earth pressure.

3. *Passive earth pressures.* These take place when the wall moves towards the soil. In a sense with passive earth pressure the vertical and horizontal pressures reverse themselves, that is, the lateral movement of the wall induces vertical stress in a soil rather than the vertical stress inducing a lateral stress. The wall movement that induces passive earth pressures is shown in Figure 10.13.

Although it may seem contradictory, active and passive earth pressures can and do take place in the same wall. If a sheet piling wall, for example, rotates, the side of the wall that rotates away from the soil (active pressures) above the pivot point is opposite from the side that rotates away from the soil below the wall.

Excavation supports are associated with cuts. As a cut is made, the soil at the face tends to expand or move into the cut area. This expansion mobilizes some of the soil's strength. If a support is placed against the excavation surface in such a manner as to prevent all movement, then the stress existing before excavation is maintained. If some movement is permitted, the distribution of pressure changes. The actual distribution of pressure will depend on the mode of movement permitted. When a retaining structure is permitted to rotate a small amount so that the top of the structure moves out with respect to its base, the distribution of pressure remains linear and increasing with depth, as shown in Figure 10.14a. It is reduced to a limiting value, proportional to depth, where the constant of proportionality is K_A, the active lateral earth pressure coefficient. This may be computed from theoretical considerations.

Excavation supports tend to move outward at the base with respect to the top, depending on their method of construction. In these situations, the soil pressure at the bottom is reduced and the soil pressure at the top increases. The total force against the support is equal to or

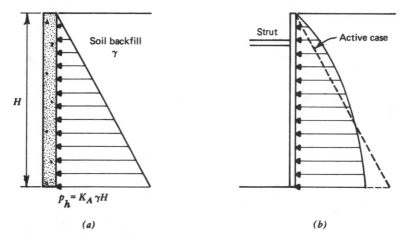

Figure 10.14   Lateral earth pressure distribution. (*a*) Active case. (*b*) Pressure on excavation bracing.

slightly greater than the total force for the active case; however, it acts higher on the structure. This is an important observation to consider in designing excavation supports. Pressure distributions similar to those shown in Figure 10.14*b* cannot be computed from theoretical considerations. Generally, we rely on empirical methods derived by observing forces on actual excavation support systems. Engineers at many locations throughout the world have measured loads in the struts supporting various types of sheeting and bracing systems. From these strut loads, back calculations of apparent earth pressures acting on the sheeting have been computed. Envelopes of maximum values of these apparent pressures have been constructed. These are shown in Figure 10.15. It is important to note that any water pressure that exists must be added to these apparent earth pressures to arrive at the total lateral pressure on a cut support. It is further important to note that these envelopes represent fictitious pressures that, when applied to an excavation support for design, will permit estimates of the maximum values of strut loads at any level. Measured individual strut loads would be expected to differ and generally would be lower than those calculated. This produces an inherent conservatism in excavation support design that is warranted by the catastrophic consequences of a cut support failure. Failure of a single strut in a long excavation results in the transfer of the load it carried to adjacent members. If these members fail, their load is in turn transferred to adjacent members and, with each load transfer, the amount of load to be carried increases. Thus, if a single strut fails, it is quite likely that the failure will lead to a collapse of the entire system.

10.5.2 Excavation Support Methods

Excavations may be supported by either cantilever structures or structures with some form of additional support. The two possibilities are schematically diagrammed in Figure 10.16. The supported system is in most common use. It consists of two essential elements: the part of the structure that is in direct contact with the soil, which might be referred to as the *soil support*, and the *bracing* structure that supports it. In the cantilever system, a single structure serves both of these functions. The resisting capability of a cantilever structure is dependent on its rigidity

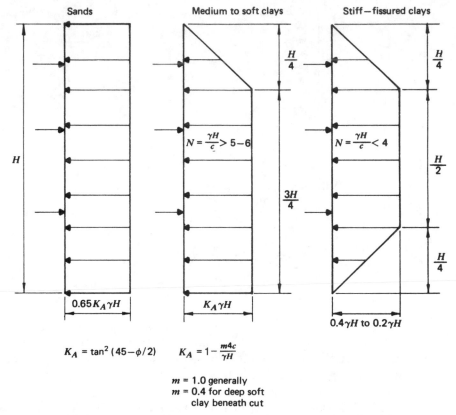

Figure 10.15 Apparent earth pressure on excavation supports. (Reprinted, by permission, from Terzaghi, K. and Peck, R.B., *Soil Mechanics in Engineering Practice*, New York, Wiley, 1967.)

and the support provided by the foundation soil below the base of the excavation. Cantilever structures are usually restricted to use in shallow excavations. The range of depth may be extended by using a rigid wall with a great embedment depth. An example of this approach is the *cylinder pile* method in which large-diameter concrete cylinders are cast adjacent to one another in augered holes to form a cantilever wall. This method has been used to successfully support excavations to 50 ft deep and more.

A number of methods for providing soil support have been developed. One of the earliest is the method employing driven *sheet piling*. In this method, either timber, concrete, steel, vinyl, or putruded fiberglass sheets are driven ahead of the excavation. These sheets are then braced from the inside as the excavation proceeds. Sheet piles can be installed by either impact or vibratory pile drivers (generally the latter), which can also remove them when the need for the temporary works is ended. A more popular method that has become highly developed is the use of the *soldier pile*, or *soldier beam, and lagging* system. In this method, H-pilings are set in predrilled holes around the periphery of the excavation. Predrilling as opposed to driving is used to provide close control of alignment and location. These piles are then grouted in place with weak concrete. As the excavation begins, the soil and concrete are carefully trimmed away from the soldier pile and the horizontal lagging is installed between adjacent pile flanges. In

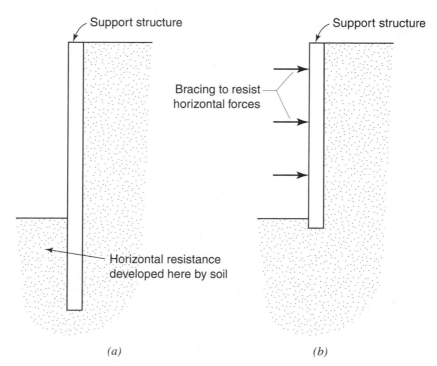

Figure 10.16 General methods of excavation support. (*a*) Cantilever systems. (*b*) Supported systems.

some cases, the lagging is eliminated when the soils are highly cohesive and are capable of arching between soldier piles. Even in these instances, however, a surface binder such as gunnite is often applied to the soil to prevent raveling. The soldier pile and lagging method obviously is inappropriate for perfectly cohesionless soil. In this type of soil, a sheeting method must be used.

For very difficult conditions in soft ground with a high water table, the *soldier pile–tremie concrete* method has been developed. In this method, soldier piles are set in predrilled holes, and the entire space between flanges of adjacent soldier piles is excavated in a slurry trench as shown in Figure 10.17. Very specialized equipment is used for this purpose. When that excavation has been completed, tremie concrete is placed and reinforcement is set, if appropriate. The result is a continuous concrete wall cast beneath the surface of the ground prior to the beginning of any excavation. When this wall is intact and complete, excavation and interior bracing can begin.

When soil support is in place, the construction of structural support or bracing begins. For narrow excavations, internal struts are often most appropriate. When the excavation level reaches the point where struts are to be installed, a horizontal member called a *waler* is placed against the soil support. Intermediate struts are then installed from waler to waler across the excavation. A typical *cross-lot strutting* system is shown in Figure 10.18. It is obvious that the presence of the struts restricts the working space within the excavation.

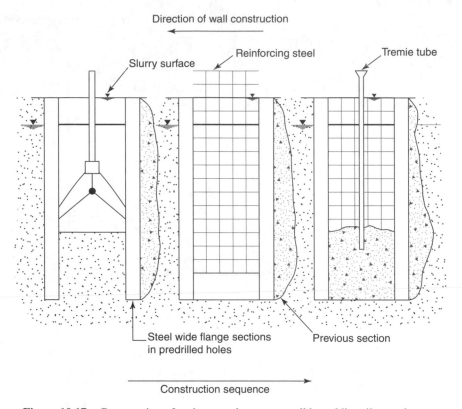

Figure 10.17 Construction of underground concrete wall by soldier pile-tremie concrete method, including excavation, placement of reinforcing steel, and tremie concreting.

Figure 10.18 A strutted excavation.

Figure 10.19   Raker bracing for soldier pile and lagging excavation supports.

For very wide excavations, it is often not feasible or reasonable to use cross-lot struts. In these circumstances, *raker bracing* such as that shown in Figure 10.19 can be used. To install a raker bracing system, the central portion of the excavation is completed by sloping the excavation boundaries. When the required depth is reached, support for the bottom of the rakers can be installed. These supports will be either driven piling or footings. Construction of the soil support and removal of the remainder of the excavation then begins. Raker bracing is installed as the room becomes available. A raker bracing system offers some advantage over cross-lot bracing in that the central portion of the work area is comparatively uncluttered.

The use of *tiebacks* is a very popular alternative to the installation of internal bracing. To install tiebacks, inclined holes are drilled through the soil support system as the excavation proceeds. A completed project is shown in Figure 10.20. Figure 10.21 shows a tied-back slurry wall constructed to support a deep cut and obviate the need for dewatering outside the cut area. The tieback installed in each hole may be a concrete prestressing tendon or simply a steel rod. In either case, it is grouted within the stratum that is to provide the support and then tensioned using hydraulic jacks pushing against the soil support system. It is general practice to pretest all tieback anchors to loads greater than the anticipated design load. The load in the anchor is then set somewhere near the anticipated design load. Because of this pretensioning, tieback systems are generally very successful in preventing movements of the excavation walls. Estimates of anchor capacity can be made using results of pullout tests from previous jobs, but are usually considered as estimates only, because installation methods affect anchor size and soil disturbance. Final design is based on pretest results.

When permanent construction inside a braced excavation is complete, it is the usual practice to leave the wall in place. Often, the wall is used as the back form for the permanent basement of the structure. It may be necessary, however, in some instances, to remove the entire

Figure 10.20   A tied-back excavation.

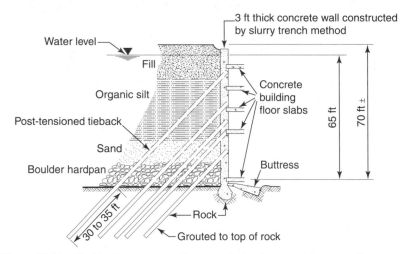

Figure 10.21   Tied-back concrete wall constructed underground for excavation support and dewatering at the original New York World Trade Center. (Reprinted, by permission, from Kapp, M.S., "Slurry Trench Construction for Basement Wall of World Trade Center," *Civil Engineering*, April, 1969.)

bracing wall. Examples include an encroachment on a public right of way or on adjacent private property might be used for subsequent subsurface construction. Tiebacks, if left in place, are always cut to relieve tension when the permanent structure can safely carry the load. In some cases, particularly with heavy stiff walls constructed using tremie methods, the temporary walls become part of the permanent construction.

10.6 PLANNING FOR EXCAVATION SUPPORTS

An outline for planning an excavation support system that has been suggested is given in Table 10.4. This outline illustrates the fact that the planning effort will encompass activities that are not necessarily part of the contractor's operation. In some cases, all of the activities may be his responsibility. In other cases, he may be asked to build an excavation support system that has been completely specified. In either case, however, the steps in planning could be essentially the same. Only the responsible parties may change. It is of primary importance to establish as accurately as possible the nature of soil and subsurface conditions. Groundwater conditions are a primary consideration. The condition of adjacent structures and utilities must also be established before excavation begins. This is to aid in planning the excavation work and the support system to preclude damage to these facilities. It also provides a very important baseline of information against which claims for damage resulting from construction may be compared. For this reason, a requirement for a preconstruction survey may be specified. During this survey, the contractor should make a complete record of any prior settlement and the damage that settlement has caused to adjacent structures. This record should include extensive photographs and video records. In this way, this damage will not be attributed to the activities of the contractor during the job.

A series of trials leads to the final establishment of a bracing and construction scheme for excavation supports. Experience, such as that documented in Figure 10.21 concerning movements, is an invaluable aid in the process. Figure 10.21 was developed from data gathered on a number of braced excavation projects. It represents the ability of contractors to control ground movements for widely varying subsurface conditions, and incorporates the fact that these abilities will vary from job to job. As such, it is an empirical, rather than a theoretical, expression of actual expectations. To estimate settlement induced adjacent to an excavation, one would first choose the point on the abscissa that represents the location at which the estimated settlement is desired.

TABLE 10.4 Engineering an Excavation

Step No.	Activity	Considerations
1.	Explore and test subsoil.	
2.	Select dimensions of excavation.	Structure size and grade requirements, depth to good soil, depth to float structure, stability requirements.
3.	Survey adjacent structures and utilities.	Size, type, age, location, condition.
4.	Establish permissible movements.	
5.	Select bracing (if needed) and construction scheme.	Local experience, cost, time available, depth of wall, type of wall, type and spacing of braces, dewater excavation sequences, prestress.
6.	Predict movements caused by excavation and dewatering.	
7.	Compare predicted with permissible movements.	
8.	Alter bracing and construction scheme, if needed.	
9.	Instrument-monitor construction and alter bracing and construction as needed.	

Source: T.W. Lambe. "Braced Excavations," *Proceedings, Specialty Conference on Lateral Stresses in the Ground and Design of Earth Retaining Structures*, American Society of Civil Engineers, 1970.

EXAMPLE   Estimate of Bracing Movement

Suppose that we have a 30-ft-deep excavation in soft clay and that the settlement 15 ft from the bracing wall must be estimated. At the point 0.5 on the abscissa, and for the line dividing regions I and II on the plot, we read, on the ordinate, about 0.6. This would indicate that the expected settlement would be about $(0.006)(30)$ or 0.18 ft. It has also been shown that horizontal movements toward the excavation of approximately equal magnitude should be planned for.

Aside from subsurface conditions and excavation dimensions, other factors, including dewatering, sequence of bracing installation, construction of the support wall, underpinning, prestressing of struts, and foundation construction, influence the magnitude of deformations. For this reason, Figure 10.22 cannot be used reliably for every job. It represents only a guide and a starting point for more detailed engineering planning. Theoretical methods are available and can be very helpful in estimating deformations caused by excavations. The methods cannot, however, simulate all of the construction activities involved. They are, therefore, best used as a supplement to informed judgment.

Like dewatering systems, excavation support systems are often redesigned during construction to accommodate unanticipated conditions. For that reason, on major excavation work, support systems are frequently monitored by measuring deflections and forces in key members. If the performance of the support system is not within the requirements of the work, it is then necessary to alter the design. Performance observations serve this very important function. They also permit us to better plan future excavations by providing the data base needed to quantify our experiences.

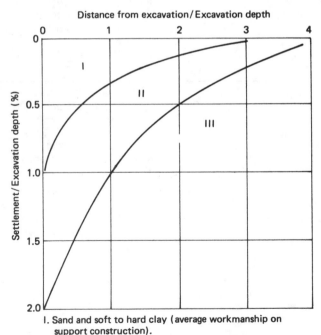

I. Sand and soft to hard clay (average workmanship on support construction).

II. Very soft to soft clay – Condition I (construction difficulties).

III. Very soft to soft clay – Significant depth below excavation bottom.

Figure 10.22   Movement limits associated with braced excavation supports. (Reprinted, by permission, from Peck, R.B., "Deep Excavations and Tunneling in Soft Ground," *Proceedings, 7th International Conference on Soil Mechanics and Foundation Engineering*, State of the Art Volume, 1969.)

10.7   SUMMARY

This section has discussed matters of considerable importance related to the stability of slopes for excavations and to methods of preventing the movements of excavation walls. The discussion given has been incomplete in the engineering sense, but has provided a look at some of the basics involved in these operations. It must be remembered that the planning of major excavation work should be left to those with extensive experience in these matters, for the consequences of failure can be heavy in terms of loss of life and property damage.

Following study of this chapter and solving of the suggested problems, the reader should

1. Appreciate the responsibilities of the contractor in projects where excavation supports are required.

2. Understand the fundamental concepts of simple slope stability analysis.

3. Be able to calculate lateral pressures in a soil deposit.

4. Know the requirements for planning each of the steps involved in constructing excavation supports.

5. Be able to describe basic methods for supporting walls of open excavations.

REFERENCES

Darragh, R.D., "Tiebacks, Type of Installation—Analysis," *Seminar on Design and Construction of Deep Retained Excavations,* American Society of Civil Engineers, Structural Engineers Association of Northern California, 1970.

Department of the Army. *Retaining and Flood Walls.* EM 1110-2-2502. Washington, DC: U.S. Army Corps of Engineers.

Fang, H.Y., "Stability of Earth Slopes," in *Foundation Engineering Handbook,* Hans F. Winterkorn and H.Y. Fang, Editors. New York; Van Nostrand Reinhold, 1975.

Gerwick, B.C., "Slurry Trench Techniques for Diaphragm Walls in Deep Foundation Construction," *Civil Engineering,* December 1967.

Gould, J.P., "Lateral Pressures on Rigid Permanent Structures," *Proceedings, Specialty Conference on Lateral Stresses in the Ground and Design of Earth-Retaining Structures,* American Society of Civil Engineers, 1970.

Lambe, T.W., "Braced Excavations," *Proceedings, Specialty Conference on Lateral Stresses in the Ground and Design of Earth-Retaining Structures,* American Society of Civil Engineers, 1970.

Occupational Safety and Health Administration. *OSHA Technical Manual.* TED 1-0.15A. Washington, DC: Occupational Safety and Health Administration, 1999.

O'Rourke, T.D., "Ground Movements Caused by Braced Excavations," *Journal, Geotechnical Engineering Division,* American Society of Civil Engineers, 107, No. GT9, 1981.

Peck, R.B., "Deep Excavations and Tunneling in Soft Ground," *Proceedings, 7th International Conference on Soil Mechanics and Foundation Engineering,* State of the Art Volume, 1969.

Taylor, D.W., *Fundamentals of Soil Mechanics.* New York: Wiley, 1948.

Terzaghi, K., and Peck, R.B., *Soil Mechanics in Engineering Practice.* New York: Wiley, 1967.

PROBLEMS

1. A 20-ft-deep cut for a basement excavation is specified to be sloped at 1 : 2 (one horizontal to two vertical). The upper 40 ft of soil at the site consists of a homogeneous clay with a

density of 90 pcf and an unconfined compressive strength averaging 2400 psf. The water table is at great depth. Can the cut be made? If so, what is the factor of safety for the slope? Is the slope safe enough for normal work to progress in the excavation without danger?

2. A vertical-sided trench was being cut in a very soft clay soil with a density of about 110 pcf. A depth of 8 ft had been reached when the sides of the trench caved in. Could the trench be completed to a depth of 10 ft by sloping the sides before installing bracing? If so, how steeply may the sides be cut? Show your calculations and sketch the procedure you would follow to complete the job. Apply both the design criteria and the OSHA requirements to this problem.

3. Assuming the soils at the site described in Figure 5.17 are normally consolidated, estimate the maximum values of the lateral total, water, and effective pressures in the natural ground. Present your results in a sketch showing variations with depth.

4. A 25-ft-diameter concrete caisson is to be sunk to the rock surface shown in Figure 5.17 at a depth of 50 ft. Describe, using sketches where appropriate, how you would do it, and what problems you might encounter.

5. A large (200-ft-square) structure is to be built on a downtown site where subsurface conditions are as shown in Figure 10.23. The structure foundations are to be placed on the surface of the dense sand. A 4-ft-deep gravel fill is then to be placed above the sand to support the basement floor. Streets and sidewalks surround the property. The structure walls proposed extend to within 5t of the sidewalk line. Propose a method for supporting the excavation walls. Use sketches and justify your assumptions.

6. Repeat Problem 5, assuming that the structure is to be founded on the rock.

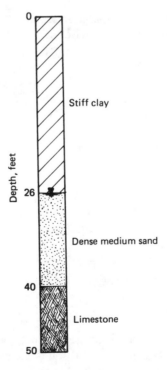

Figure 10.23 Soil profile for Problem 5.

CHAPTER 11

Foundation Construction

Foundations for buildings are generally divided into two broad classifications: shallow foundations and deep foundations.

Shallow foundations are simply large footings that either are installed below the structural loads of the building (spread footings) or cover the structure's entire base (mat or raft foundations.) Most structures are built on shallow foundations. In a structure that has been designed by an engineer, the foundation sizes have been selected to control settlement and provide an adequate margin of safety against shear failure in the foundation soils. Careful attention to the specified details of construction is necessary to ensure that foundation performance is in accord with design assumptions.

This chapter will show

1. How settlement estimates and bearing-capacity calculations are made.

2. Why the usual provisions of specifications for shallow foundation construction are important.

Deep foundations (piles and piers) are used to bypass unsatisfactory soils and transfer structure loads to a suitable bearing stratum. Due to the many variables involved and the specialized nature of the work, driven pile construction has unique uncertainties, and because of this it has become associated with a certain amount of mystery. Some of the mystery is imagined and some is very real. Controversy surrounds many aspects of pile construction, probably more so than any other type of foundation work. This chapter will consider the fundamentals of the subject in an attempt to dispel some of the misconceptions that seem to be part of many such controversies. It will also deal with the construction of deep foundations with the use of drilled piers. The reader will learn

1. Why piling and piers are used and how they carry loads.

2. Basic pile types and materials and the advantages and disadvantages of each.

3. The principal types of pile-driving and pier-drilling equipment and their applicability to certain job conditions.

4. Methods of determining deep-foundation load-carrying capacity.

5. The usual provisions in a pile-driving contract.

6. About some common problems arising in pile and pier construction and how they are dealt with.

11.1 SETTLEMENT ANALYSIS

Settlement forecasts are always a part of foundation design analysis. Understanding how these analyses are made contributes to an understanding of certain provisions in construction contracts. It may also be useful to know the basics of settlement calculations in those instances where temporary foundations must be installed.

11.1.1 Footings on Cohesive Soils

Equation (5.21) provides the basis for consolidation settlement calculations on cohesive soils. Consolidation settlements result from a decrease in the soil's void ratio under load, and they occur at a rate governed by soil permeability and drainage. On all soils, immediate settlements occur because of elastic distortion of the area influenced by the footing load. These settlements are usually quite small on overconsolidated cohesive materials. To use Equation (5.21) the compression index of the soil must be known, along with the effective pressures in the soil before and after loading. If the soil is layered, the equation must be applied to determine the consolidation for each layer and then summed up for the total settlement. With a homogeneous soil, it is necessary to divide the soil into layers so that the varying stress levels in the soil—and thus the varying settlement amounts—are properly taken into consideration. As a general rule, these layers should be no greater than 5–15 feet in depth. The next example illustrates these calculations.

EXAMPLE Settlement Analysis in Cohesive Soils

Assume that the square footing in Figure 11.1 rests on the surface of an overconsolidated clay soil for which tests have shown that $\gamma = 110$ pcf, $e = 0.76$, p'_c is 10,000 psf, and the slope of the recompression curve (see Figure 5.9) is 0.043. The first step in the settlement analysis is to determine the initial stresses and final stress in the soil profile for the depth of significant influence beneath the footing. The calculations are made for a number of layers of soil of increasing size. For our illustration, we have chosen layers designated 1 to 4. The pressure calculations are made at the middepth of each layer. The initial vertical effective stresses are

$$\text{Level } a \quad p'_{1a} = 110 \times 2 = 220 \text{ psf,}$$
$$\text{Level } b \quad p'_{1b} = (110 \times 6) - (62.4 \times 2) = 535 \text{ psf,}$$
$$\text{Level } c \quad p'_{1c} = (110 \times 12) - (62.4 \times 8) = 821 \text{ psf,}$$

and

$$\text{Level } d \quad p'_{1d} = (110 \times 20) - (62.4 \times 16) = 1202 \text{ psf.}$$

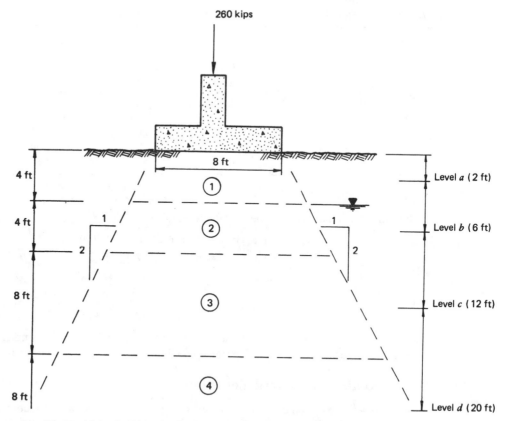

Figure 11.1   Problem for settlement analysis illustration.

The contact pressure at the footing base is the column load divided by the footing area. The pressure resulting from the column load decreases with depth beneath the footing. Reasonable estimates of the pressure at any depth may be made by dividing the column load by an area, derived by increasing the dimensions of the area influenced by the depth in the soil beneath footing grade. (Although there are more precise methods of computing the stress in the soil induced by a foundation.) This is commonly known as the 2:1 rule. Application to the footing shown yields

$$\text{Level } a \quad \Delta p'_a = 260{,}000 \div 10^2 = 2600 \text{ psf,}$$
$$\text{Level } b \quad \Delta p'_b = 260{,}000 \div 14^2 = 1327 \text{ psf,}$$
$$\text{Level } c \quad \Delta p'_c = 260{,}000 \div 20^2 = 650 \text{ psf,}$$

and

$$\text{Level } d \quad \Delta p'_d = 260{,}000 \div 28^2 = 332 \text{ psf.}$$

TABLE 11.1   Results of Settlement Analysis Illustration

Soil Layer	h_1 Layer Thickness (in.)	$\dfrac{C_c}{1 + e_1}$	p'_1 (psf)	p'_2 (psf)	Δh (in.)
1	48	0.0244	220	2820	1.3
2	48	0.0244	535	1862	0.6
3	96	0.0244	821	1471	0.6
4	96	0.0244	1202	1534	0.2
				Total estimated settlement	2.7

Adding the stress increment to the initial stresses,

$$\text{Level } a \quad p'_{2a} = 220 + 2600 = 2820 \text{ psf,}$$
$$\text{Level } b \quad p'_{2b} = 535 + 1327 = 1862 \text{ psf,}$$
$$\text{Level } c \quad p'_{2c} = 821 + 650 = 1471 \text{ psf,}$$

and

$$\text{Level } d \quad p'_{2d} = 1202 + 332 = 1534 \text{ psf.}$$

The second step is to make the settlement calculations for each layer by using Equation (5.21) and sum those results to obtain the settlement estimate. The results are shown in Table 11.1.

11.1.2   Footings on Cohesionless Soils

Settlements of footings on cohesionless soils generally occur as soon as the load is applied and result from elastic distortion of the ground without significant change in the soil's void ratio. Estimates of immediate settlement in cohesionless soils are made from empirical correlations among settlement, bearing pressure, standard penetration resistance, and footing size derived from load tests, and settlement observations and field testing. If N is the average corrected standard penetration resistance within a depth below the footing equal to its width, then

$$s = \frac{8q}{N} \qquad (B < 4 \text{ ft)}, \tag{11.1}$$

and

$$s = \frac{12}{N}\left(\frac{B}{B+1}\right)^2 \qquad (B > 4 \text{ ft)}. \tag{11.2}$$

In these expressions, B is the footing width in feet, q is the bearing pressure beneath the footing in tons per square foot, and s is the estimated settlement in inches. For design purposes, the settlements calculated from Equations (11.1) and (11.2) are normally doubled if the foundation soils are submerged. Observations have shown that the use of these equations results in very conservative (smaller than actual) settlement predictions and that actual settlements are about one-half of the calculated values, even without the correction just indicated for the presence of groundwater beneath the footing. Thus, if the footing in Figure 11.1 were on sand for which $N = 18$ instead of the clay indicated, design settlement analysis would indicate that

$$s = \frac{12 \times 260,000}{18 \times 64} \times \frac{1}{2000}\left(\frac{8}{9}\right) \times 2 = 2.1 \text{ in.}$$

Whereas, in reality, the actual settlement would probably be on the order of

$$s = \frac{12 \times 260,000}{18 \times 64} \times \frac{1}{2000}\left(\frac{8}{9}\right) \times \frac{1}{2} = 0.5 \text{ in.}$$

It is possible to obtain more reliable settlement estimates for foundations on sands than is indicated by the preceding procedures. The methods for doing so, however, involve extensive laboratory testing and analysis and are not warranted for design purposes in all but very special cases.

11.2 FOUNDATION BEARING FAILURE

While foundations may settle excessively under working load, there is also the possibility that they may be so heavily loaded that the strength of the supporting ground is inadequate to resist a punching type of sudden failure. An example is shown in Figure 11.2, where the mudsills supporting the form shoring on a bridge project failed during a concrete pour. Similar *bearing*

Figure 11.2 Failure of shoring foundations during a concrete pour.

failures often occur beneath fills on soft ground. Design analyses for these situations are directed to providing an adequate factor of safety against bearing failures. The fundamentals of the methods employed are outlined here.

11.2.1 Shallow Foundations

Figure 11.3 illustrates a generalized situation in which an area of width B is loaded by a pressure of intensity q at depth D_f in the ground. The pressure is presumed to result from a foundation load.

If the pressure intensity is increased to the point where a ground failure is produced, it will have a value q_d that may be estimated from

$$q_d = \frac{\gamma B}{2} N_\gamma + \gamma D_f N_q + c N_c \tag{11.3}$$

if the loaded area is very long ("continous foundation") and

$$q_d = 0.4 \gamma B N_\gamma + \gamma D_f N_q + 1.3 c N_c \tag{11.4}$$

if the loaded area is square. The intensity q_d is the ultimate bearing capacity, γ is the soil unit weight, and c is its cohesive strength below a-a. The terms N_γ, N_q, and N_c are referred to as the bearing capacity factors. These factors are derived theoretically and are functions of the soil's internal friction angle ϕ. In the form shown, the equations assume that the soil above the base of the foundation is uniform. These equations are referred to as the Terzaghi bearing capacity equations. Although there are newer formulations of these equations that take into account more factors and can analyze more footing shapes, virtually all of them have the same format as that of Equations (11.3) and (11.4), and the foregoing equations shown are suitable for estimating purposes. For square footings on clay for which $\phi = 0$, $N_\gamma = 0$, $N_q = 1$, and $N_c = 5.7$, and Equation (11.4) reduces to

$$q_d = \gamma D_f + 7.4c \tag{11.5}$$

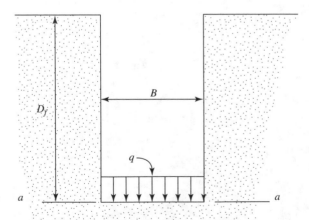

Figure 11.3 Terminology for bearing-capacity analysis.

TABLE 11.2   Bearing Capacity Factors for Shallow Foundation Design (Terzaghi Method) (after EM 1110–1-1905)

ϕ', Degrees	N_q	N_c	N_γ
0	1.00	5.70	0.0
2	1.22	6.30	0.2
4	1.49	6.97	0.4
6	1.81	7.73	0.6
8	2.21	8.60	0.9
10	2.69	9.60	1.2
12	3.29	10.76	1.7
14	4.02	12.11	2.3
16	4.92	13.68	3.0
18	6.04	15.52	3.9
20	7.44	17.69	4.9
22	9.19	20.27	5.8
24	11.40	23.36	7.8
26	14.21	27.09	11.7
28	17.81	31.61	15.7
30	22.46	37.16	19.7
32	28.52	44.04	27.9
34	36.50	52.64	36.0
35	41.44	57.75	42.4
36	47.16	63.53	52.0
38	61.55	77.50	80.0
40	81.27	95.66	100.4
42	108.75	119.67	180.0
44	147.74	151.95	257.0
45	173.29	172.29	297.5
46	204.19	196.22	420.0
48	287.85	258.29	780.1
50	415.15	347.51	1153.2

and for footings on sands with $c = 0$,

$$q_d = 0.4\gamma B N_\gamma + \gamma D_f. \tag{11.6}$$

The bearing capacity factors may be estimated based on standard penetration test results or laboratory tests. They are tabulated in Table 11.2.

The allowable pressure on the foundation q_a is selected by dividing q_d by a factor of safety selected with consideration for the reliability of the information available for analysis, including soil properties and probable loads. The value of the safety factor is usually about 3.

EXAMPLE   Bearing Capacity Analysis

To illustrate the application of these methods, consider the footing shown on Figure 11.1. The soil is purely cohesive; the shear strength of the soil $c = 3000$ psf. Since we are dealing with a cohesive soil, we can use Equation (11.5). However, because the base of the footing is at the surface of the soil, $D_f = 0$, and the first term drops out. The bearing capacity can then be computed as

$$q_d = 7.4 \times 3000 = 22,200 \text{ psf},$$

and, since the actual bearing pressure under the load shown is 4062 psf, the factor of safety against bearing failure is

$$\text{FS} = \frac{q_d}{q_a} = \frac{22,200}{4062} = 5.47,$$

which would normally be considered adequate. On the other hand, let us assume that the footing is on sand with $\gamma = 110$ pcf and $N_{60} = 18$, as previously assumed for the settlement analysis illustration in Section 11.1.2. Because $c = 0$, we can use Equation (11.6). From Table 7.1, we find that $\phi \approx 32°$, and from Table 11.2, we find that $N_q = 28.52$. As was the case with the cohesive soil example, since $D_f = 0$, the N_γ term (here the second term of Equation (11.6) drops out, and the ultimate bearing capacity is computed as

$$q_d = 0.4 \times (110 - 62.4) \times 8 \times 28.52 = 4344 \text{ psf}.$$

Note that, in this application, the soil unit weight below the foundation is taken as the *effective unit weight*, $\gamma - \gamma_w$, provided that the water table is relatively near the footing base. Hence, the results indicate that the footing is very near failure under the load shown.

11.2.2   Deep Foundations

Axially loaded piling and piers are generally supported by soil resistance at the tip and along the sides of the embedded portion. Such distribution of load is indicated schematically in Figure 11.4. The total capacity of the piling Q_d is the sum of the capacities that result from side friction and point resistance, as indicated here:

$$Q_d = Q_s + Q_p. \tag{11.7}$$

Both the side friction and the point resistance are computed by multiplying unit values of resistance to the area that the soil and pile interface. In the case of the side friction, it is generally the perimeter of the pile multiplied by the length of the pile (or the layer of soil where the soil properties are relatively consistent), and for the point resistance, the cross-sectional area of the pile at the toe (the toe plate projected area for closed-ended piles) multiplied by the unit point resistance. Expanding Equation (11.7) along these lines, we have

$$Q_d = Q_s + Q_p = A_s f_s + A_p q_p, \tag{11.8}$$

where f_s and q_p are unit values of side friction and point resistance, respectively. A_s and A_p are side and point areas. Q_d is then is divided by a suitable factor of safety to arrive at an allowable load. This factor of safety is necessary to account for uncertainties in the analytical method and uncertainties in loads that may be placed on the piling.

In some situations, support may be derived either principally from side friction or principally from point resistance. Here the structures are referred to as being either *friction* or *point-bearing* elements, respectively. Point-bearing piles or piers are usually economical where surface soils are soft and a firm-supporting stratum exists at a reasonable distance from the ground surface. Depths in excess of 100 ft are common. Friction piles are used when a satisfac-

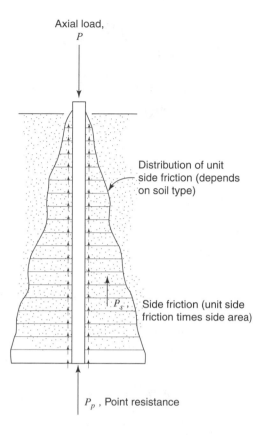

Axial load, P

Distribution of unit side friction (depends on soil type)

P_s , Side friction (unit side friction times side area)

P_p , Point resistance

Figure 11.4   Soil resistance on an axially loaded pile or pier.

tory tip-bearing stratum cannot be economically reached. In the usual application, friction piles are installed to comparatively shallow depths.

In some applications, deep foundations must carry lateral load from a structure. Two methods of installation are used in these applications. They are illustrated in Figure 11.5. Batter piles are inclined so that the resultant force on the pile head acts along the pile axis. The lateral force and the vertical force on the pile are in the same proportion as the lateral-to-vertical pile batter. In situations where the lateral design loads are moderate, analyses may be made for the lateral capacity and deflection of a vertical pile or pier. From the contractor's viewpoint, piles driven vertically are, of course, preferable to piles driven on a batter, because of the ease of construction involved. Piers are usually installed plumb (vertically).

Equation (11.8) shows that deep foundations carry load due to their capability to develop point resistance and skin friction. How much resistance is available may be estimated by several methods. These methods will be different, depending upon whether the soil is cohesionless or cohesive; they may also depend upon the job specifications from the owner as well. The methods presented here for both types of soil are those of Dennis and Olson (1983). Such methods were developed for steel pipe piles, but can be used with other types of piles. They will be outlined and illustrated in the next examples. Examples of these methods are outlined in this section, not to present a state of the art treatment of the subject, but to acquaint the reader with the fundamental considerations in deep foundation design.

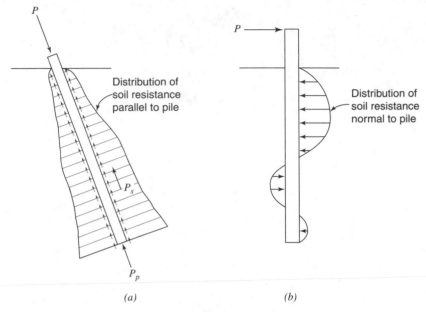

Figure 11.5 Soil resistance on laterally loaded piles and piers. (*a*) Batter piles. (*b*) Horizontal load on vertical pile or pier.

EXAMPLE   Capacity of Driven Pile in Clay

Suppose that a 12-in.-square prestressed concrete pile is to be driven 40 ft into a uniform deposit of clay, having an unconfined compressive strength q_u of 2000 psf. The unit skin friction is given by the equation

$$f_s = \alpha c F_c F_L \tag{11.9}$$

where α = adhesion factor, F_c = correction factor for strength, and F_L = correction factor for pile penetration. The pile length factor is determined by the equation

$$
\begin{aligned}
F_L &= 1 \ (\text{L} < 100 \text{ ft}) \\
F_L &= 0.01067L - 0.067 \ (100 < \text{L} < 175 \text{ ft}) \\
F_L &= 1.8 \ (\text{L} > 175 \text{ ft}),
\end{aligned}
\tag{11.10}
$$

where L is the length of the pile in feet. The adhesion factor is determined by

$$
\begin{aligned}
\alpha &= 1 \ (cF_c < 600 \text{ psf}) \\
\alpha &= -0.00083cF_c + 1.5 \ (600 < cF_c < 1200 \text{ psf}) \\
\alpha &= -0.0000526cF_c + 0.563 \ (1200 < cF_c < 5000 \text{ psf}) \\
\alpha &= 0.3 \ (cF_c > 5000 \text{ psf}),
\end{aligned}
\tag{11.11}
$$

The strength correction factor is given in Table 11.3. These values should be used if local experience is unavailable. For this problem, $F_c = 1.8$.

TABLE 11.3   Strength Correction Factors for Piles in Clay

Test Method for Soil Properties	Strength Correction Factor F_c
Unconfined compression tests on samples of high quality	1.1
Unconfined compression tests on samples taken with typical driven samplers	1.8
In situ vane shear tests	0.7

Having defined the equations, the first thing we should note is that we are given the un-confined compression strength, not the cohesion or shear strength. The cohesion is computed by dividing the unconfined compression strength by two, and thus

$$c = \frac{q_u}{2} = \frac{2000}{2} = 1000 \text{ psf.}$$

For a 40-ft-long pile, Equation (11.10) gives the value of $F_L = 1$. Since $cF_c = (1000)(1.8) = 1800$ psf, from Equation (11.11),

$$\alpha = -0.0000526cF_c + 0.563 = -0.0000526 \times 1800 + 0.563 = 0.47.$$

From Equation (11.9), the unit skin friction of the pile is thus

$$f_s = 0.47 \times 1000 \times 1.8 \times 1 = 846 \text{ psf.}$$

The unit point resistance is given by the equation

$$q_p = 9cF_c \tag{11.12}$$

and for this problem is

$$q_p = 9 \times 1000 \times 0.47 = 4230 \text{ psf.}$$

From Equation (11.8), the ultimate pile capacity is

$$Q_d = Q_s + Q_p = A_s f_s + A_p q_p = 4 \times 1 \times 40 \times 846 + 1^2 \times 4230 = 139{,}590 \text{ lbs.}$$

With a factor of safety of 3, the allowable pile axial load would be about 46.5 kips, or 23.3 tons.

EXAMPLE   Capacity of Driven Pile in Sand

Let us consider the same 12-inch-wide square, 40-ft-long concrete pile as before, but this time the pile is driven into sand with $\phi = 30°$ and $\gamma = 120$ pcf. The water table is at the surface. Compute the capacity of the pile.

The unit skin friction for piles in sand is

$$f_s = F_{SD}Kp'_{va} \tan \delta, \tag{11.13}$$

where p'_{va} = average effective stress along the side of the pile, K = earth pressure coefficient, δ = frictional angle between pile and soil, and

$$F_{SD} = \frac{5}{3}e^{-\frac{L}{60D}}, \tag{11.14}$$

TABLE 11.4   Pile-Soil Friction Angles and Bearing Capacity Factors for Piles Driven in Sand (After Dennis and Olson, 1983.)

Soil Type	δ, degrees	N_q
Silt	15	8
Sandy Silt	20	12
Silty Sand	25	20
Sand	30	40

in which D = diameter of the pile, ft. Values of K depend on the type of pile and whether it is being loaded in tension or compression. Full displacement piles (such as the one in this example) should have a $K = 1$ in compression and 0.7 in tension. Partial displacement piles should have a $K = 0.7$ in compression and 0.5 in tension. H-piles should have a $K = 0.8$. Values for δ are given in Table 11.4.

For this problem, $F_{SD} = 0.856$, $K = 1$, $p'_{va} = (120 - 62.4)(40)/2 = 1152$ psf, and $\delta = 30°$, we substitute and obtain the result

$$f_s = F_{SD} K p'_{vo} \tan \delta = 0.856 \times 1 \times 1152 \times \tan(30°) = 570 \text{ psf.}$$

The unit tip resistance is given by

$$q_p = F_D p'_{vo} N_q, \tag{11.15}$$

where p'_{vo} = effective stress at the pile tip, values for N_q are given in Table 11.4 and

$$F_D = \frac{1}{0.15 + 0.008L}. \tag{11.16}$$

It is important that L be given in feet for Equation (11.16). For this problem, $p'_{vo} = (120 - 62.4)$ $(40) = 2304$ psf, $F_D = 2.13$, and $N_q = 40$. The unit tip resistance for this pile is

$$q_p = F_D p'_{vo} N_q = 2.13 \times 2304 \times 40 = 196,301 \text{ psf.}$$

The ultimate axial capacity of the pile is thus

$$Q_d = Q_s + Q_p = A_s f_s + A_p q_p = 4 \times 1 \times 40 \times 570 + 1^2 \times 196,301 = 287,421 \text{ lbs.}$$

Again with a factor of safety of 3, the allowable axial capacity for the pile is 96 kips or 48 tons. There are several observations that need to be made about these methods:

- Both of these examples assumed a uniform soil. For layered soils, including those which have both clay and sand layers, the pile has to be broken up into segments, the capacities solved, and the results summed. For sand layers whose lower limit is above the pile tip, the value of p'_{vo} is best taken as the value at the bottom of the layer being analyzed, not at the pile tip.

- It is interesting to note that, not only in this case does the pile in sand have a higher capacity, but also most of this capacity is at the pile tip, while with the clay pile it is along the pile side. This is a typical difference between piles in sand and piles in clay.

- The capacity of the pile in clay is independent of the effective stress, whereas the capacity of the pile in sand is very much dependent upon the effective stress. This was predicted by the discussion of soil strength in Chapter 5.

The most reliable method of determining capacity of a deep foundation is by static load testing. Test loads are often required in a construction contract to overcome the uncertainties associated with the determination of allowable capacity. Sometimes load tests are justified, but usually only on large jobs. The load-test method involves placing a load on the pile or pier to be tested by stacking weights on a platform attached thereto, or by jacking, using adjacent foundations for reactions. The latter method is illustrated in Figure 11.6 and is the most common method of static load testing in the United States. A standard test method is specified in ASTM D1143. Since incorporating this information into the project may require some lead time, tests to failure generally will be done before the beginning of a project and may be a separate contract unto themselves.

Load tests are generally of two types: proof tests and tests to failure. Proof tests are used simply to confirm a design assumption. For instance, the design may call for piles having a minimum allowable bearing capacity of 200 kips with a factor of safety of 2. Therefore, the required minimum ultimate capacity of the piling would be 400 kips. A proof test would involve measuring the deflection of the pile head at the 400-kip load. If this deflection were satisfactory, the pile would be acceptable.

A test load to failure is used as a means of obtaining actual ultimate capacity or unit values of soil side friction and point resistance. These values obtained from actual load tests may be used to design piling of greater or lesser length or of different cross section than the actual test piling. Once such a test is run, it must be interpreted properly for the results to be meaningful. Several techniques are used, but the most common is the Davisson's method, which is illustrated in the next example.

Figure 11.6 Pile load test conducted by jacking against reaction piles.

EXAMPLE Davisson's Method for Static Load Tests

An 18-in.-diameter, 30-foot-long solid round concrete pile has been statically load tested and the results are plotted as the curve in Figure 11.7. The concrete has a modulus of elasticity $E_p = 3000$ ksi. Determine the ultimate capacity of the pile for failure.

Davisson's method is an "offset yield" method in that, instead of attempting to find the "break" in the load-deflection curve between the straight (elastic) and curved (plastic) portions of the curve, it seeks to find the intersection of the load-deflection curve with an elastic load-displacement line offset by a given amount. For Davisson's method, that amount is 0.15 in. $+ B/120$, where B is the diameter or basic dimension of the pile in inches. The slope of the line is given by

$$m = \frac{A_p E_p}{L_p},\qquad(11.17)$$

where A_p, E_p, and L_p are the cross-sectional area, the modulus of elasticity, and the length of the pile, respectively. This line, passing through the origin, is the load-deflection curve of the pile if it were a purely compression member with a rigid tip and no column buckling. The slope computed is also the slope of the offset line as well. For this pile, the slope is

$$m = \frac{\left(\dfrac{\pi B^2}{4}\right) E_p}{L_p} = \frac{\left(\dfrac{\pi \times 18^2}{4}\right) 3000}{30 \times 12} = 2121 \text{ kips/in.}$$

Figure 11.7 shows a straight line with this slope drawn from the origin. It can be seen that the actual load-deflection curve near the origin is identical with the line and bends away from it. The bending away represents the plastic yielding of the soil.

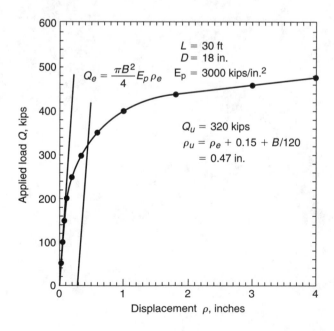

Figure 11.7 Pile load test results showing load against settlement of pile head. (After EI 02C097.)

Another line with the same slope as the first is drawn parallel to the first, but to the right by

$$\text{Offset} = 0.15 + \frac{B}{120} = 0.15 + \frac{18}{120} = 0.3 \text{ in.}$$

The intersection of the second line with the pile's load-deflection curve takes place at the load $Q_d = 320$ kips, which is the ultimate capacity of the pile.

11.3 FOOTING AND RAFT CONSTRUCTION

Specifications usually require that groundwater be controlled in an excavation where shallow foundations are to be built and that the footings be placed on undisturbed earth. This is done so that the as-built foundations bear on soils that have the same properties as those used in the design. In some cases, foundation soils will be improved by compaction or by replacement with better quality material.

Machine excavation usually is done with a mechanical shovel, backhoe, or tracked vehicle. Grading with such equipment loosens the upper few inches of supporting soil. The example in Section 11.1.1 shows that the major portion of the settlement beneath a footing is seated in the soil closest to the footing base. The loosened soil, which is more compressible than the natural soil, must, therefore, be removed. Hand excavation is often required to do so. Foundation construction in progress is shown in Figure 11.8.

Uncontrolled water in an excavation on either the surface or seeping upward because of inadequate dewatering will also loosen the surface soils. Water must therefore be controlled by sumping and dewatering. In the event that softening takes place, the unsuitable soil should be removed before construction of footings begins. The practice of dumping imported granular materials on the base of an excavation to provide trafficability for equipment is most often unacceptable, since the softened soil usually fills the voids in the coarser materials, resulting in a base that has nearly the same undesirable high compressibility.

It is a common practice to require that foundations and slabs be constructed on a thin lift of compacted sand or gravel. The usual purpose is to provide a highly pervious material interfacing with a foundation drain system. In other cases, this is done to provide material with limited compressibility immediately beneath footings. In both cases, the result desired is a clean, dense imported layer. Excavation in preparation for placing the layer must usually be done with the same care as that for foundations. Loosened material not removed will contaminate the coarser imported soil more easily than the undisturbed subgrade, particularly during compaction. If very coarse gravels are to be used, a thin layer of well-graded sand is sometimes placed first, to act as a filter and prevent intrusion of the natural soil into the voids of the coarse gravel.

11.4 FOUNDATION PILING

The principal use of driven piling in the context of this chapter is as load-bearing structural elements. They are used to transmit load either vertically or laterally to a satisfactory soil-bearing stratum. On occasion, piles are also used to compact loose soils. In harbors and around ship berths, piles are used for fendering systems and dolphins to protect structures that are more permanent.

Figure 11.8 Footing foundation construction for a multistory building.

The use of compaction piles has been discussed in Chapter 8. Such piles are driven in loose cohesionless soil deposits to occupy void space and thereby densify the material. They serve no structural purpose whatsoever. Since compaction piles are designed to carry no load, and since they are usually driven to a predetermined depth, the specifications are not particularly restrictive. Therefore, many of the problems associated with foundation pile installation do not arise.

11.4.1 Pile Types

Depending on the basis of classification used, it would be possible to identify scores of different types of foundation piling. Considering the materials used, however, there are only four principal types. These are illustrated in Table 11.5.

The three principal materials involved are timber, concrete, and steel. Composite piles are made by combining two of these materials. No single pile type is universally accepted as being the best. Each has its own particular advantages and disadvantages that determine its suitability for a particular project. From a design viewpoint, steel piles and concrete piles generally are used to carry the heaviest load. Timber piles are limited in their carrying capability by the natural size of the tree from which the pile was cut. It has become an increasingly prevalent practice to permit design loads on timber piles up to 100 kips. The actual maximum will be based on the allowable stress for the wood and the pile size and load-carrying mode. The load-carrying capacity of steel and concrete piles is limited only by the size of section selected and the allowable stress used, which presumes, of course, that there is adequate soil support to develop this capacity.

It is a well-known fact that untreated timber piling installed full length below the permanent water table may be considered as permanent. Untreated piles above the water table are prone to decay because of both bacterial action and attack from marine parasites (the result is similar in effect to damage caused by termites). Creosote and other types of treatment for timber piles alleviate the problem; however, due to environmental concerns and restrictions, treated wood is becoming more difficult to use as a piling material. Steel piles are susceptible to corrosion in a harsh chemical environment. An abundant supply of oxygen and dissolved salts contributes to their deterioration. Studies have shown that the oxidation rate of steel piles driven full length in the ground is most rapid in that portion of the ground that is made up of recent fill. Concrete piles are subject to attack by acids that may be dissolved in the ground. Certain acid-resistant cements may be used to retard acid attack. Overall, when selecting a particular pile type for a given application, one must consider not only the load to be carried, but also the environment in which the pile is to be installed.

From the contractor's viewpoint, the drivability of a piling and the ease with which its length may be adjusted are probably more important considerations than merely its capability to carry load. Piles that displace a large volume of soil as they are installed are more difficult to drive than *nondisplacement piles*. One could compare, for instance, a steel H-piling and a

TABLE 11.5 Basic Piling Types

Material	Advantages	Limitations
Timber	Cutting, handling ease.	Damage during hard driving.
Precast concrete	Availability.	Length changes difficult.
		Handling care required.
		Harder driving than other high-capacity piles.
Cast in-place concrete (steel shells)	Adjustable length.	Two-step construction sequence.
Steel H-section or steel pipe	Adjustable length.	Economy.
	Easy driving.	Availability.
	Easy handling.	
Composite	Advantages of each material.	Disadvantages of each material.
		Three-step construction operation.

steel pipe piling driven with a closed end. The former will drive much more easily than the latter if the same hammer is used, even considering the possibility that the piles may be made of the same material and have the same material cross-sectional area. Concrete piles and timber piles are also considered *displacement piles*. Displacement piling then is particularly appropriate for use as friction piles. Driving displacement piling to great depth for point bearing may be difficult.

In many applications, the length of a driven piling cannot be determined prior to installation. In those cases where the bedrock surface has not been well defined by the exploratory drilling program or when the bedrock itself is variable in depth (as is the case with Karst topography), the length of point-bearing piles may vary considerably from estimates. In this situation, steel piles offer the advantage of relative ease in the adjustment of length. Extra sections may be added or extra lengths may be removed readily in the field. To alter the lengths of concrete piles by either splicing or cutting is a difficult field operation. The possibility that these adjustments in length may be required in a particular application should be considered by the contractor in preparing his bid. Traditionally, it has been the practice with timber piles to order extra length and after driving was completed, to cut off the remaining portion. This was permitted by the relative ease of making the cutoff and the comparative low cost of the timber material. Because material costs have risen, such practice is no longer a desirable alternative to accurate predetermination of driven lengths, particularly on large jobs.

Concrete piles may be cast in practically any size, according to the particular application for a job. Steel piles are available in standard rolled sections such as those indicated in Figures 11.9 and 11.10. Timber pile sections are standard within ranges. These ranges are specified in ASTM D25 and are shown in Figure 11.11.

11.4.2 Pile Hammers

A pile-driving rig consists essentially of a set of leads, the hammer, and a crane as shown in Figure 11.12. The function of the leads is to guide the hammer and the pile during driving. In Figure 11.12, the leads are fixed (attached at the tip of the crane's boom), and there is a horizontal brace or strut connecting the leads and crane. In this case, the brace can be extended or shortened to incline the leads so that piles can be driven on a batter. In some cases, the attachment of the leads at the boom is such that the leads can be raised or lowered by sliding at the point of attachment. For side batters, the brace may be designed to move laterally or it may have a moonbeam (upward curved beam supporting the lower end of the leads) at its end. Swinging leads are suspended from, rather than attached to, the boom. An example of swinging leads is shown in Figure 11.13. In this case, the leads are tilted to enable a pile to be driven on a batter (at an angle) without having to used fixed leads. Such practice is common, but requires special care on the part of the contractor. For offshore applications, the pile may be supported by an external template and driven by a suspended hammer. A special "stub" type leader is used in this case. Figure 11.14 shows offshore pile driving that use this setup.

The essentials of a hammer-pile system are shown in Figure 11.15. A driving cap or helmet rests on the top of the pile to keep the hammer and pile together during driving. In driving concrete piles, a cushion, usually wood, is inserted between the pile head and the cap to prevent pile damage. In some circumstances, a cap block is used to cushion the hammer blow on the cap. The cap block frequently consists of a coiled length of wire rope, although other materials are used.

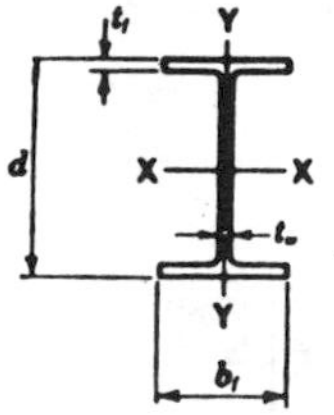

HP SHAPES
PROPERTIES FOR DESIGNING

			Flange		Web	Elastic Properties					
						Axis X-X			Axis Y-Y		
Designation	Area A	Depth d	Width b_f	Thickness t_f	Thickness t_w	I	S	r	I	S	r
	(In.²)	(In.)	(In.)	(In.)	(In.)	(In.⁴)	(In.³)	(In.)	(In.⁴)	(In.³)	(In.)
HP 14 × 117	34.4	14.21	14.885	0.805	0.805	1220	172	5.96	443	59.5	3.59
× 102	30.0	14.01	14.785	0.705	0.705	1050	150	5.92	380	51.4	3.56
× 89	26.1	13.83	14.695	0.615	0.615	904	131	5.88	326	44.3	3.53
× 73	21.4	13.61	14.585	0.505	0.505	729	107	5.84	261	35.8	3.49
HP 13 × 100	29.4	13.15	13.205	0.765	0.765	886	135	5.49	294	44.5	3.16
× 87	25.5	12.95	13.105	0.665	0.665	755	117	5.45	250	38.1	3.13
× 73	21.6	12.75	13.005	0.565	0.565	630	98.8	5.40	207	31.9	3.10
× 60	17.5	12.54	12.900	0.460	0.460	503	80.3	5.36	165	25.5	3.07
HP 12 × 84	24.6	12.28	12.295	0.685	0.685	650	106	5.14	213	34.6	2.94
× 74	21.8	12.12	12.217	0.607	0.607	566	93.4	5.10	185	30.2	2.91
× 63	18.4	11.94	12.125	0.515	0.515	472	79.1	5.06	153	25.3	2.88
× 53	15.5	11.78	12.045	0.435	0.435	393	66.8	5.03	127	21.1	2.86
HP 10 × 57	16.8	9.99	10.225	0.565	0.565	294	58.8	4.18	101	19.7	2.45
× 42	12.4	9.70	10.075	0.420	0.415	210	43.4	4.13	71.7	14.2	2.41
HP 8 × 36	10.6	8.02	8.155	0.445	0.445	119	29.8	3.36	40.3	9.88	1.95

Figure 11.9 Standard steel H-pile sections. (Reprinted, by permission, from American Institute of Steel Construction.)

Size O.D.	Wall Thickness (in.)	Weight per linear ft (lb)	I Moment of Inertia (in.^{4})	r Radius of Gyration (in.)
8⅝″	.172	15.52	40.81	2.99
	.188	16.90	44.36	2.98
	.203	18.26	47.65	2.98
	.219	19.64	51.12	2.97
9″	.172	16.20	46.49	3.12
	.188	17.64	50.54	3.12
	.203	19.07	54.30	3.11
	.219	20.51	58.26	3.11
10″	.172	18.04	64.14	3.48
	.188	19.70	69.77	3.47
	.203	21.24	74.99	3.46
	.219	22.88	80.51	3.46
	.230	24.00	84.28	3.46
	.250	26.03	91.05	3.45
10¾″	.172	19.42	79.97	3.74
	.188	21.15	87.01	3.73
	.219	24.60	100.48	3.72
	.250	28.04	113.71	3.71
11″	.172	19.87	85.77	3.83
	.188	21.65	93.34	3.82
	.203	23.40	100.37	3.82
	.219	25.18	107.81	3.81
	.230	26.45	112.88	3.81
	.250	28.70	122.03	3.80
12″	.172	21.71	111.79	4.18
	.188	23.72	121.70	4.18
	.203	25.58	130.92	4.17
	.219	27.56	140.67	4.17
	.230	28.91	147.33	4.16
	.250	31.37	159.33	4.16
	.281	35.17	177.70	4.14
	.312	38.95	195.77	4.13
12¾″	.172	23.09	134.43	4.45
	.188	25.16	146.38	4.44
	.219	29.31	169.27	4.43
	.250	33.38	191.82	4.42
	.281	37.45	214.03	4.41
	.312	41.51	235.90	4.40
13″	.172	23.54	142.61	4.54
	.188	25.65	155.30	4.53
	.203	27.74	167.10	4.52
	.219	29.86	179.61	4.52
	.230	31.36	188.15	4.52
	.250	34.04	203.56	4.51
	.281	38.20	227.16	4.50
	.312	42.34	250.41	4.49

Figure 11.10 Dimensions and properties for standard steel-pipe piling. Additional sizes are available. (Courtesy of L.B. Foster Company.)

Size O.D.	Wall Thickness (in.)	Weight per linear ft (lb)	*I* Moment of Inertia (in.⁴)	*r* Radius of Gyration (in.)
14″	.172	25.38	178.62	4.89
	.188	27.66	194.57	4.88
	.219	32.20	225.14	4.87
	.250	36.71	255.30	4.86
	.281	41.21	285.04	4.85
	.312	45.68	314.38	4.84
16″	.172	29.06	267.86	5.60
	.188	31.66	291.90	5.59
	.219	36.87	338.06	5.58
	.250	42.05	383.66	5.57
	.281	47.22	428.72	5.56
	.312	52.36	473.24	5.55
	.375	62.58	562.08	5.53
18″	.219	41.54	483.55	6.29
	.250	47.39	549.13	6.28
	.281	53.22	614.05	6.27
	.312	59.03	678.25	6.25
	.375	70.59	806.63	6.23

Figure 11.10   (continued)

Specified Butt Diameters (Inches)	7	8	9	10	11	12	13	14	15	16	18
Required Minimum Circumference 3 ft from Butt	22	25	28	31	35	38	41	44	47	50	57
Length (Feet)	Minimum Tip Circumferences (in.) and Corresponding Diameter in Parentheses										
20	16.0 (5.0)	16.0 (5.0)	16.0 (5.0)	18.0 (5.7)	22.0 (7.0)	25.0 (8.0)	28.0 (8.9)				
30	16.0 (5.0)	16.0 (5.0)	16.0 (5.0)	16.0 (6.0)	19.0 (6.0)	22.0 (7.0)	25.0 (8.0)	28.0 (8.9)			
40				16.0 (5.0)	17.0 (5.4)	20.0 (6.4)	23.0 (7.3)	26.0 (8.3)	29.0 (9.2)		
50					16.0 (5.0)	17.0 (5.5)	19.0 (6.0)	22.0 (7.0)	25.0 (8.0)	28.0 (8.9)	
60						16.0 (5.0)	16.0 (5.0)	18.6 (5.9)	21.6 (6.9)	24.6 (7.8)	31.6 (10.0)
70						16.0 (5.0)	16.0 (5.0)	16.0 (5.0)	16.2 (5.1)	19.2 (6.1)	26.2 (8.3)
80							16.0 (5.0)	16.0 (5.0)	16.0 (5.0)	16.0 (5.0)	21.8 (6.9)
90							16.0 (5.0)	16.0 (5.0)	16.0 (5.0)	16.0 (5.0)	19.5 (6.2)
100							16.0 (5.0)	16.0 (5.0)	16.0 (5.0)	16.0 (5.0)	18.0 (5.8)
110										16.0 (5.0)	16.0 (5.0)
120											16.0 (5.0)

Figure 11.11   Standard timber pile sections. Select table according to specification requirement for butt or tip. (ASTM D25-70, reprinted, with permission, from the *Annual Book of ASTM Standards*. Copyright © 1970, American Society for Testing and Materials, 1916 Race Street, Philadelphia, PA, 19103.)

Specified Tip Diameter (Inches)	5	6	7	8	9	10	11	12
Tip Circumference, Required Minimum	16	19	22	25	28	31	35	38
Length (Feet)	Minimum Circumferences 3 ft from Butt (in.) with Diameter in Parentheses							
20	22.0 (7.0)	24.0 (7.6)	27.0 (8.6)	30.0 (9.5)	33.0 (10.5)	36.0 (11.5)	40.0 (12.7)	43.0 (13.7)
30	23.5 (7.5)	26.5 (8.4)	29.5 (9.4)	32.5 (10.3)	35.5 (11.3)	38.5 (12.2)	42.5 (13.5)	45.5 (14.5)
40	26.0 (8.3)	29.0 (9.2)	32.0 (10.2)	35.0 (11.1)	38.0 (12.1)	41.0 (13.0)	45.0 (14.3)	48.0 (15.3)
50	28.5 (9.0)	31.5 (10.0)	34.5 (11.0)	37.5 (11.9)	40.5 (12.9)	43.5 (13.8)	47.5 (15.1)	50.5 (16.0)
60	31.0 (9.8)	34.0 (10.8)	37.0 (11.8)	40.0 (12.7)	43.0 (13.7)	46.0 (14.6)	50.0 (15.9)	53.0 (16.8)
70	33.5 (10.6)	36.5 (11.6)	39.5 (12.6)	42.5 (13.5)	45.5 (14.4)	48.5 (15.4)	52.5 (16.7)	55.5 (17.7)
80	36.0 (11.4)	39.0 (12.4)	42.0 (13.4)	45.0 (14.3)	48.0 (15.3)	51.0 (16.2)	55.0 (17.5)	58.0 (18.4)
90	38.6 (12.2)	41.6 (13.2)	44.6 (14.2)	47.6 (15.1)	50.6 (16.0)	53.6 (17.0)	57.6 (18.3)	60.5 (19.2)
100	41.0 (13.0)	44.0 (14.0)	47.0 (15.0)	50.0 (15.9)	53.0 (16.8)	56.0 (17.8)	60.0 (19.0)	
110	43.6 (13.8)	46.6 (14.8)	49.6 (15.7)	52.6 (16.7)	55.6 (17.7)	61.0 (19.4)		
120	46.0 (14.6)	49.0 (15.6)	52.0 (16.6)	55.0 (17.5)	58.0 (18.4)			

Figure 11.11 (*continued*)

Figure 11.12 Driving concrete piles with a differential acting hammer. (Courtesy of Vulcan Foundation Equipment.)

Figure 11.13 Single-acting hammer in use with swinging leaders. (Courtesy of Vulcan Foundation Equipment.)

Figure 11.14 Driving offshore pile with stub leaders and using the platform as a template. (Courtesy of Vulcan Foundation Equipment.)

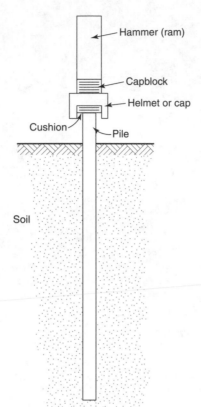

Hammer (ram)

Capblock

Helmet or cap

Cushion

Pile

Soil

Figure 11.15 Schematic of hammer-pile system.

The falling weight of the hammer strikes an anvil. The anvil, the cap block, and the cushion may or may not be present according to the type of hammer and the piling being driven. There are five basic types of hammers. They are illustrated conceptually in Figure 11.16. Detailed specifications for a number of hammers are shown in Appendix B.

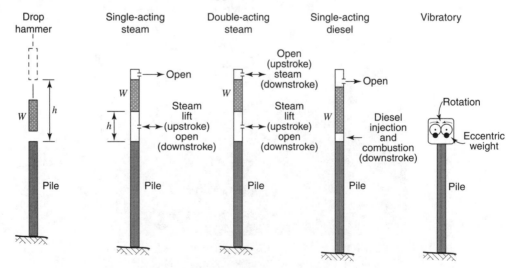

Figure 11.16 Conceptual representation of pile hammers.

Figure 11.17   A lightweight drop hammer used for installation of timber piles. Helmet is shown suspended on hammer above pile head.

The drop hammer (see Figure 11.17) is the simplest hammer type and, yet, because of the low available energy for driving and the high impact stresses it produces in piling, is not the most satisfactory hammer to use. The concept of a drop hammer, however, is a useful background for the discussion of other hammer types. All hammers have a rated energy. The energy of a drop hammer is simply the potential energy of the raised weight above the pile head. The theoretical maximum amount of energy produced by a drop hammer is the product of the weight and the height of drop. Due to energy losses, the rated energy is multiplied by an efficiency factor to arrive at the energy available for driving.

A drop hammer is operated by lifting the weight with a cable and allowing it to fall freely, striking the head of the pile. This operation is cumbersome, time consuming, and not suitable for high-production operations. A single-acting steam or air hammer operates on the same principle, except that steam or air is used to raise the ram before it is dropped. At the top of a stroke, the chamber pressure raising the ram is exhausted and the ram falls. Single-acting hammers are widely used and particularly suited to driving heavy piles.

The rams for double-acting or differential-acting steam or air hammers are driven on both the up and down strokes. The ram is raised to the maximum height in much the same manner as a single-acting hammer. On the downstroke, steam or air is injected, which accelerates the downward movement of the ram. Accordingly, the maximum energy available is the sum of the potential energy resulting from the weight of the ram and the additional energy imparted by the injection at the top of the downstroke. Double-acting hammers are widely used for driving light piling in relatively easy driving conditions. They operate rapidly. It should be pointed out that a double-acting hammer might develop the same rated energy as that of a comparable single-acting hammer in one of two ways. If the weights of the rams in the two hammers are the same, they must have the same velocity at impact in order to have the same kinetic energy. This requires that the height of fall of the double-acting hammer be lower than for the comparable single-acting hammer of the same weight and that the additional energy to accelerate the ram be made up by injection of gas at the top of the stroke. However, if the ram weight for the double-acting hammer is less than that of the corresponding single-acting hammer, the ram of the double-acting hammer must travel at a higher velocity at impact in order to achieve the same energy. This additional velocity must be achieved by either an increased height of drop, an increased speed due to gas injection, or a combination of both. These matters are particularly important when considering potential damaging effects of hammers on piling. It can be shown, for instance, that the stresses produced in a pile by driving are proportional to the velocity of the hammer at impact. Therefore, single-acting hammers and double- or differential-acting hammers of the same rated energy are not necessarily equally suited to the driving of a particular pile.

Diesel hammers (see Figure 11.18) are very popular. They offer the advantage of being entirely self-contained, requiring no external boiler or compressor system for operation. Combustion of diesel is produced as the result of the compression of a fuel–air mixture in the space below the falling ram. Therefore, a diesel hammer must be started manually on the first stroke. For each succeeding stroke, the hammer runs itself, since the explosion raises the ram and permits it to fall, compressing the gas for the next ignition. The rated energy of a diesel hammer is a combination of the potential energy produced by the height of drop of the ram and the additional energy imparted by the resulting fuel combustion. Double-acting diesel hammers are also available. Diesel hammers operate at rated energy generally when hard driving resistance is encountered. In softer driving, the lower resistance does not permit the ram to recoil for its full height of drop, and the operation is therefore slower and results in less energy. In very soft ground conditions, diesel hammers will sometimes fail to operate properly since the pile does not offer enough resistance to penetration to permit the needed compression for ignition to occur.

Hydraulic impact hammers (Figure 11.19) are a relatively new type of pile hammer. Instead of using steam or air, the ram is moved up and down using pressurized hydraulic fluid, which is furnished for the smallest hammers from the crane's hydraulic system and for most hammers from a separate power pack. As is the case with air compressors or boilers, the hydraulic power pack can be mounted on the rear of the crane and acts as part of the counterweight the crane requires. Hydraulic hammers have high rated efficiencies and many instrumentation and operational features, but they tend to be expensive and require a higher caliber of maintenance than air, steam, or diesel hammers.

Figure 11.18 Diesel pile hammer.

Figure 11.19 Hydraulic impact hammer. (Courtesy of Vulcan Foundation Equipment.)

Figure 11.20   A vibratory pile hammer used to drive H-beams. (Reprinted, by permission, from Warrington, Don C., "Vibratory Pile Driving Equipment," *Vulcan Iron Works Inc.* © Copyright 1989.)

Vibratory hammers (see Figure 11.20) are fundamentally different from the other types of hammer previously discussed. These hammers transmit a sinusoidally varying force to the top of the pile because of the centrifugal forces produced by a pair of eccentric weights operated in synchronization by electric motors in the hammer. The vibrations produce changes in the soil properties that enable the pile to sink into the ground by its own weight. Vibratory hammers are particularly suited for driving nondisplacement piling and sheet piling in sandy soils.

A number of factors bear on the selection of a particular hammer type for a particular job, not the least of which is ownership or availability of the equipment. The pile size and type must be considered, as well as the driving energy required. In addition, the speed of the driving operation is an important factor in preparing the bid. For operation in adverse weather conditions or underwater, special hammer types are required. Generally, it is advantageous to drive piling with a hammer that is large enough to produce the maximum penetration rate and, at the same time, not damage the piles. Traditionally, hammer selection has been largely a matter of the experience of a particular contractor within the limits imposed on him or her by the contract specifications. Analytical approaches to hammer selection are in continuing stages of development. They are based on techniques for computing pile capacity or driving stresses. These concepts are introduced in later sections of this chapter.

11.4.3   Jetting, Spudding, and Predrilling

In some circumstances, piling must be driven through dense or hard materials to bearing at a greater depth. Several driving aids have been developed for this purpose. The principal functions of these driving aids are to speed the driving operation and to prevent damage to piling that results from heavy driving. *Jetting* is applicable to those situations where piling must be driven through cohesionless soil materials to greater depths. Jetting is accomplished by pumping water through pipes attached to the side of the pile as it is driven. The flow of water creates a *quick* condition and thereby reduces skin friction along the sides of the pile. The result is that the pile drives more easily. Where piles have been designed to carry a large portion of their load in skin friction, jetting may not be permitted, since it loosens the soil and thereby could reduce pile capacity.

Spudding and *predrilling* are both operations that produce a hole into which a driven piling may be inserted. Spudding is accomplished by driving a heavy steel mandrel that is subsequently removed. Spudding is particularly applicable to the installation of piling through materials containing debris and large rocks. Predrilling accomplishes the same results except that an auger, rather than a mandrel, is used. In both cases, the stratum to be spudded or predrilled must be cohesive enough so that the resulting hole remains open until pile driving begins.

11.4.4   Pile Dynamics

Most geotechnical engineering applications involve design considerations that are purely static in nature. Pile driving, on the other hand, is a very dynamic process, involving high velocities and impact stresses. The art or science of analyzing the movement of piles during driving and relating these observations to the performance of driven piles in use is referred to as *pile dynamics*. Most of the difficulties with driven piles stem from problems in pile dynamics. A contractor who wants successful driven piles must understand these considerations completely.

An impact hammer ram strikes the pile cushion or driving accessory with an impact energy given by the formula

$$E_r = \frac{WV_h^2}{2g},$$ (11.18)

where W is the weight of the ram, V_h is the impact velocity of the ram, and g is the acceleration due to gravity. This energy can be spent in a number of ways in the driving system, such as

- plastic deformation of the hammer and pile cushion, which results in heat generation;
- losses due to poor alignment of the hammer, driving accessory, and pile;
- plastic deformation of the pile (usually not a large amount); and
- plastic deformation and energy radiation into the soil.

The last point is the most important, because plastic deformation of the soil is the object of pile driving; this deformation produces the permanent set for each blow of the hammer.

As the pile is driven into the soil, the resistance of the soil increases and the set per blow decreases. The point at which further driving would result in very small movement of the pile and very large potential damage to the hammer and the pile is referred to as the refusal point. The value of the ultimate resistance at which this point occurs depends upon the properties of the soil and its interaction with the pile. It also depends upon the distribution of the energy losses described previously.

Since the ultimate capacity of the soil also depends upon the interaction of the pile and the soil, it makes sense that there is some kind of relation between the behavior of the pile during driving and its performance during use. Quantifying this relationship is the most important objective of pile dynamics.

The first method used was the *dynamic formulae*. The dynamic formulae assume the pile is a solid mass and that the impact between ram and pile could be modeled using Newtonian impact mechanics. Many of these formulae have been set forth; the most important one in the United States is the *Engineering News* formula, set forth in the late 1880s. This formula computes the allowable pile capacity with the equation

$$Q_a = \frac{2Wh}{s + c} = \frac{2E_r}{s + c}, \tag{11.19}$$

where Q_a is the allowable load of the pile in pounds, h is the drop height of the ram (for single-acting hammers), E_r is the rated striking energy of the hammer in ft-lbs, s is the permanent set per hammer blow of the pile in inches, and c is a constant that depends on the type of hammer being used. For drop hammers, $c = 1$; for air and steam hammers, $c = 0.1$. It should be noted that it is incorrect to apply a hammer efficiency to this formula, as the factor of safety of 6 already includes this value. Values of s depend upon the hammer manufacturer and the contract specifications. A common value for refusal used is $s = 0.1$ inch, which can also be inverted to be expressed as 120 blows per foot, or 120 BPF.

This formula, along with the other dynamic formulae, became very popular to use and was written into many driven-pile specifications; even today, many specifications still use these formulae. Unfortunately, two limitations forced reconsideration of the use of these formulae. The first was the erratic relationship between the results of the formulae and the results of load tests. The second concerned problems with precast (as opposed to prestressed) concrete piles breaking in the middle during easy driving. The *Engineering News* formula came into use when only timber piles were being driven; the rigid-body assumption could not explain this breakage.

It remained for the Australian civil engineer D.V. Isaacs in 1931 to propose the use of wave mechanics to model pile driving, to estimate stresses during driving, and to estimate the set of the pile for a given resistance to driving. The pile was modeled as a rod with distributed mass and elasticity; waves traveled down the pile from impact and were reflected from the pile tip. The breakage in concrete piles was due to compression waves being reflected as tension waves at low tip resistance and thus breaking the piles.

Implementation of wave mechanics was beyond practical computational limits in the 1930s. By the late 1950s the computer became available, and E.A.L. Smith of the Raymond Concrete Pile Company, at the time the greatest pile-driving organization in the world, developed a numerical solution to the problem. This solution involved dividing the pile into discrete masses connected by springs, with the soil modeled by a combination of springs and dashpots, as shown in Figure 11.21. The solution is commonly referred to as the *wave equation* for piling and, with modern developments, is the preferred method of dynamically analyzing drivability and capacity of driven piles today.

Wave-equation programs are used to model the displacement of the pile during driving; in this way, the expected pile sets can be correlated with various pile capacities, and job-site

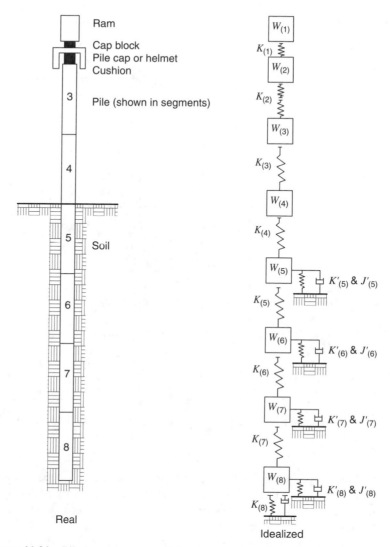

Ram

Cap block
Pile cap or helmet
Cushion

Pile (shown in segments)

Soil

Real

Idealized

Figure 11.21   Pile-hammer representation for wave-equation analysis.

control can be based on this correlation. Moreover, as shown in Figure 11.22, the stresses in the pile during driving can be predicted and compared with the maximum allowable driving stresses for a given pile material, which is important, since a driven pile must be sized for both its service stresses and its driving stresses.

Field implementation of the wave equation is not limited to running wave-equation programs; a device called the pile-driving analyzer is used both to monitor the driving stresses as they occur and to predict the capacity of a given pile, using the stress wave from the hammer impact. A schematic of the pile-driving analyzer is shown in Figure 11.23. Strain gauges and accelerometers are mounted at the pile top, and these, in turn, transmit data to the analyzer. Depending upon the capabilities of the analyzer, it will both output the force–time and veloci-

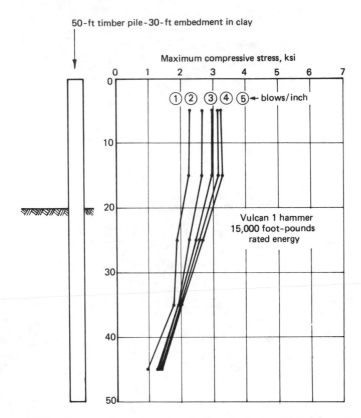

50-ft timber pile-30-ft embedment in clay

Figure 11.22 Compressive stresses in a timber pile during driving from wave-equation analysis.

ty–time curves for each blow and plot them on a screen or paper, as shown in Figure 11.24. As this figure shows, the analyzer can also estimate the capacity of the pile by taking the measured force–time and velocity–time curves and running a wave-equation analysis, varying the soil configuration until the computed curves match the measured ones. Plots of penetration resistance against depth for piles driven in sands and clays are shown in Figure 11.25.

The wave equation, along with the pile-driving analyzer, is the preferred method both of prediction of driving performance of both hammer and pile and of jobsite control of pile driving. The specialized nature of the analysis and the lack of methods have given a "black box" mystique to the entire process, which leads to uncritical acceptance of the results. Such uncritical acceptance can have serious consequences for the proper execution of a pile-driving project, for both the owner and contractor alike.

11.4.5 Contract Provisions

Although both static and dynamic methods of prediction of pile capacity have been advanced, there is still considerable uncertainty in the accuracy of the predictions. There are also problems associated with the external effects of the pile-driving operation. All of these uncertainties contribute to further uncertainties in bidding and in the payment for work actually accomplished.

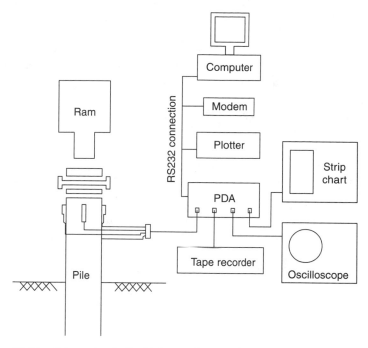

Figure 11.23 Schematic of pile-driving analyzer. (After EI 02C097.)

The objective of the owner is for the pile to carry a certain axial or lateral load. To accomplish this objective, it is necessary to (a) select a pile type and (b) drive it to a sufficient length to support the axial or lateral load. Depending upon the structure of the specifications, the contractor can have input on one or both of these parameters. How these parameters are specified can also influence how the contractor will bid the job.

Concerning the pile type and material, the particular pile material to be used is usually selected after careful consideration by the design engineer; nevertheless, the contractor in some instances is permitted the latitude of offering a substitute. For example, a contractor may be offered the alternative of driving prestressed concrete piling to tip bearing, instead of steel H-piling in the case where the concrete piles may be easier to obtain. The contractor must recognize, however, that the concrete piles will likely drive harder, being displacement piles, and that he or she will probably be required to bear the additional expense associated with splicing or cutting.

Turning to the issue of pile length, in theory, the method described earlier to estimate the capacity of the pile (or another method) should allow the engineer or contractor to determine the length of the pile required to support the design load. But, in reality, these methods are not accurate enough without other forms of verification. A static load test is one way to make the proper estimate, but if there are variations in soil conditions in different parts of the job, the results of the load test may not be representative. Here is where pile dynamics become significant for the contractor, because such methods can allow an estimate of pile capacity by using the pile hammer as a measuring tool.

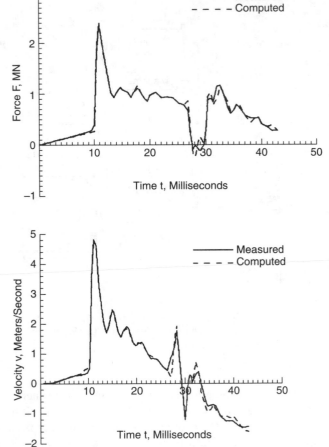

Figure 11.24 Comparison of measured and predicted force–time and velocity–time curves, using a pile-driving analyzer. (From EI 02C097.)

Inspection of Equation 11.19 shows that, for a given hammer, the set per blow (or blows per foot or meter of pile penetration) will determine the capacity of the pile. This has led to a *blow-count specification*, in which the contractor is required to drive the pile until a certain blow count is achieved. If this blow count takes place, the pile can be accepted. Although the wave equation has complicated the determination of the blow count, blow-count specifications are still common for driven piles. As we noted before, the contractor can check this method by using a *drivability analysis* with a wave equation program; for some jobs, a drivability analysis is required in the bid submission. A wave-equation analysis also enables the contractor to deal with issues such as hammer selection, pile type, and pile stresses during driving.

One difficulty in blow-count specifications is the issue of inaccurate "as furnished" pile lengths. The top of the piles must be at the proper elevation to interface with the pile caps and the structure they are supporting. The contractor must have sufficient material to achieve proper top elevation, but not too much to end up having to pay for unnecessary material. Any type of pile (including concrete piles) can be cut with modern equipment. Splicing is more difficult with concrete piles than with steel and is generally unadvisable to use with wood piles.

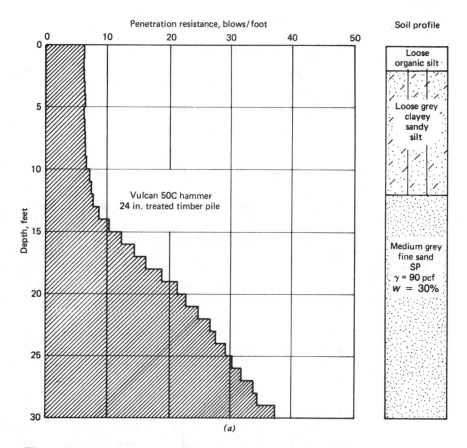

Figure 11.25 Example of penetration-resistance against depth plots for piles. (*a*) Driven in sand. (*b*) Driven in clay.

The alternative to a blow-count specification is a *tip-elevation specification*, in which the pile must be driven to a specific tip elevation and thus a length. These are very common with laterally loaded piles, as the ability to verify the lateral load capacity of the pile is more difficult than with axially loaded piles. The major difficulty with tip-elevation specifications occurs when the pile cannot be driven to the specified depth by the equipment on the job, either at all or in a reasonable period of time. In bid preparation, the contractor must therefore make a thorough drivability study to insure that the piles can be driven to the desired tip elevation.

In some cases, specifications will call for both blow-count and tip-elevation standards. This type of specification is especially difficult to deal with, and the contractor needs to get such a specification clarified.

Both of these specifications relate to the methods of payment for driven piles. They can be broken down into two basic types: fixed price and per unit length of driven pile. A fixed-price contract requires the contractor to drive the piles to the desired tip-elevation or blow-count specification for a fixed price, forcing the contractor to include provisions in his or her bids for uncertainties that are inherent in this kind of work. By using a contract that pays the

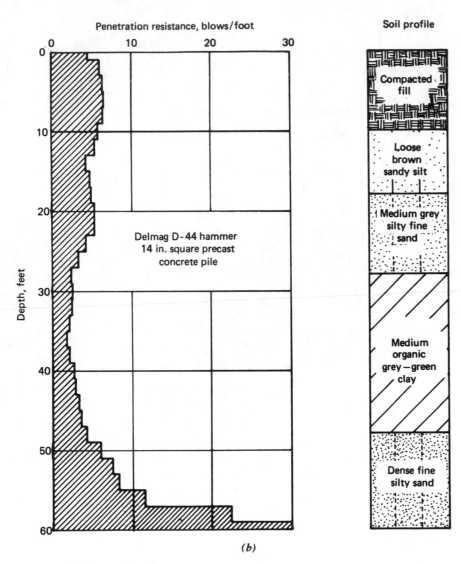

Penetration resistance, blows/foot

Soil profile

Delmag D-44 hammer
14 in. square precast
concrete pile

Depth, feet

Compacted fill

Loose brown sandy silt

Medium grey silty fine sand

Medium organic grey—green clay

Dense fine silty sand

(b)

Figure 11.25   (continued)

contractor on the basis of a per unit length of driven pile, the risk is placed on the owner, rather than the contractor. If the owner has a clear idea of how far the piles need to penetrate, both owner and contractor can benefit from a fixed-price contract. There are also contracts that combine both features (i.e., a fixed-price contract with possible additions for unanticipated work).

The test-pile program is one method used (especially on larger projects) to minimize the uncertainties of pile driving. It involves driving one or more piles at various places in the job-site and then testing them either statically, dynamically, or both. This method enables the owner, the contractor, and the engineer to get a better idea of the course of the work, as well as make both bidding and construction more realistic.

EXAMPLE Drivability Analysis

The best way to illustrate some of the issues involved in the drivability of piles is to consider the 12″, 40′ long concrete pile driven in sand whose capacity we estimated earlier. We need to select a pile hammer that will drive the pile successfully. The specifications call for a refusal blow count of 120 blows per foot, which translates into a pile set of 0.1 inches. The Dennis and Olson method estimates an ultimate capacity of 287 kips and an allowable capacity of 96 kips. The specifications also require a wave equation analysis for drivability.

Although the *Engineering News* formula (Equation 11.19) gives an allowable capacity, as a result of the variations in dynamic formulae, it is better practice to use an ultimate capacity for hammer selection purposes. Substituting the ultimate capacity and pile set just given yields a hammer rated striking energy of 28,700 ft-lbs allowing the use of a Vulcan 010 or Raymond 2/0 to drive these piles.

Since the specifications called for a wave-equation analysis, one was run and the results are shown in Table 11.6. The important assumptions behind the analysis are as follows:

- Standard values were used for all material and equipment properties, including the pile concrete, hammer and pile cushion, and hammer properties.

- A 6″-thick plywood cushion was used between the hammer driving accessory and the pile. (Generally, with concrete pile one cushion is used with each pile.)

- The skin friction is assumed to increase linearly with depth. This is the "triangular" distribution assumption; given that cohesionless soils increase their strength and friction with depth, the "triangular" distribution is a sensible assumption.

As most of the output shows the hammer, pile, and soil properties, special attention is directed to the very last table of figures at the end of Table 11.6. For a complete analysis, we not only analyzed drivability at the ultimate capacity of the pile, but also at a range of ultimate capacities. The data columns are as follows:

- Rult: Ultimate capacity, kips.

- Bl Ct: Blow count, blows per foot

- Stroke: Hammer stroke, ft.

- Minstr I,J: Minimum stress (maximum tensile stress), ksi, and the element at which it appears

- Maxstr I,J: Maximum stress (maximum compressive stress), ksi, and the element at which it appears

- Enthru: Energy from the hammer actually transmitted to the pile, ft/kips

Examination of the results of the wave equation analysis reveals the following:

- The Vulcan 08 can drive the pile at the estimated capacity of 287 kips at a blow count of 113 blows/foot. This is close to the specification but is acceptable. The hammer is slightly smaller than the Vulcan 010 that the *Engineering News* formula recommends.

- The range of blow counts and capacities shows the increase of capacity with increasing blow count. How it affects construction depends upon the structure of the specifications:

  - If the specification is for a certain blow count, then the contractor could stop driving when the desired blow count—113 BPF in this case—was achieved, even if the blow count was at, for instance, 35′ tip elevation.

TABLE 11.6 **Results of Wave-Equation Analysis for Example Problem**

1 weap87: wave-equation analysis of pile foundations 1987, version 3.00 soils in construction example hammer model of: vul 08 made by: vulcan

element	weight (kips)	stiffness (k/in)	coeff. of restitution	d-nl ft	cap dampg (k/ft/s)
1	8.000				
cap/ram	2.620	5935.2	.800	.0100	5.3
cushion		720.0	.500	.0100	

assembly	weight (kips)	stiffness (k/in)	coeff. of restitution	d-nl. ft
1	4.940	30280.5		
2	4.940	30280.5	.800	.0100

hammer options:

hammer no.	fuel settg.	stroke opt.	hammer type	dampng-hamr
207	1	0	3	2

hammer performance data

ram weight (kips)	ram length (in)	max stroke (ft)	stroke (ft)	efficiency
8.00	50.00	3.25	3.25	.670

rtd press. (psi)	act press. (psi)	eff. area (in2)	impact vel. (ft/s)
.00	.00	.00	11.84

hammer cushion	area (in2)	e-modulus (ksi)	thickness (in)	stiffness (kips/in)
	148.49	350.0	8.500	6114.3

pile cushion	area (in2)	e-modulus (ksi)	thickness (in)	stiffness (kips/in)
	144.00	30.0	6.000	720.0

pile profile:

lbt (ft)	area (in2)	e-mod (ksi)	sp.w. (lb/ft3)	wave sp (ft/s)	ea/c (k/ft/s)
.00	144.0	5000.	150.000	12426.4	57.9
40.00	144.0	5000.	150.000	12426.4	57.9

(continued)

TABLE 11.6 (*continued*)

WAVE TRAVEL TIME $-$ 2L/C $-$ = 6.438 MS
PILE AND SOIL MODEL FOR RULT = 50.0 KIPS

no	weight (kips)	stiffness (k/in)	d-nl (ft)	splice (ft)	cor	soil-s (kips)	soil-d (s/ft)	quake (in)	l bt (ft)	area (in**2)
1	.750	12000.	.010	.000	.850	.2	.050	.100	5.00	144.0
2	.750	12000.	.000	.000	1.000	.7	.050	.100	10.00	144.0
3	.750	12000.	.000	.000	1.000	1.2	.050	.100	15.00	144.0
4	.750	12000.	.000	.000	1.000	1.7	.050	.100	20.00	144.0
5	.750	12000.	.000	.000	1.000	2.3	.050	.100	25.00	144.0
6	.750	12000.	.000	.000	1.000	2.8	.050	.100	30.00	144.0
7	.750	12000.	.000	.000	1.000	3.3	.050	.100	35.00	144.0
8	.750	12000.	.000	.000	1.000	3.8	.050	.100	40.00	144.0
toe						34.0	.150	.100		

pile options:

n/uniform	auto s.g.	splices	dampng-p	d-p value (k/ft/s)
0	0	0	1	1.159

soil options:

% skin fr	% end bg	dis. no.	s damping
32	68	1	smith-1

analysis/output options:

iteratns	dtcr/dt(%)	res stress	iout	auto sgmnt	outpt incr max t(ms)
0	160	0	0	0	1 0

rult kips	bl ct bpf	stroke (eq.) ft	minstr i,j ksi	maxstr i,j ksi	enthru ft-kip
50.0	8.5	3.25	−.25 (6, 47)	2.04 (1, 29)	9.8
100.0	17.3	3.25	.00 (1, 0)	2.05 (2, 30)	9.6
150.0	27.3	3.25	.00 (1, 0)	2.06 (2, 31)	9.2
200.0	41.7	3.25	.00 (1, 0)	2.07 (2, 33)	8.1
225.0	53.3	3.25	.00 (1, 0)	2.08 (2, 33)	7.7
250.0	71.1	3.25	.00 (1, 0)	2.19 (7, 44)	7.2
287.0	113.6	3.25	.00 (1, 0)	2.34 (7, 46)	6.5
300.0	127.0	3.25	.00 (1, 0)	2.38 (7, 47)	6.3
325.0	159.8	3.25	.00 (1, 0)	2.45 (7, 48)	6.0
350.0	206.1	3.25	.00 (1, 0)	2.52 (7, 50)	5.9

- If the specification is for a certain tip elevation, then the contractor must drive the pile to the tip elevation specified. If the actual capacity is lower than anticipated, then the tip elevation can be achieved faster than estimated. If the actual capacity is higher than the estimate, the contractor will either be forced to overdrive hammer and pile—risking damage to both—or mobilizing larger equipment to finish driving, which adds to the contractor's expense. Given, in this case, that the estimated blow count at 40′ is so close to the 120 BPF specification, the contractor might want to use a larger hammer as a "factor of safety" of success.

- If the blow counts are significantly lower than anticipated at the required tip elevation, the owner and engineer may want the piles to be driven further. The contractor's view is obviously conditioned by the provisions of the specifications relative to payment of additional pile length driven.

- The wave-equation analysis estimates the stresses—tensile and compressive—in the pile during driving. Allowable stresses depend upon the strength of the concrete and job specifications; the values shown at Table 11.6 are reasonable for most concrete piles. Note also that tension stresses only appear at the lowest resistance and disappear for the higher ones.

- The "enthru" decreases with increasing capacity. Pile driving in higher blow counts is a "diminishing returns" phenomenon, in that, the higher the blow count goes, the less hammer energy is transmitted to the pile and thus the incremental increase in pile capacity is lessened. In many cases the contractor may be instructed to simply drive the piles at high blow counts until the tip elevation is achieved. Higher blow counts, however, are frequently not the answer for such problems.

- The example we are using is for cohesionless soil. With cohesive soils, generally, the increase in pore water pressure during driving decreases the resistance the hammer-pile system experiences during driving. Such resistance will return in time, making restriking the pile (whether in a test pile program or during production) result in a higher blow count than during the original driving. The amount of this kind of setup varies with different clay types and the contractor needs to obtain estimates set for drivability studies and construction control.

11.4.6   Driving Effects

Pile-driving operations (especially when applying excessive hammer energy) can be damaging to the piling and to the structures in the surrounding area. The damage may manifest in broken or damaged pile tops or in a sudden change in the rate of pile penetration.

In recent years, there has been a tendency for increased damage of timber piles during installation when these piles have been designed to take large loads. In the past, it has been the practice to permit loads on timber piling that range up to a maximum capacity of 50 kips. More recently, this maximum capacity has been extended to about 100 kips. Timber piles are very often driven with control by using the *Engineering News* formula. Considering that the formula has a theoretical factor of safety of 6, we can see that, while in the past, the ultimate capacity of a timber piling was theoretically up to 300 kips, present practice may cause this to approach 600 kips. Thus, it is not surprising that when this capacity is actually approached, we see an increased frequency of breakage as a result from pile driving.

Concrete piles may be damaged by either hard or soft driving. When very little tip resistance is encountered and a high-capacity hammer is being used, high tensile stresses may be

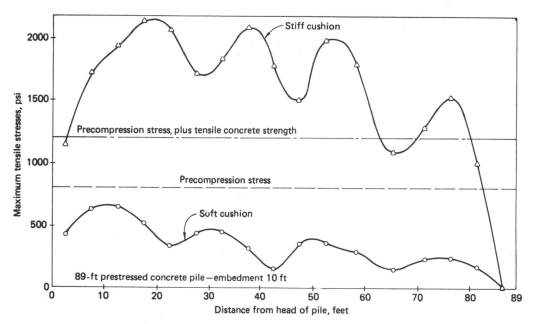

Figure 11.26   Calculated tensile stresses in a concrete pile during driving. Cracking of the pile should occur where computed tensile stress exceeds the sum of the stress that results from prestressing and the concrete tensile strength.

produced in a pile through tension cracks caused by wave propagation in the piles. If these tensile stresses exceed the sum of the prestressing force in a concrete pile and the tensile strength of the concrete, transverse tension cracks may be produced. These cracks may be observed to open and close with each hammer blow. As we saw in the example, tension cracking is predictable and might be overcome by adequate cushioning during driving or by the use of reduced hammer energy until hard driving is encountered. Requiring both of these methods is a common contract provision. A sample prediction by using wave analysis is illustrated in Figure 11.26.

Pile driving produces vibrations of the surrounding ground of varying magnitudes. The effects of this vibration may be damaging to adjacent structures. It has been suggested that, when driving in clay soils, the vibrations will be within tolerable limits at a distance of about three pile lengths from the driving operation. In sandy soils, this limit is reduced to one pile length. These are approximate relationships only and, in sensitive situations, it would be prudent to conduct special studies for potential damage that may result from driving. Pile-driving vibrations can also produce volume changes in soils. In cohesionless soils, the effect of vibrations is to produce densification, resulting in settlement of the surface of the surrounding area. Clay soils do not compact because of pile driving. On the contrary, when displacement piles are driven, the volume occupied by the driven piles is reflected in heave either laterally or vertically, depending on which way it is easiest for the soil to move. One consequence of heave may be to raise adjacent piles that have been previously driven. For this reason, driving contracts will usually contain some provision for ascertaining the amount of the heave and for requiring

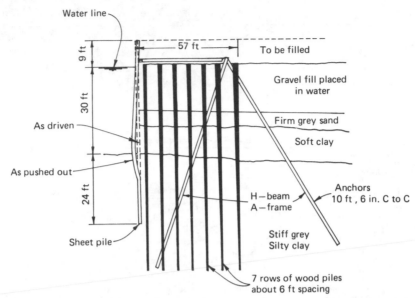

Figure 11.27 Lateral heave of sheet pile bulkhead, resulting from displacement of clay soil by pile driving. (Reprinted, by permission, from Sowers, G.B. and Sowers, G.F., "Failures of Bulkhead and Excavation Bracing," *Civil Engineering,* January, 1967.)

subsequent redriving if it exceeds a specified limit. Lateral displacement of a clay soil during driving in one instance produced large displacements in a retaining structure that had been previously constructed. This case is illustrated in Figure 11.27.

11.5 DRILLED PIERS

Drilled piers are sometimes referred to as caissons or cast-in-place concrete piles, both because of local custom and because of the similarities between methods of construction and between the resulting foundations. The true caisson method of construction was discussed in Chapter 9 and is usually employed for larger-diameter structures and structures that include usable constructed space within the caisson, below ground elevation. Cast-in-place piles are usually of comparatively small diameter and formed by other methods. Drilled piers, then, are those deep foundations consisting of a concrete shaft from about 6 in. up to 10 ft or more in diameter, which may be equipped with an expanded tip, cast in a machine-drilled hole. Examples are shown in Figure 11.28.

11.5.1 Applications

Piers are constructed to transmit structure load to a suitable bearing stratum. One obvious application is when the surface soils are compressible and where a suitable bearing stratum exists at reasonable depth. Another widespread application is in the control of the effects of expansive soils, where piers are used to bypass the zone of seasonal moisture change and arrangements are made to provide a void space between the grade beam and ground surface. They are an alternative to foundation piling in some cases and may be selected over piling for a number of reasons. For instance, by using piers, it may be possible to provide very high-capacity foundations

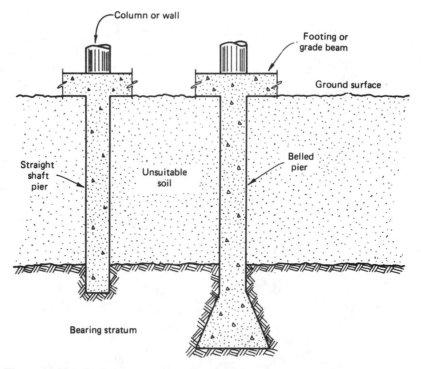

Figure 11.28 Configurations of machine-drilled piers.

without the heave or vibration problems associated with heavy pile driving. It may also be possible to establish very economical foundations of comparatively low capacity. Usually, if geologic conditions common to a geographical area are favorable for construction, piers will be a very popular foundation type. On the other hand, certain geologic conditions make pier construction difficult and may favor the installation of driven piles.

11.5.2 Construction

Installing a pier involves excavation of the shaft and bell (drilling and underreaming), setting casing when required, setting reinforcing steel, placing concrete, and removing casing. Each of these operations must be done with care to produce a continuous pier with the intended shape and concrete with the specified properties.

Augers for pier drilling include both the flight and bucket types. Examples are shown in Figure 11.29. For large, deep piers the augers are mounted on a square shaft or *kelly bar*, which is driven by a ring gear. On the most modern equipment, the kelly is equipped with a device to provide downward force and speed drilling, which can telescope for deep work. The boring is advanced by filling the flights or bucket of the auger, raising it to the surface to discharge the soil, and reentering the hole to repeat the operation. In strongly cohesive materials, an open hole may be drilled. In very soft soils, or cohesionless soils, it may be necessary to case the hole or drill through a clay slurry.

(a)   *(b)*

Figure 11.29   Auger types for pier drilling. (*a*) Flight auger. (*b*) A bucket auger.

If a belled end is to be formed, a special bucket, such as that shown in Figure 11.30, is substituted for the auger. The belling bucket underreams the hole as its expandable sides extend. The tip diameter formed in this manner is about three times that of the main shaft.

Depending on foundation conditions, a casing may be set in the completed hole to keep it open and to aid in cleaning the bottom. In some cases, the casing is equipped with a cutting edge so that it may be seated in the bearing stratum. In a cased hole where a seal can be effected at the tip, and in a stable, open hole, the end of the pier can be seated on a material that has been thoroughly cleaned and visually inspected.

When the hole is ready for construction of the pier itself, the reinforcement is placed first if any is to be used. The steel reinforcement is usually arranged in a circular pattern of vertical bars attached to a spiral cage. All elements are connected firmly, so that the reinforcement may be positioned accurately and maintained in position in the hole during the remainder of the construction operation.

In an open hole, concrete is placed down the center of the hole, so that it does not impact the steel cage. In a hole filled with slurry or water, concrete is placed by the tremie method. The

Figure 11.30 Belling bucket for pier-drilling machine.

casing, if one is used, may be withdrawn or left in place. If it is withdrawn, the level of the fresh concrete inside must be very carefully controlled, and the casing must be kept vertical at all times. A frame for pulling the casing for piers is shown in Figure 11.31.

11.5.3 Problems in Construction

During drilling of comparatively large-diameter piers in soft ground, unless casing or fluid is used, there is a possibility of loss of ground by squeezing toward the hole, as well as subsequent settlement of the surrounding ground surface. If the water level in a pier hole is lower than in the surrounding area, either because of pumping or simply due to a lag during drilling, cohesionless materials may be carried to the hole with similar results. The latter situation is illustrated in Figure 11.32a and is analogous to the problems associated with sumping in cohesionless soils. (See Chapter 9.) Loss of ground may be prevented by casing and by maintaining the fluid level in the hole above the surrounding groundwater level at all times.

Unless the stratum in which a belled end is to be formed is strongly cohesive, the excavation may not support itself and may cave, as shown in Figure 11.32b. If this problem is not discovered by subsequent inspection and concrete for the pier is placed, the tip-bearing area is considerably reduced, with obvious consequences. Belling should therefore be attempted only within a suitable stratum, the location of which is confirmed by competent inspection at each pier location. No pier should ever be completed without inspection to provide assurances that the tip dimensions and conditions are as provided for in the plans.

Removal of casing is an operation that must be performed with great attention to detail in order to ensure that a situation like that in Figure 11.32c does not develop. If a casing is to be pulled, the head of concrete must always be great enough so that it will flow as the casing is withdrawn, but not so great that the flow rate is slower than the rate of casing withdrawal. Continuous inspection of the concrete surface is required to confirm that these conditions are met and comparisons between the concrete volume and hole volume should always be made. In cases where the withdrawal of casings presents great difficulties, they should be left in place.

Figure 11.31 Frame for withdrawing pier casings.

11.6 SUMMARY

This chapter has included a discussion of some basic concepts of foundation design and construction. The subject has been presented as an introduction to the scope of problems associated with the construction of shallow and deep foundations. Complete coverage of the state of the art in these matters would require many volumes. Consistent with the scope of this text, the reader should

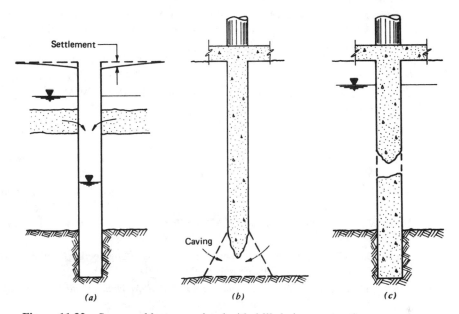

Figure 11.32 Some problems associated with drilled-pier construction. (a) Settlement resulting from loss at ground associated with seepage toward hole. (b) Caving of excavated bell before concreting. (c) Discontinuous shaft caused by poor control during extraction of casing.

1. Have learned how foundations transmit load to the soil.

2. Be able to describe the basic foundation types and their advantages and disadvantages, both from the standpoint of design and construction.

3. Have obtained knowledge of how the basic types of foundations are built and the special equipment required for construction.

4. Understand the fundamentals of analysis of foundations for bearing capacity and settlement.

5. Be aware of some of the problems arising during foundation construction and how to solve them.

REFERENCES

American Petroleum Institute, *Recommended Practice for Planning, Designing, and Constructing Fixed Offshore Platforms*, April 1977.

American Society for Testing and Materials, *1974 Book of ASTM Standards*, Philadelphia: ASTM, 1974.

Barrero-Quiroga, Franklin, *Pile Hammer Selection by Driving Stress Analysis*, Unpublished M.S. Thesis, Oregon State University, 1974.

Bethlehem Steel Corporation, *Steel H-Piles*. Bethlehem, PA: Bethlehem Steel Corporation.

Chellis, R.D., *Pile Foundations*. New York: McGraw-Hill, 1961.

Dennis, N.D., and Olson, R.E. "Axial Capacity of Steel Pipe Piles in Clay." *Proceedings of the Conference on Geotechnical Practice in Offshore Engineering*. New York: American Society of Civil Engineers, 1983.

Dennis, N.D., and Olson, R.E., "Axial Capacity of Steel Pipe Piles in Sand." *Proceedings of the Conference on Geotechnical Practice in Offshore Engineering*. New York: American Society of Civil Engineers, 1983.

Department of the Army. *Bearing Capacity of Soils*. EM 1110-1-1905. Washington, DC: U.S. Army Corps of Engineers, 1992.

Department of the Army. *Design of Deep Foundations*. EI 02C097. Washington, DC: U.S. Army Corps of Engineers, 1996.

Hagerty, D.J., and Peck, R.B., "Heave and Lateral Movements Due to Pile Driving," *Journal, Soil Mechanics and Foundations Division, American Society of Civil Engineers*, 97, No. SM11, 1971.

Hirsch, T.J., Lowery, L.L., Coyle, H.M., and Samson, C.H., "Pile-driving Analysis by One Dimensional Wave Theory: State of the Art," *Highway Research Record 333*, 1970.

Issacs, D.V., "Reinforced Concrete Pile Formula," *Transactions of the Institute Engineers Australia*, Vol. XII, pp. 305–323, 1931.

Luna, William A., "Ground Vibrations Due to Pile Driving," *Foundation Facts, Raymond International*, 3, No. 2, 1967.

Mayo, David F., "Development of Nonpneumatic Caisson Engineering," *Journal, Construction Division, American Society of Civil Engineers*, 99, No. C01, 1973.

Meyerhof, G.G., "Shallow Foundations," *Journal, Soil Mechanics and Foundations Division, American Society of Civil Engineers*, 91, No. SM2, 1965.

Oregon State Highway Division, *Standard Specifications for Highway Construction*, 1974.

Peck, R.B., Hanson, W.E., and Thornburn, T.H., *Foundation Engineering*. New York: Wiley, 1974.

Peurifoy, R.L., *Construction Planning, Equipment, and Methods*. New York: McGraw-Hill, 1970.

Romanoff, M., *Corrosion of Steel Pilings in Soils*, United States Bureau of Commerce, National Bureau of Standards, Monograph 58, 1962.

Smith, E.A.L., "Pile-driving Analysis by the Wave Equation," *Journal, Soil Mechanics and Foundations Division, American Society of Civil Engineers*, 86, No. SM4, 1960.

Sowers, G.B., and Sowers, G.F., "Failures of Bulkhead and Excavation Bracing," *Civil Engineering*, January 1967.

State of California, Division of Highways, *Standard Specifications*, 1973.

Terzaghi, K., and Peck, R.B., *Soil Mechanics in Engineering Practice*. New York: Wiley, 1967.

U.S. Bureau of Reclamation, *Concrete Manual*, 1966.

Warrington, D.C., "A new type of wave equation analysis program." *Proceedings of the Third International Conference on the Application of Stress-Wave Theory to Piles*, Ottawa, Ontario, 25–27 May 1988.

Warrington, D.C., *Vibratory Pile Driving Equipment*. Chattanooga, Tennessee: Vulcan Iron Works, Inc., 1989.

PROBLEMS

1. Footings for a structure are to be built beneath a basement slab on a clay soil at a depth of 20 ft below present ground level. Representative consolidation tests on the soil produce results typified by Figure 5.9. The groundwater table is 15 ft below the ground surface. The soil-void ratio is about 0.87. If the average bearing pressure on the footings is 2,600 psf, estimate the settlement of 10-ft and 5-ft-square footings.

2. What is the factor of safety for the footings in Problem 1 against bearing failure if the clay soil has an unconfined compressive strength of 3,600 psf?

3. What is the maximum column load in kips that could be placed on a 4-ft square temporary footing on a sand for which the standard penetration resistance is about $N = 15$ and the water table is at the surface? A factor of safety of 2 is to be maintained against bearing failure.

4. Solve Problem 3 if settlement of the footing must be limited to $1/2$ in.

5. Six hundred timber piles are to be driven to a depth of 40 ft in sandy silt below the water table. The soil has a wet unit weight of 105 pcf. Compute a representative ultimate pile capacity if minimum tip diameter is 8 in. and minimum butt diameter is 20 in. Suggest a type and size of pile-driving hammer, using the *Engineering News* formula. The pile design load is 50 kips. Justify your choice and describe what problems you might encounter.

6. According to the provisions of a contract you are preparing a bid for, prestressed concrete friction piles, 12 in. square, are to be driven to an allowable design load of 80 kips in a soft to medium clay deposit. Capacity is to be determined by the *Engineering News* formula and lengths are not specified. Select a tentative hammer type and size, justifying your choice. How long should the piles you furnish be? Describe problems you may encounter.

7. Steel H-piles are to be driven to tip bearing through about 150 ft of soft clay and plastic silt. The ultimate capacity of the piles is 200 kips. What type of hammer would you use? What special problems might you encounter that should be considered in bidding the work?

8. Specifications require that timber piles be driven to a depth of 50 ft in a stiff to very stiff clay soil. Minimum tip diameter is 8 in. and minimum butt diameter is 20 in. The specifications further require a minimum, demonstrated allowable capacity of 80 kips, with a factor of safety of at least 3. Calculated capacities from driving records for the first few piles according to the *Engineering News* formula are less than 25 tons. You have all 55-ft piles on hand. How would you proceed? Justify your answer.

9. Figure 10.15 shows the soil profile at a building site. Pier foundations are to be used and design requirements dictate two alternatives. The first employs belled piers with 6-ft-diameter tips at 26 ft in depth. The second requires 2-ft-diameter straight shafts to 42 ft in depth. Which alternative would you prefer to build? Justify your answer, giving the advantages and disadvantages of each alternative.

CHAPTER 12

Construction Access and Haul Roads

On earth-moving projects and many other jobs, temporary roads are built to move materials to the project site. In some cases, existing roads are used, and often they are upgraded for this purpose. Sometimes such roads become part of the permanent works. In other cases, haul roads are temporary only. Public roads are sometimes used for access, and occasionally must accommodate very heavy traffic and loads. Access and haul roads may be surfaced or unsurfaced, according to project requirements.

The purposes of this chapter are to

1. Give the reader an introduction to pavement and subgrade structural performance requirements.

2. Show how subgrade strength and bearing capacity analysis may be employed to design or evaluate construction roads.

3. Provide an introduction to the rationale for commonly employed road maintenance and improvement procedures.

12.1 PAVEMENT COMPONENTS AND THEIR FUNCTIONS

Figure 12.1 shows a cross section through a pavement. The pavement consists of the higher quality (usually imported) materials above the subgrade, including the wearing surface, the base course, and the subbase. The function of these materials is to protect the subgrade from traffic loads and weather. Their quality and thickness requirements will be determined by subgrade conditions and traffic loadings. In many instances, particularly depending on subgrade conditions and the weather, some construction roads may be built to be used without a pavement.

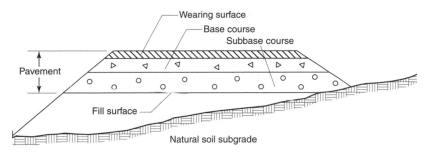

Figure 12.1   Pavement section.

The wearing surface on any road is the highest quality material in the pavement structure, except in those instances when the subgrade consists of hard, durable intact rock. On permanent roads, it usually consists of asphalt concrete, Portland cement concrete, or stabilized aggregate. It must be sufficiently strong to resist compressive stresses from wheel loads and cohesive enough to withstand traffic abrasion.

The base course is usually the major structural element in a pavement, except for those surfaced with Portland cement concrete. It consists of a comparatively thick, highly stable layer of well-graded clean, coarse, crushed aggregate. Uncrushed aggregate is sometimes used when suitable natural gradations are available. Subbases are employed in permanent works but not often in construction pavements. The quality of the subbase is intermediate between base course and subgrade.

12.2   SUBGRADE MATERIALS AND STRENGTH REQUIREMENTS

The general suitability of various soils as subgrade materials may be inferred from soil classifications. Such correlations are shown in Figure 4.11 for the AASHTO system and in Figure 4.2 for the Unified System. Beyond general suitability, subgrade strength is evaluated by methods used in conjunction with various pavement design procedures. Most of the methods used for pavement design are empirical or semiempirical. Each requires complex equipment and tests to determine subgrade strength, and each employs very specialized analytical procedures. These pavement design methods are beyond the scope of this text. A reasonably good understanding of pavement analysis methods and subgrade material strength requirements can be acquired, however, by considering the general bearing capacity and stress analysis methods discussed in Section 11.2.1 and by applying these theoretical procedures to the particular problem of designing a pavement to support a loaded wheel. Figure 12.2 provides a summary of the necessary definitions. Table 12.1 gives some examples of wheel loads for typical construction vehicles.

For usual wheel loads a circular contact "print" of the tire with diameter d_p is presumed. The contact stress q_p is equal to the tire pressure p. The diameter of the "print" or loaded area is thus derived from

$$P = q_p \frac{\pi d_p^2}{4} = p \frac{\pi d_p^2}{4}. \tag{12.1}$$

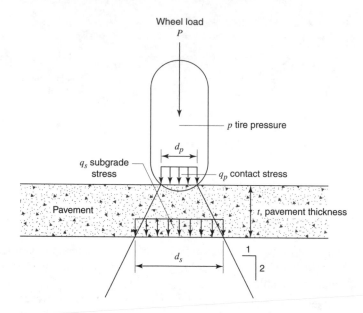

Figure 12.2 Definition of terms describing wheel load transfer through pavement to subgrade.

TABLE 12.1 Typical Wheel Loads for Construction Vehicles

Vehicle	Maximum Axle Load (kips)	Maximum Wheel Load (kips)	Tire Pressure[a] (psi)
Highway truck, legal, single axle, dual tires	18–20	9–10	60–90
Highway truck, maximum, single axle, dual tires	32	16	60–90
Front loader, light	29	15	50–90
Front loader, heavy	118	59	50–90
Off highway truck, light	66	33	50–90
Off highway truck, heavy	86	43	50–90
Scraper, light	30	15	50–90
Scraper, medium	57	28	50–90
Scraper, heavy	94	47	50–90

[a] Tire pressures are varied to optimize performance for given loadings and road conditions.

We then solve for the diameter of the print

$$d_p = \sqrt{\frac{4P}{\pi p}}. \tag{12.2}$$

For multiple-wheel configurations, loads may be superimposed to calculate a pressure distribution and "print" on the subgrade surface, using the method described in Section 11.1.1. If we now assume that stresses from a single wheel are distributed beneath the loaded area in Figure 12.2 over the width

$$d_s = d_p + t, \tag{12.3}$$

then, from Equation (12.2), the width of the loaded area on the subgrade is

$$d_s = 2\sqrt{\frac{P}{\pi p}} + t. \tag{12.4}$$

Equation 11.4 relates the ultimate bearing capacity of the soil beneath a loaded area to the strength of the soil and the size of the area. If, in Figure 11.3, the area above the loaded surface a–a were filled with the same material as exists outside the width B, the net ultimate bearing capacity of the subgrade, from Equation (11.4), would be

$$q_d - \gamma D_f = q_{d(\text{net})} = 0.4\gamma B N_\gamma + \gamma D_f N_q + 1.3 c N_c - \gamma D_f. \tag{12.5}$$

For the terminology shown in Figure 12.2, from Equation (12.5) we find that the ultimate strength of the subgrade q_{sd} will be

$$q_{sd} = 0.4\gamma B N\gamma + \gamma D_f (N_q - 1) + 1.3 c N_c. \tag{12.6}$$

When this strength is compared with the subgrade stress q_s, we have an expression for the factor of safety against bearing failure beneath the pavement:

$$\text{FS} = \frac{q_{sd}}{q_s}. \tag{12.7}$$

Normally, the value of the factor of safety should exceed about 2, to prevent plastic deformation in the subgrade.

EXAMPLE   Pavement Subgrade Strength for Cohesive Soils

To illustrate the use of Equation (12.6), suppose we wish to know whether a pavement consisting of 12 in. of crushed rock over a medium ($c = 750$ psf) clay subgrade is adequate to support a 10,000-lb single-wheel load. Tire pressure is 80 psi.

Although we could work our way through the preceding equations to obtain the answer, a more useful result would be to derive a relationship for the necessary thickness of the pavement. To begin with, for cohesive soils ($\phi = 0$), from Table 11.2, $N_\gamma = 0$; We also usually neglect the N_q term as well for problems with this kind of soil. Equation (12.6) thus reduces to

$$q_{sd} = 1.3 c N_c. \tag{12.8}$$

Since the wheel load is transferred from the surface through the subgrade, in addition to Equation (12.1), we can also state

$$P = q_s \frac{\pi d_s^2}{4}. \tag{12.9}$$

Substituting the value for d_s obtained in Equation (12.4) into Equation (12.9) and solving for q_s, we have

$$q_s = \frac{4Pp}{4P + 4\sqrt{\pi p P t} + \pi p t^2}. \tag{12.10}$$

Substituting q_s from Equation (12.10) and q_{sd} from Equation (12.7) and substituting and solving for t, we have, at last,

$$t = \frac{2p\sqrt{130\pi c N_c P(\text{FS})} - 26cN_c\sqrt{\pi pP}}{13\pi c N_c p}.$$

(12.11)

We thus have an expression that we can use to directly solve for the thickness of the pavement. For purely cohesive soils, $N_c = 5.7$. The problem states that $P = 10,000$ lbs., $p = 80$ psi, and $c = 750$ psf $= 5.208$ psi. Also, FS $= 2$. Substituting these values into Equation (12.11) and solving yields $t = 13.1$ in. A 12-in. pavement thickness is inadequate for this wheel load.

This example was for a cohesive soil. Since we neglect N_q, we do not include any strength from the subgrade to support the wheel load. Bases and subbases should be made from cohesionless materials to insure proper drainage.

The foregoing procedure considers only a single application of the wheel load. More widely used procedures consider the effects of numbers of load applications on pavement performance. Thus, it is possible that one can utilize a "pavement life" concept in arriving at a design, by evaluating the expected amount of traffic and the magnitude of the wheel loads.

12.3   SUBGRADE IMPROVEMENT

The previous section displayed how theoretical methods may be used to evaluate the suitability of a pavement for a single passage of a wheel load. Should analysis by these or other pavement analysis methods show that the pavement is unsuitable, two options are available to us. The first is to use a thicker pavement, and the second is to improve or strengthen the subgrade. As indicated in Chapter 8, subgrade improvement can be accomplished by compaction or by chemical stabilization. These procedures would of course be applicable only to newly built roadways. For example, it has been shown that, for cohesionless sands, the angle of internal friction can be related approximately to relative density by

$$\phi' = 25° + 0.15D_r.$$

(12.12)

Thus, by going from $D_r = 30$ percent to $D_r = 80$ percent as a result of compaction, we obtain an angle of internal friction increase from about 30° to about 37°. Referring to Table 11.2, one can see that the bearing capacity factors N_γ and N_q to be applied in Equation (12.6) are more than doubled as a result. Obviously, compaction has a beneficial effect on subgrade strength.

Chapter 8 also provides some indication of how the cohesive strength of some fine soils may be increased by mixing with chemicals and how cohesion may be developed in sands and gravels by the addition of cement. For both soil types, in real applications, studies are usually conducted with on-site soils and proposed stabilizing agents to determine what strength improvements might be available.

Laboratory studies have shown that the relationships in Figure 12.3 represent the 28-day unconfined compressive strengths to be expected when mixing cement with various types of soils. Mixtures of soil and cement will continue to gain strength over an extended period, and the gain would be expected to be approximated by the relation

$$q_{ud} = q_{ud0} + K \log_{10}\left(\frac{d}{d_o}\right),$$

(12.13)

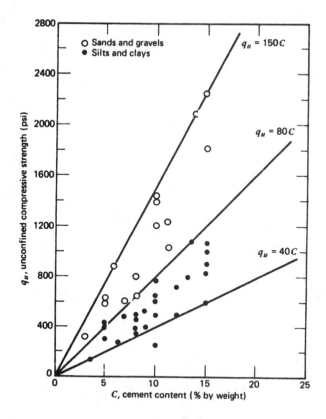

Figure 12.3 Relationships between cement content and 28-day unconfined compressive strength for soil and cement mixtures. (Adapted from FHWA-1P-80-2, October 1979.)

where

$$q_{ud} = \text{unconfined compressive strength at an age of } d \text{ days, in psi}$$

$$q_{ud0} = \text{unconfined compressive strength at an age of } d_0 \text{ days, in psi}$$

$$K = 70C \text{ for granular soils and } 10C \text{ for fine-grained soils}$$

$$C = \text{cement content, in percent by weight.}$$

EXAMPLE   Soil-Cement Mixture

For a silty soil with a 6 percent cement content that has a 7-day strength of 200 psi, we would expect that the strength at 60 days would be

$$q_{u60} = 200 + 10 \times 6 \times \log_{10}\!\left(\frac{60}{7}\right) = 256 \text{ psi.}$$

The cohesive strength to be used in Equation (12.6) is, of course, by definition one-half the unconfined compressive strength.

For lime treatment of road subgrades, selected laboratory and field results are presented in Table 12.2. Note that, in general, the field strength values shown are greater than the laboratory values. This is not usually the case since there is less thorough mixing in the field operation.

TABLE 12.2 Unconfined Compressive Strength Summary

| | | | | Unstabilized Subgrade | | Lime Stabilized Subgrade | | | | | | | |
| | | | | | | Field Testing Program | | | | Laboratory Testing Program | | | |
Test Site	Classification	Age (yr)	Lime (%)	q_u (psi)	Average q_u	w (%)	γ_d (pcf)	q_u (psi)	Average q_u	w (%)	γ_d (pcf)	q_u (psi)	Average q_u
				LEAN CLAY SUBGRADES									
Perry Co., Missouri Location #1/Test Site 2-66	A-7-6(18)/CL	14	4.5	36 21	29					21.1 15.4	105.4 105.8	86 51	68
Ft. Hood, Kileen, Texas Test Site 4-66	A-5(11)/CL	4	4.0	28 35 27 26 41 33	32					13.2 12.7	111.9 110.6	133 152	142
Frederick Co., Virginia Test Site 7-67	A-6(16)/CL	5	7.0			16.2	109.5	267	267	16.5 15.9 15.4	111.2 106.5 107.4	61 60 64	62
				FAT CLAY SUBGRADES									
Bergstrom AFB Austin, Texas	A-7-5(19)/CH	9	4.0	7 13	10					18.0 17.7	97.1 102.5	66 101	83
Giles Co., Virginia Location #1/Test Site 6-67	A-7-5(18)/CH	6	7.0			27.3	89.6	171	171				
Location #2	A-7-5(18)/CH	6	7.0	22	22	20.7	99.2	268	268	21.4 21.8	105.5 92.0	116 154	135

Source: From R. E. Aufmuth, "Strength and Durability of Stabilized Layers under Existing Pavements," *Technical Report M-4, Construction Engineering Research Laboratory*, Champaign, Illinois, 1970.

The field data, in this case, however, represents subgrades in place for considerable time periods. In practice, field strengths are usually taken as equal to laboratory values, and extra lime is added to offset mixing problems. No simplified relationships for the effect of lime content on soil strengths are known.

Recently, a number of geotextiles, or engineering fabrics, have been developed, and these fabrics have a number of useful applications in the construction field. One of these applications is strengthening roads on weak fine soils. In practice, geotextiles may be laid directly on the subgrade and covered with an aggregate surfacing. In general, for geotextile-reinforced roads on fine soils, we can use Equation (12.8) and, because the presence of the fabric restricts plastic deformation, may be applied with a factor of safety as low as 1. Various methods are available for analyzing the degree of improvement in specific cases, and it has been shown that not all available fabrics provide the same beneficial effects. Road reinforcement using fabrics on soft subgrades has, however, been proven in practice as a means of reducing the pavement thickness required and it is worthy of consideration where significant savings may be realized.

12.4 MAINTENANCE REQUIREMENTS

During a project, it may be necessary to restore a road to its original condition to improve its riding qualities. When a project is finished, it is sometimes required that the condition of preexisting roads be reestablished.

For paved roads with a high-quality wearing surface, patching and repairs at damaged locations may be all that is required. Alternatively, overlays or new surfaces may be necessary. Whether for the contractor's own benefit or for postproject service, the need for such improvements should be evaluated in the estimating and bidding process and the appropriate costs included in the proposal. In making such an evaluation, one may consider the feasibility of using light hauling equipment, which will not adversely impact the road, versus the option of using heavy equipment and repairing the consequent damage.

For aggregate surfaced or unsurfaced roads, maintenance during construction usually involves watering, regrading, and compaction. Light watering is principally used to control dust. Maintenance of the subgrade at near optimum water contact, however, results in better control of rutting and deformation by improving subgrade strength. The value of maintaining an even, dense road surface can be assessed by considering the rolling resistance of equipment. Rolling resistance may be defined as the force required to move a given unit of weight over the surface in question. It is typically expressed in pounds per ton and depends not only on road surface condition, but also on equipment characteristics. Soft, deformable road surfaces have higher rolling resistance because greater energy must be exerted to cause a wheel to climb out of the depression created by the load it carries. Underinflation of tires results in higher rolling resistances on uniform, nonyielding surfaces. Some example values of rolling resistance are provided in Table 12.3.

The selection of haul roads for construction equipment, especially for long distances or heavy loads, depends on a number of factors, including road grade and machinery available. Table 12.3 shows that just the maintenance of haul roads on large projects can be a wise investment that considerably reduces power requirements and produces significant fuel savings.

TABLE 12.3 Representative Rolling Resistances for Various Types of Wheels and Surfaces (in Pounds Per Ton of Gross Load)

Type of Surface	Steel Tires, Plain Bearings	Crawler-Type Track and Wheel	Rubber Tire, Antifriction Bearings	
			High Pressure	Low Pressure
Smooth concrete	40	55	35	45
Good asphalt	50–70	60–70	40–65	50–60
Earth, compacted and well maintained	60–100	60–80	40–70	50–70
Earth, poorly maintained, rutted	100–150	80–110	100–140	70–100
Earth, rutted, muddy, no maintenance	200–250	140–180	180–220	150–200
Loose sand and gravel	280–320	160–200	260–290	220–260
Earth, very muddy, rutted, soft	350–400	200–240	300–400	280–340

Source: From R.L. Peurifoy, *Construction Planning, Equipment and Methods*, New York: McGraw-Hill, 1970.

12.5 SUMMARY

Access and haul roads can be a significant consideration on many construction projects. Requirements for road construction, improvement, and maintenance are affected by site conditions, weather, equipment type, and expected traffic. Pavement design and analysis procedures can be used to evaluate the different factors of a particular job site. The reader should

1. Understand terminology associated with pavement components.
2. Be able to make a theoretical evaluation of the suitability of a given pavement for a given wheel load.
3. Be aware that more detailed analyses are required to account for effects of nonhomogeneous paving materials and repeated traffic loads.
4. Appreciate that substantial improvements in subgrade behavior may be achieved by chemical stabilization, compaction, or the use of geotextiles.
5. Recognize the importance of a planned maintenance program for haul roads.

REFERENCES

American Association of State Highway Officials, *Standard Specifications for Highway Bridges*, Washington, DC, 1965.

Aufmuth, R.E., "Strength and Durability of Stabilized Layers under Existing Pavements," *Technical Report M-4, Construction Engineering Research Laboratory*, Champaign, Illinois, 1970.

Giroud, J.P., and Noiray, L., "Geotextile-Reinforced Unpaved Road Design," *Journal, Geotechnical Engineering Division, American Society of Civil Engineers*, 107, No. GT9, 1981.

Peurifoy, R.L., *Construction Planning, Equipment and Methods*, McGraw-Hill, New York, 1970.

U.S. Department of Transportation, *Soil Stabilization in Pavement Structures—A User's Manual, Vol. 2*, FHWA-1P-80-2, 1979.

Yoder, E.J., *Principles of Pavement Design*, New York: Wiley, 1959.

PROBLEMS

1. A front loader will operate in an aggregate crushing operation on river bottomland. The soil is saturated with an unconfined compressive strength averaging about 900 psf. The maximum wheel load for the loader proposed is 40 kips. Tire pressure is 70 psi. Calculate the thickness of aggregate that should be placed over the natural soil to provide a suitable working surface.

2. Off-highway trucks with maximum 80-kip axle loads and 60-psi tire pressures will operate on a sand fill. The single rear axle has dual tires 2 ft apart. Compare the suitability of the fill to support the trucks after it has been compacted ($D_r = 85$ percent) to that after it was placed ($D_r = 55$ percent).

3. A private farm road is to be used for hauling aggregate from a borrow pit to a construction project. The road is constructed with a 4-in. asphalt stabilized aggregate surface over a clay subgrade with an unconfined compressive strength of about 3000 psf. Legal highway dump trucks will be used. The farmer has stipulated that the road must be restored to its original condition after construction. Is it likely that significant repairs will be required?

4. Compare the estimated thicknesses of gravel required to support an overloaded highway truck (32-kip axle load) on a soft subgrade ($q_u = 750$ psf) with and without a geotextile at the fabric-soil interface.

APPENDIX A

Laboratory Testing Exercises*

A.1 CALIBRATION OF SAND-CONE DEVICE AND DENSITY-TESTING SAND, AND DETERMINATION OF FIELD UNIT WEIGHT OF SOILS

Purpose. The sand cone field density testing procedure is a widely used, standardized method for determining the unit weight (density) of soils, in the field, in place. The standard method is described in ASTM D1556. To use the method, the sand-cone device must be calibrated, and the unit weight of the sand used to make field volume determinations must be established. Equipment required for this exercise is shown in Figure A.1.1. Example data sheets are shown

Figure A.1.1 Equipment required for the calibration and use of the sand-cone apparatus.

*The laboratory exercises described in this chapter are presented in an order that has been found to be useful in supporting the course for which *Soils in Construction* was designed. Testing methods are similar to those specified by the American Society for Testing and Materials and the U.S. Army Corps of Engineers, EM 1110-2-1906, "Laboratory Soils Testing," which are referenced herein. However, the tests presented here are intended to be used in classroom demonstrations and exercises only. Published standard methods should always be used for actual construction control.

in Figures A.1.2 and A.1.3. The purpose of this exercise is threefold: (1) to calibrate a sand-cone device and density-testing sand for later use in field testing; (2) to provide the student with experience early in the course of study, related to determination of weights and volumes; and (3) to prepare the student for performing unit weight determinations in the field.

Reference Standard. ASTM D1556—Density of Soil in Place by the Sand-Cone Method. Equipment (Refer to Figure A.1.1)

CALIBRATION OF SAND-CONE AND DENSITY SAND

FOR _____

BY _____

DATE _____   APPARATUS NUMBER _____

SAND DESCRIPTION _____

JAR VOLUME DETERMINATION

	Trial 1	Trial 2	Trial 3
Weight of Jar + Cone			
Weight of Jar + Cone + Water			
Weight of Water			
Water Temperature			
Water Density			
Volume of Jar			
Average Jar Volume			

DENSITY OF SAND CALIBRATION

	Trial 1	Trial 2	Trial 3
Weight of Sand + Jar + Cone			
Weight of Jar + Cone			
Weight of Sand			
Density of Sand			
Average Sand Density			
Weight of Sand + Jar + Cone			
Weight of Residue + Jar + Cone			
Weight of Sand in Cone + Plate			
Average Weight of Sand in Cone + Plate			

Figure A.1.2   Example data sheet for calibrating sand-cone device and density-testing sand.

1. Sand-cone apparatus. This consists of a jar and the metal cone/valve assembly.
2. The sand-cone base plate and an additional plate to be placed under the base plate during calibration.
3. Uniform, medium-grained sand.
4. Thermometer.

FIELD DENSITY – SAND-CONE METHOD

TEST FOR _____

TEST BY _____   DATE _____

SAMPLE DESCRIPTION_____

TEST LOCATION _____

FIELD DENSITY	CYLINDER
1. Wt. Sand + Jar	1. ASTM Designation
2. Wt. Residue + Jar	2. Wt. Soil + Mold
3. Wt. Sand Used (1) - (2)	3. Wt. Mold
4. Wt. Sand in Cone & Plate	4. Wt. Soil
5. Wt. Sand in Hole (3) - (4)	5. Wet Density, pcf (4) ÷ Vol. Mold
6. Density of Sand	6. Moisture Content
7. Wt. Container + Soil	7. Dry Density, pcf (5) ÷ [1 + (6)]
8. Tare Wt. Container	MOISTURE CONTENT PAN NO. _____
9. Wt. Soil (7) - (8)	WW _____ DW _____
10. Vol. of Hole (5) ÷ (6)	DW _____ TW _____
11. Wet Density, pcf (9) ÷ (10)	WATER _____ SOIL _____
12. Moisture Content, %	PER CENT MOISTURE _____
13. Dry Density, pcf (11) ÷ [1 + (12)/100]	NOTES:
Representative Curve No.	
14. Optimum Moisture, %	
15. Maximum Dry Density, pcf	
16. Relative Compaction, % (13) ÷ (15)	
FIELD–MOISTURE DETERMINATION	
Method _____	
Percent Moisture _____	

Figure A.1.3   Data and calculations for field-density test by sand-cone method.

5. Balance sensitive to 0.1 g.
6. Tools to dig and clean a small hole in the field.
7. Air-tight container to store excavated soil.
8. Appropriate apparatus for water-content determination (not shown; see A.4).

A.1.1 Determine Volume of Sand-Cone Jar

1. Assemble the jar and sand cone. Make sure that the gasket on the cone is in place so that a water-tight seal is provided. Ensure that the jar, sand cone, and plate are identifiable, separately from others in the laboratory. This is usually best accomplished by numbering sets of equipment.
2. Weigh the assembled jar and cone and record that weight.
3. Open the valve on the sand cone and fill the jar and part of the cone with water.
4. Close the valve and remove the water above it. Wipe moisture quickly from the cone and the outside of the device.
5. Weigh the water-filled device and determine the water temperature. Convert the weight of water to volume using the information presented in Table A.1.1

TABLE A.1.1   Water-Density–Volume Temperature Relationship

Temperature (°C)	Water Density (g/cc)
10	0.999700
11	0.999605
12	0.999498
13	0.999377
14	0.999244
15	0.999099
16	0.998943
17	0.998774
18	0.998595
19	0.998405
20	0.998203
21	0.997992
22	0.997770
23	0.997538
24	0.997296
25	0.997044
26	0.996783
27	0.996512
28	0.996232
29	0.995944
30	0.995646

Bureau of Reclamation, U.S. Department of the Interior, *Earth Manual*, 2d edition 1974, p. 458.

6. Complete Steps 3 through 5 for at least three trials producing a maximum volume variation of 3 ml. The average of these three trials shall be taken as the volume of the jar. Note that the jar and cone should be marked so that, in the event they are separated, they may be reassembled without altering the volume.

A.1.2 Calibration of Density Testing Sand

1. Select a relatively uniform (poorly graded) dry sand of medium grain size. Bagged sands meeting the requirements are commercially available.
2. Ensure that the jar and cone are dry. Open the valve in the sand-cone device and place the device on a firm surface. Fill the jar with sand to above the valve, taking care to avoid vibration. Close the valve and remove the remaining sand from the cone.
3. Weigh the apparatus and sand and determine the weight of the sand.
4. Determine the unit weight (density) of the sand by dividing its weight by the volume of the jar.

A.1.3 Calibration of Sand Cone and Plate

1. Fill the jar with sand and determine the sand's weight.
2. Place the plate so that the bottom bears directly on a clean level surface. A piece of plywood or plastic, larger than the hole in the plate, that will fit within any protrusions on the plate bottom and is thicker than the height of the protrusions is needed.
3. With the valve closed, invert the sand cone and jar, placing the edge of the cone in the plate recess.
4. Open the valve. Close the valve when the sand has stopped flowing.
5. Determine the weight of sand needed to fill the cone and plate.

A.1.4 Determination of Field Unit Weight

1. Determination of the total unit weight of a soil in the field is made by excavating a small quantity of soil, determining its weight and also the volume of density-testing sand required to backfill the sample hole. The first step involves placing the base plate flush against an even portion of the site.
2. Using the center hole of the base plate as a template, excavate soil to a depth of at least 4 in. below the base plate. The hole should not extend laterally under the base plate. Retain the excavated soil in an air-tight container (e.g., sample jar or can, heavy zip-tight plastic bag). The moist and dry weights of a portion of the excavated soil will be required for subsequent calculations of moisture content, moist unit weight, and dry unit weight.
3. Once the hole has been cleared of all loose soil, place the sand-cone device on the base plate, with the edge of the cone in the plate recess.
4. Open the valve. Close the valve when the sand has stopped flowing.
5. Determine the weight of sand needed to fill the cone and hole. The unit weight (density) of the soil can be calculated by following the steps in the data sheet in Figure A.1.3. Determination of relative compaction and optimum moisture requires a laboratory compaction test. (See A.5.)

A.2   PREPARATION OF SOIL SAMPLES AND PERFORMANCE OF PARTICLE-SIZE ANALYSES

Purpose.   The suitability of a soil for a particular use in construction is often dependent on the distribution of grain sizes in the soil mass. There are two tests used to analyze the particle-size distribution in a soil. One of these methods is the sieve analysis. This test is a fundamental requirement for identification and for specification compliance testing for coarse soils. The other is the hydrometer test, which is used for fine-grained soils. This exercise is intended to provide experience performing sieve analyses, constructing gradation curves, and developing familiarity with the Unified Soil Classification System (ASTM D2487).

Reference Standards: ASTM D421—Dry Preparation of Soil Samples for Particle-Size Analysis and Determination of Soil Constants.
ASTM D422—Particle-Size Analysis of Soils.
ASTM D1140—Amount of Material in Soils Finer than the No. 200 (75-mm) Sieve.
EM 1110-2-1906, Appendix V: Grain-Size Analysis

A.2.1   Sieve Analysis

Equipment (Refer to Figures A.2.1 and A.2.2)

1. Sieves, bottom pan, and a cover lid.
2. Balance.
3. Mortar and rubber-covered pestle.
4. Sample splitter (if necessary).
5. Mechanical sieve shaker.
6. Drying oven.
7. Evaporating dish.
8. Sieve brush(es).

Figure A.2.1   Equipment necessary for the performance of a soil grain-size analysis.

Figure A.2.2 Mechanical sieve shaker.

Sample Preparation and Testing. Several methods exist for determining the particle-size distribution of soils. Mechanical separation by sieving is used to classify *coarse-grained soils* (i.e., those with minimum dimensions greater than 0.075 mm (No. 200 sieve)). Soils passing the No. 200 sieve are classified as *fine-grained soils.* Particle-size analysis of these soils is based on the settling rate of soil grains in water and performed with either a hydrometer or centrifuge. The laboratory procedures described in this section are specific to predominantly coarse-grained soils for situations when it is not necessary to determine the grain-size distribution of the minus-No. 200 fraction.

Two methods of sieve analysis are commonly used for soils. The specification of one or the other will depend on the size of the coarsest constituents of the soil. Coarse soils passing the No. 4 (4.74 mm) sieve require relatively little preparation and are sieved in a simple, straightforward manner. Soils including gravel or cobbles require several additional steps during preparation for what is termed the *combined-sieve analysis.* Both of these analyses techniques will be described, proceeding first with the method for minus-No. 4 soils.

A.2.1.1 *Procedure for Minus-No. 4 Coarse Soils*

1. Obtain the following sieves; Nos. 4, 10, 20, 40, 60, 140, and 200. This series of sieves is generally sufficient for most projects that use minus-No. 4 soils; however, sieves of intermediate sizes can be added if necessary. Clean the sieves with the appropriate brushes in order to dislodge soil grains from previous tests.

2. Weigh the sieves to 0.1 g.

3. Obtain a representative sample of dry (preferably air-dried) soil. Use the mortar and pestle to break down the aggregations into individual particles. The pressure required to achieve this disaggregation should be minimized to avoid breaking the solid particles themselves.

4. Weigh to 0.1 g an appropriate amount of the prepared soil. For soils passing the No. 4 sieve (4.75 mm) this corresponds to roughly 500 g of the dry soil. For samples containing a significant percentage of particles coarser than the No. 4 sieve, larger weights are needed. The data sheet shown in Figure A.2.3 can be used to record all pertinent data.

SIEVE ANALYSIS DATA AND GRAIN-SIZE CURVE

TEST FOR _____

DATE _____

TEST BY _____

SAMPLE DESCRIPTION _____

SAMPLE WEIGHT _____

U.S. STANDARD SIEVE SIZE	WEIGHT RETAINED	TOTAL WEIGHT RETAINED	TOTAL WEIGHT PASSING	PERCENT FINER
3"				
1½"				
¾"				
NO. 4				
NO. 10				
NO. 20				
NO. 40				
NO. 60				
NO. 100				
NO. 140				
NO. 200				
PAN (TOTAL)				

FINE FRACTION	NOTES
NO. 200 UNWASHED WEIGHT _____	
NO. 200 WASHED WEIGHT _____	
(Enter above)	
FINES BY WASHING _____	
FINES BY SIEVING _____	
FINES Enter (TOTAL) above	
_____	

Figure A.2.3   Data and results from laboratory sieve analysis.

5. Assemble the stack of sieves such that the sieves grade from the coarsest mesh on the top to finest mesh at the bottom of the stack. A bottom pan must be placed beneath the finest mesh (No. 200) to ensure that all of the sample can be accounted for after sieving.

6. Place the lid on the topmost sieve and place the stack of sieves in a mechanical shaker. The soil sample should be subjected to approximately 10 minutes of shaking.

7. Weigh to 0.1 g the bottom pan and individual sieves with the soil retained on them.

8. If a significant portion (~5–10%) of the sample is retained on the No. 200 sieve, it must be washed.

9. Washing, or wet sieving, is achieved by taking the No. 200 sieve with the soil retained on it and pouring water through the sieve. The slurry of fined-grained soil particles may be stirred or spread (using a minimum amount of pressure against the delicate mesh) to facilitate sieving. It is advisable to allow the slurry to sit for a minimum of two hours to allow the aggregated particles to hydrate and break down. Once this has been achieved, continue washing the soil on the No. 200 sieve until the water passing through the sieve is free of silt and clay-sized particles. Obtain the weight of the soil retained on the No. 200 sieve by washing the material into an evaporating dish, oven drying the sample, and weighing to 0.1 g.

10. Now that the weights of soil retained on each sieve are known, the percentages finer than each sieve size can be determined and plotted on the graph in Figure A.2.4. Once the gradation of the soil is plotted, the USCS designation for the soil is easily established.

A.2.1.2 *Procedure for Coarse Soils with Gravel or Cobbles*

Laboratory procedures are slightly modified for soils containing a significant percentage of particles greater than the No. 4 sieve. This procedure, the combined sieve analysis, is specified for projects involving gravelly soils. In order for samples of gravelly soils to be considered representative of the material in the field, large volumes of material must be sampled. For ease of testing, sample splitters are used to divide the bulk sample into smaller, representative batches. Data for this sieve analysis can be recorded on the sheet in Figure A.2.5.

1. Use a sample splitter to obtain a test sample of roughly 1000 to 3000 g of soil.

2. To the sieves listed in Section A.2.1 add the 3-in., 1 1/2-in., and 3/4-in. sieves or others, as appropriate.

3. Break up the soil aggregates and weigh the test sample.

4. Separate the test sample by sieving the soil with the No. 10 sieve and the pan. Set the material passing the No. 10 sieve aside for subsequent testing.

5. The soil retained on the No. 10 sieve is now ground again by using the mortar and pestle to eliminate any remaining aggregations. Resieve this fraction on the No. 10 sieve.

6. Mix the material passing the No. 10 sieve during Step 5 with the soil that passed the No. 10 sieve in Step 4. Weigh this mixture and record as "Pass No. 10."

7. The soil retained on the No. 10 sieve after the second sieving is now washed to remove any fine-grained particles, dried, and weighed to 0.1 g. Record this mass as "Retained No. 10."

8. Sieve the coarser fraction by using the 3-in., 1 1/2-in., 3/4-in., No. 4, and No. 10 sieves. Weigh each of the sieves plus the retained soil. Calculate the percentages passing each of these sieves by using both the mass of the total sample and the mass of the "Retained No. 10" fraction as indicated in the example data reduction shown in Figure 3.3 of the text (Section 3.1).

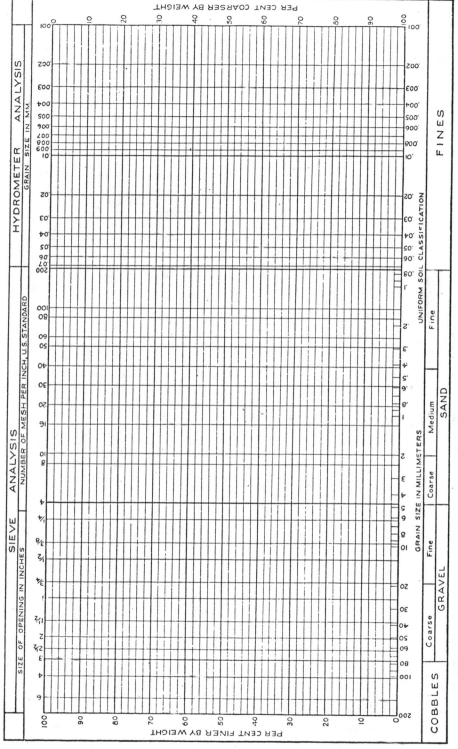

Figure A.2.4 Grain-size chart and ASTM grain-size scale.

COMBINED-SIEVE ANALYSIS

TEST FOR _____

TEST BY _____ DATE _____

SAMPLE DESCRIPTION _____

_____

Total Sample Weight After Splitting, W_1 _____

Total Weight Retained on No. 10 Sieve, W_2 _____

Weight of Plus No. 10 Portion After Splitting, W_3 _____

Total Weight Passing No. 10 Sieve, W_4 _____

Weight of Minus No. 10 Portion After Splitting, W_5 _____

	Percent of Total (3)											Percentages Passing
		3"	1½"	¾"	No. 4	No. 10	No. 40	No. 80	No. 100	No. 200	Pan	
Retained No. 10	$\left[\dfrac{W_2}{W_1}\right]$											
Pass No. 10	$\left[\dfrac{W_4}{W_1}\right]$												
Total Sample	100											
Alternate Sieve Sizes												

KEY

(2) ◄——— Percent of Combined Sample

(1) ◄——— Percent of Split Portion Sieved

$$\frac{W}{W_3} \text{ or } \frac{W}{W_5}$$

NOTES

W = Total Weight Passing Any Sieve

(2) = (3) × (1)

Figure A.2.5 Data-collection and reduction form for the combined-sieve analysis.

9. Obtain a representative sample of the "Pass No. 10" fraction from Step 7 weighing approximately 100 to 150 g. Use a sample splitter if necessary.

10. Sieve this portion of the fine fraction by using the No. 40, 60, 140, and 200 sieves. The No. 80 and 100 sieves can also be used if desired.

11. Record the weights of the soil retained on each of the sieves and calculate the percentages passing while using both the mass of the "Pass No. 10" fraction and the mass of the sample obtained by splitting in Step 9.

12. The results of the calculations made in Steps 8 and 11 are combined, as shown in Figure 3.3, to obtain the percentages passing for the total sample.

A.2.2 Hydrometer Analysis

The hydrometer analysis is based on Stokes' law, which relates the terminal velocity of a sphere falling freely through a fluid to the diameter of the sphere. The theory assumes laminar flow around the sphere. According to Stokes' law, the larger spheres will have a higher terminal velocity, and, thus, in a suspension of particles, will fall to the bottom sooner. The hydrometer is used to determine the percentage of dispersed soil particles remaining in suspension at a given time. Although the hydrometer analysis can be used with soils passing the No. 10 sieve, this analysis is best when used for particles passing the No. 200 sieve that can no longer be analyzed by using the sieve. In a combined analysis, these particles are taken after sieve analysis and used for the hydrometer test.

Equipment

1. Hydrometer.
2. Dispersion apparatus, either a mechanically operated device, in which a rotating paddle is inserted into a special cup with the supension, or an air dispersion device.
3. Sedimentation cylinder.
4. Centigrade thermometer.
5. Timing device.
6. Balance.
7. Oven.

Procedure

1. Record all identifying information for the sample, such as project, boring number, or other pertinent data, on a data sheet such as is shown in Figure A.2.6.
2. Determine the dispersing agent correction and the meniscus correction, unless they have already been given. Record these on the data sheet.
3. Determine or estimate the specific gravity of the solids. (See A.3.)
4. Oven dry the sample, allow to cool, and then weigh and record the dry weight. Place the sample in a submerged dish, and add distilled or demineralized water until the sample is submerged. Add the dispersing agent, and allow the sample to soak overnight.

HYDROMETER ANALYSIS	DATE

PROJECT	

BORING NO.:

SAMPLE OR SPECIMEN NO.	CLASSIFICATION

DISH NO.	GRADUATE NO.	HYDROMETER NO.

DISPERSING AGENT USED	QUANTIY

DISPERSING AGENT CORRECTION, C_d = _____   MENISCUS CORRECTION, C_m = _____

TIME	ELASPED TIME MIN	TEMP °C	HYDRO. READING (R')	CORRECTED READING (R)	PARTICLE DIAMETER (D), MM	TEMP CORRECTION (m)	$R - C_d + m$	PERCENT FINER	
								PARTIAL	TOTAL

WEIGHT IN GRAMS	DISH PLUS DRY SOIL		
	DISH		
	DRY SOIL		W_0

Specific gravity of solids, G_S =

Corrected hydrometer reading (R) = hydrometer reading (R') + C_m

The particle diameter (D) is calculated from Stokes' equation using corrected hydrometer reading. Use nomographic chart for solution of Stokes' equation.

Hydrometer graduated in specific gravity W_s = total oven-dry wt of sample used for combined analysis

Partial percent finer = $\dfrac{G_S}{G_S - 1} \times \dfrac{100}{W_0} (R - C_d + m)$ W_0 = oven-dry wt in grams of soil used for hydrometer analysis

Hydrometer graduated in grams per liter W_1 = oven-dry wt of sample retained on No. 200 sieve

Partial percent finer = $\dfrac{100}{W_0} (R - C_d + m)$

Total percent finer = partial percent finer $\times \dfrac{W_S - W_1}{W_S}$

REMARKS _____

_____

TECHNICIAN	COMPUTED BY	CHECKED BY

Figure A.2.6   Data and results from hydrometer analysis. (From EM 1110-2-1906.)

5. Transfer the soil-water slurry from the dish to a dispersion cup. Add distilled water to the dispersion cup, until the water surface is 2–3 in. below the top of the cup. Place the cup in the dispersion machine and disperse the solution from 1–10 minutes.

6. Transfer the suspension into a 1000 ml sedimentation cylinder and add distilled or demineralized water until the volume of the suspension equals 1000 ml.

7. One minute before starting the test, take the cylinder in one hand and, using a suitable stopper, shake the suspension vigorously for a few seconds to transfer the sediment into a uniform suspension.

8. Set the cylinder on the table. Slowly immerse the hydrometer into the liquid 20–25 seconds before each reading.

9. Observe and record the hydrometer reading on the data sheet after 1 and 2 minutes have elapsed from the time the cylinder is placed on the table. Immediately after the two minute reading, carefully remove the hydrometer from the suspension and place it in clean water. Repeat Steps (8) and (9) after elapsed times of 4, 15, 30, 60, 120, 240, and 1440 minutes.

10. For each hydrometer reading, compute the particle diameter based on Stokes' equation, and record these diameters on the data sheet. To assist in this, you can use the nomograph in Figure A.2.7. A key showing the steps to follow in computing D for various values of R is shown on the chart.

11. The percent finer for each hydrometer reading can be computed for a hydrometer calibrated in grams per liter by dividing the corrected hydrometer height reading R (with temperature and dispersing agent corrections included) by the weight of the solids in the hydrometer test and multiplying the result by 100.

12. A gradation curve for a soil sample tested with both sieve and hydrometer tests can be constructed by first plotting the results of the sieve-analyzed portion of the sample, which was described in the previous section. Afterward, take the results from Step (11) of the hydrometer tests and multiply them by $(W_s - W_l)/W_s$, where W_s is the total dry weight of the sample used for the combined analysis and W_1 is the dry weight of the sample retained on the No. 200 sieve.

A.3   SPECIFIC GRAVITY OF SOILS

Purpose.   The specific gravity of a soil is the ratio of the weight in air of a given volume of soil particles to the weight in air of an equal volume of distilled water at a temperature of 4°C. We are, therefore, determining the specific gravity of the mineral grains that comprise the solid phase of the soil. The specific gravity of the grains (Gs) is incorporated in many of the mass–density relationships used in geotechnical engineering.

Reference Standard.   ASTM D854—Specific Gravity of Soils.
    EM 1110-2-1906, Appendix IV: Specific Gravity

Equipment (Refer to Figure A.3.1)

1. Volumetric flask (500 ml).
2. Thermometer.
3. Balance or scale.

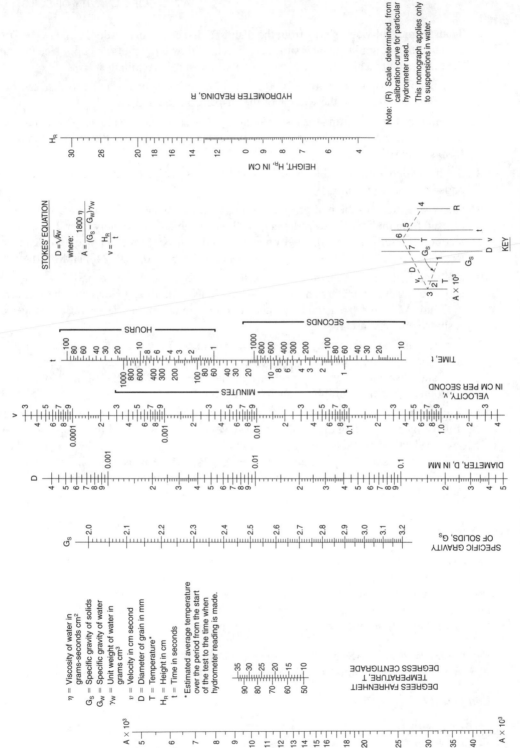

Figure A.2.7 Nomographic chart for hydrometer analysis. (From EM 1110-2-1906.)

Note: (R) Scale determined from calibration curve for particular hydrometer used.

This nomograph applies only to suspensions in water.

HYDROMETER READING, R

HEIGHT, H_R, IN CM

STOKES' EQUATION

$$D = \sqrt{Av}$$

where:

$$A = \frac{1800\,\eta}{(G_S - G_W)\gamma_W}$$

$$v = \frac{H_R}{t}$$

KEY

$A \times 10^3$

TIME, t
HOURS
MINUTES
SECONDS

VELOCITY, v, IN CM PER SECOND

DIAMETER, D, IN MM

SPECIFIC GRAVITY OF SOLIDS, G_S

η = Viscosity of water in grams-seconds cm^2
G_S = Specific gravity of solids
G_W = Specific gravity of water
γ_W = Unit weight of water in grams cm^3
v = Velocity in cm second
D = Diameter of grain in mm
T = Temperature*
H_R = Height in cm
t = Time in seconds

* Estimated average temperature over the period from the start of the test to the time when hydrometer reading is made.

DEGREES FAHRENHEIT
TEMPERATURE, T
DEGREES CENTIGRADE

$A \times 10^3$

284

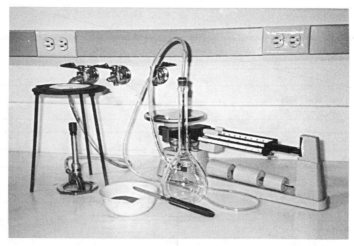

Figure A.3.1   Equipment required for the determination of specific gravity of soil.

4. Distilled water.
5. Heat (or vacuum) source.
6. Evaporating dishes.
7. Dropper or plastic squeeze bottle.
8. Drying oven.

Procedure

1. To a clean flask add deaired, distilled water up to the calibration (500 ml) mark.
2. Determine the weight of the flask, plus water to 0.01 g. Record measurements on the table in Figure A.3.2.
3. Measure and record the temperature of the water to 0.1°C. Obtain the water temperature at several depths in the flask to check the uniformity of the temperature. Once consistent, uniform temperatures are obtained, pour out roughly half of the water.
4. Put approximately 150 g of oven-dried soil into an evaporating dish.
5. If the soil is composed of coarse particles, then the sample may be transferred directly to a volumetric flask that is roughly half full of water. Cohesive (fine-grained) soils must be hydrated with deaired, distilled water to the consistency of a smooth paste. Allow the soil to soak for at least one hour in the evaporating dish. Transfer the paste to the flask, making sure that no soil grains are lost.
6. Remove all of the air that is entrapped in the soil–water mixture. This can be achieved by one of two methods:

   (a) Gently boiling the flask containing the soil–water mixture for at least 10 minutes while occasionally rolling the flask to assist in the removal of air.

   (b) Subjecting the contents of the flask to a partial vacuum.

SPECIFIC GRAVITY TEST

DETERMINATION NO	1	2	3	4
FLASK NO.				
WT. FLASK + WATER + SOIL (W_1)				
TEST TEMPERATURE t, °C				
WT. FLASK + WATER (W_2)				
EVAPORATING DISH NO.				
WT. DISH + DRY SOIL				
WT. DISH				
WT. DRY SOIL (W_S)				
SPECIFIC GRAVITY OF WATER AT t, (G_t)				
SPECIFIC GRAVITY OF SOIL (G_S)				

TEST FOR _____

TEST BY _____ DATE _____

SAMPLE DESCRIPTION _____

REMARKS _____

$$G_S = \frac{G_t W_S}{W_S + W_2 - W_1}$$

Figure A.3.2 Results and calculations of specific gravity test.

7. Once all of the entrapped air is removed, allow the soil–water mixture to cool to room temperature (if the heating technique was used). Measure the temperature of the fluid to ensure that it has cooled to within 61°C of the temperature recorded in Step 3.

8. Fill the volumetric flask to the calibration mark with deaired, distilled water.

9. Determine the weight of the flask plus contents to 0.01 g.

10. Pour the soil–water mixture into an evaporating dish. Rinse the flask to make sure that all of the soil has been transferred to the dish.

11. Dry the soil in the oven.

12. Determine the weight of the dry soil in the evaporating dish. The grain specific gravity can be calculated by filling out the data reduction sheet in Figure A.3.2 and referring to Table A.1.1.

A.4   SOIL CONSISTENCY TESTS

Purpose.   The engineering behavior of soils containing an appreciable percentage of fine-grained constituents is influenced by the mineralogy of the clay fraction, the percentage of clay minerals in the soil, and the water content of the soil. All other factors being equal, the consistency and workability of a clayey soil is a function of its water content. A series of index tests has been devised for determining the water contents (decreasing in the order following) at which a clayey soil passes from a viscous liquid to plastic material (liquid limit), from a plastic material to a brittle semisolid (plastic limit), and from a semisolid to a solid (shrinkage limit). These tests are known collectively as the *Atterberg limits tests,* and they indicate the water content of the soils that are the boundaries of these relative states of consistency.

The Atterberg limits and associated plasticity index reflect the type and amount of clay minerals in a soil. For this reason the Atterberg limits are used as the basis for classifying fine-grained soils. In addition, numerous relationships that correlate the Atterberg limits with other engineering index properties for fine-grained soils have been adopted in engineering practice. This laboratory exercise is intended to present the methods for obtaining the liquid and plastic limits of soils and to familiarize the student with the Unified Soil Classification System for fine-grained soils.

Reference Standards.   ASTM D2216—Laboratory Determination of Water (Moisture) Content of Soil, Rock, and Soil-Aggregate Mixtures.
EM 1110-2-1906, Appendix I: Water Content–General.
ASTM D4318—Liquid Limit, Plastic Limit, and Plasticity Index of Soils.
EM 1110-2-1906, Appendix III: Liquid and Plastic Limits.

A.4.1   Plastic Limit Test

Equipment (Refer to Figure A.4.1)

1. Evaporating dish.
2. Spatula.

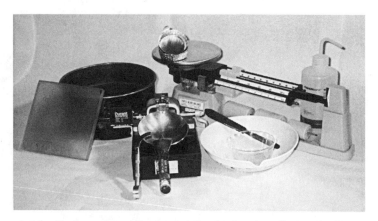

Figure A.4.1   Equipment used for determining the Atterberg limits of soil.

3. Plastic squeeze bottle with distilled water.

4. Moisture content container with cover.

5. Ground glass plate.

6. 1/8-in. diameter rod.

7. Balance sensitive to 0.01 g.

8. Drying oven.

Procedure

1. The first step in preparing soil samples for the limit tests involves separating the soil by using the No. 40 (0.425 mm) sieve. Only the material passing the No. 40 sieve is used for determining the Atterberg limits of a soil. Soils containing a fraction greater than the No. 40 sieve are air-dried and then sieved. For soils in which all of the grains are smaller than 0.425 mm, it is preferable to use samples initially at their natural water content; therefore, drying should be avoided. In both cases, approximately 250 g of the minus-No. 40 soil should be prepared for testing. It is much easier to increase the moisture content of a soil than to reduce it; hence, it is advantageous to perform the plastic-limit test prior to the liquid-limit test for air-dried samples. Most of the 250-g sample will be used in the liquid-limit test that follows.

2. To approximately 20 g of the dry soil, add small quantities of distilled water and mix thoroughly. Continue to add water and mix the soil until threads about 1/8 in. in diameter can be rolled on the glass plate.

3. Record the weight of the moisture can on the data sheet in Figure A.4.3.

4. With your fingers mold several balls of soil. Roll one of these soil samples into a 1/8-in.-diameter thread on the glass plate. (Refer to Figure A.4.2.) If the sample begins to crumble before the diameter of the thread can be reduced to 1/8 in., remold the ball and add several drops of distilled water. If the sample can be rolled into a thread having a diameter of 1/8 in. or less without cracking, then allow some of the soil moisture to evaporate by

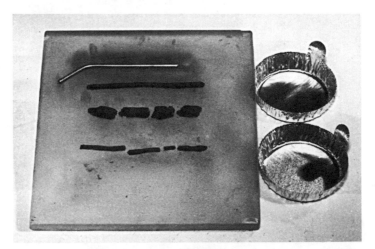

Figure A.4.2 Soil threads rolled from specimens at three different water contents.

ATTERBERG LIMITS AND WATER CONTENT

TEST FOR _____

TEST BY _____ DATE _____

SAMPLE DESCRIPTION _____

_____

PLASTIC LIMIT AND WATER CONTENT

TYPE OF TEST						
CONTAINER NO.						
WET WT. + CONTAINER						
DRY WT. + CONTAINER						
WT. OF WATER						
WT. OF CONTAINER						
DRY WT. OF SOIL						
WATER CONTENT, %						

LIQUID LIMIT

CONTAINER NO.						
NO. OF BLOWS						
WET WT. + CONTAINER						
DRY WT. + CONTAINER						
WT. OF WATER						
WT. OF CONTAINER						
DRY WT. OF SOIL						
WATER CONTENT, %						

W _____

LL _____

PL _____

PI _____

Water content (%) vs Number of drops (N): 10, 20, 25, 30, 40, 50

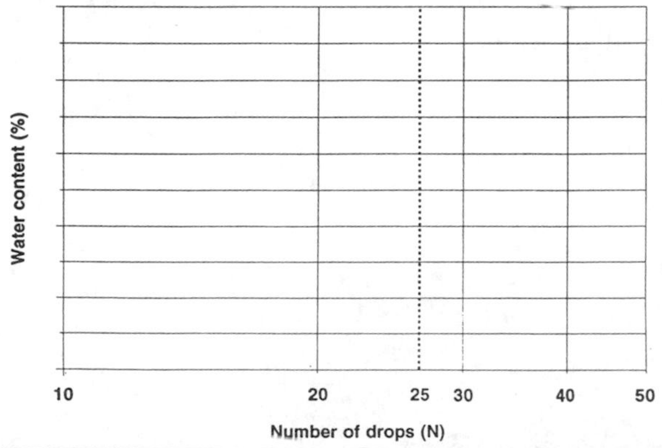

Figure A.4.3 Data and calculations for water-content and Atterberg limits test.

(a) continued remolding by hand or (b) leaving the sample exposed on the glass plate for a short duration. Towel or microwave drying is not recommended.

5. Repeat Step 4 until the thread begins to crumble into several pieces when it reaches a diameter of 1/8 in. Quickly transfer the crumbled threads into the weighing cup and seal. Note that as the plastic limit is approached, it is common for a sample that may have been slightly wet of the limit during one determination to begin to crumble before reaching a diameter of 1/8 in. on the subsequent attempt. This sample is considered acceptable, provided that the soil was not allowed to dry appreciably between tests.

6. Repeat Steps 4 and 5 for two or three of the balls of soil molded from the original sample in Step 4. Retain the crumbled threads from these tests in the weighing cup.

7. Obtain the weight of the cup plus soil, remove the cap from the can, and place the cup in the drying oven.

8. Once the cup and soil reach a constant weight, record the weight of the can, the dry soil, and the cap. The water content of the soil sample is the plastic limit for the soil.

A.4.2 Liquid-Limit Test

Equipment (Refer to Figure A.4.4)

1. Casagrande liquid limit device.
2. Grooving tool. (The ASTM standard flat grooving tool and the common curved tool are both shown in Figure A.4.4.)
3. Moisture cans.
4. Evaporating dish.
5. Spatula.
6. Drying oven.
7. Balance sensitive to 0.01 g.
8. Plastic squeeze bottle full of distilled water.

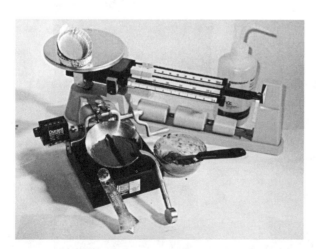

Figure A.4.4 Required equipment for the liquid-limit test.

Procedure

1. Set the fall height of the brass cup on the liquid-limit device by turning the crank of the device until the cup is at its highest position, and adjust it so that it is equal to 0.394 in. (10 mm). The end of the grooving tool is a standard 0.394-in. (10.0-mm) thickness and this is used to calibrate the device. The end of the grooving tool should just touch the base of the cup when slid along the base of the liquid-limit device.

2. Add small quantities of distilled water to the sample prepared prior to performing the plastic-limit test. Water should be added until the soil–water mixture has the consistency of a uniform stiff paste.

3. Measure the weights of the three moisture cans and record this data on the data sheet in Figure A.4.3.

4. Using the spatula, place a portion of the soil paste in the brass cup of the liquid-limit device. Smooth the surface of the soil in the cup unit so that the maximum depth of the soil is about 1/3 in. (8 mm).

5. Use the grooving tool to incise a groove along the centerline of the soil pat in the cup. (Refer to Figure A.4.4.)

6. Turn the crank of the liquid-limit device at the rate of roughly two revolutions (taps) per second. Note that the soil begins to move towards the center of the cup, gradually closing the groove, as the tapping is continued. Continue turning the crank until the soil has closed along 1/2 in. of the groove length. (Refer to Figure A.4.5.)

7. If the number of taps (N) required to close the groove 1/2 in. is between 25 and 35, extract approximately 15–20 g of the soil from the cup for a moisture-content determination. Quickly weigh the moisture can, plus the moist soil. If the sample required more than 35 taps to achieve the proper closure, then add water to the sample and repeat the test.

8. Remove the remaining soil from the brass cup, and mix thoroughly with additional prepared soil. Add small quantities of distilled water until the soil–water mixture is only

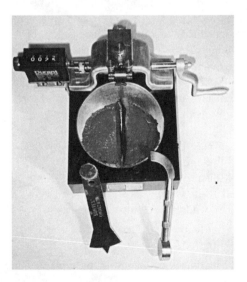

Figure A.4.5 Configuration of the soil specimen at the end of a liquid-limit trial.

slightly less stiff than the previous test sample, and repeat Steps 4 through 7. The second sample should require fewer taps to close a portion of the groove, preferably 20 to 30.

9. Obtain a water-content sample, and, again, remove the remaining soils from the liquid-limit device and add additional water to the original sample.

10. Repeat Steps 4 through 7 until the sample requires 15 to 25 taps to form the specified closure.

11. Obtain a third water-content sample and oven dry the three samples.

12. Use the graph on the example data sheet in Figure A.4.3 to plot the water content against the number of taps. The liquid limit of the soil is specified as the water content corresponding to $N = 25$.

13. Given the plastic and liquid limits, the plasticity index (PI) can be calculated. The plasticity index is the range of water content over which the soil behaves as a plastic material. It is defined as the difference between the liquid limit and the plastic limit. The PI is computed and recorded along with the liquid and plastic limits.

A.5   COMPACTION TESTS

Purpose.   The densification, or compaction, of soils by mechanical means is the most common form of soil improvement used in construction. As discussed in Chapter 8, compaction specifications are developed to (1) minimize the compressibility of the soil, (2) increase shear strength, and (3) decrease the permeability. Compaction tests are performed to evaluate the influence on unit weight of the water content of the soil at the time of compaction, which can be correlated with these soil properties. The resulting water-content-versus-unit-weight (compaction) curve is used to establish the acceptable range of water content for compaction in the field and a unit-weight specification.

It is important to realize that the moisture–density curve for a soil is not unique. The optimum water content and characteristics of the compaction curve are a function of the compactive effort utilized during the test. For this reason, it is imperative that the laboratory method used to establish the specification be annotated on the compaction curve and referenced in the construction documents.

Since the two compaction procedures outlined herein differ only in the size of the hammer, fall height of the weight, and size of the compaction mold (and therefore the compactive effort), only the standard compaction method will be thoroughly presented. This laboratory exercise will demonstrate the following: (1) laboratory compaction methods, (2) the concept of compactive effort, (3) moisture–density relationships and the zero-air voids curve, and (4) the basis for calculating relative compaction.

Reference Standards.   ASTM D698—Moisture–Density Relations of Soils and Soil–Aggregate Mixtures Using 5.5-lb Rammer and 12-in. Drop.

ASTM D1557—Moisture–Density Relations of Soils and Soil–Aggregate Mixtures Using 10-lb Rammer and 18-in. Drop.

EM 1110-2-1906, Appendix VI: Compaction Tests.

Equipment (Refer to Figure A.5.1)

1. Compaction mold (4 in. diameter), including base plate and extension collar.

Figure A.5.1   Partial assembly of equipment necessary to perform soil-compaction tests.

2. No. 4 sieve.

3. Standard compaction hammer (5.5 lb).

4. Balance sensitive to 0.01 lb.

5. Balance sensitive to 0.1 g.

6. Large flat pan.

7. Steel straight edge.

8. Moisture cans.

9. Drying oven.

10. Plastic squeeze bottle with water.

11. Sample extruder.

12. Large spoon or small hand spade.

13. Mixing pan.

Procedure.   The procedure that will be described next is applicable for soils passing the No. 4 sieve. It should be noted that correction factors are applied to the results of laboratory compaction tests for soils containing small amounts of plus-No. 4 particles. In addition, due to the size of the compaction molds, neither the standard nor the modified test equipment is appropriate for obtaining moisture-density relationships for soils composed of more than 30 percent by weight greater than 3/4-in. particles.

1. Prepare a representative batch of the soil to be tested by breaking down soil clumps into individual particles.

2. Pass enough of the prepared soil through the No. 4 sieve to yield approximately 6 to 7 lbs of minus-No. 4 material.

3. For clayey soils, it is often useful to determine the water content of the air-dried sample. The amount of moisture in air-dried soil is termed the hygroscopic moisture content. This moisture content can typically be as high as 4 to 8 percent for clayey soils (and as high as 15 to 20 percent for Na–Montmorillonite). The moisture should be accounted for when wetting the soil prior to testing. Obtain 100 to 150 g of the minus-No. 4 soil for a water-content determination. Record subsequent data as shown on the example data sheet (Figure A5.2).

MOISTURE-DENSITY RELATIONSHIP

TEST FOR _____

TEST BY _____ DATE _____

SAMPLE IDENTIFICATION_____

COMPACTIVE EFFORT _____

TRIAL NUMBER	WATER CONTENT OF AIR-DRIED SOIL	1	2	3	4	5	6
Wt. of Wet Soil & Cylinder							
Wt. of Cylinder							
Wt. of Wet Soil							
Vol. of Cylinder							
Unit Wet Weight							
Unit Dry Weight							
Pan Number							
Wt. of Wet Soil & Pan							
Wt. of Dry Soil & Pan							
Wt. of Water							
Wt. of Dish							
Wt. of Dry Soil							
Water Content, %							

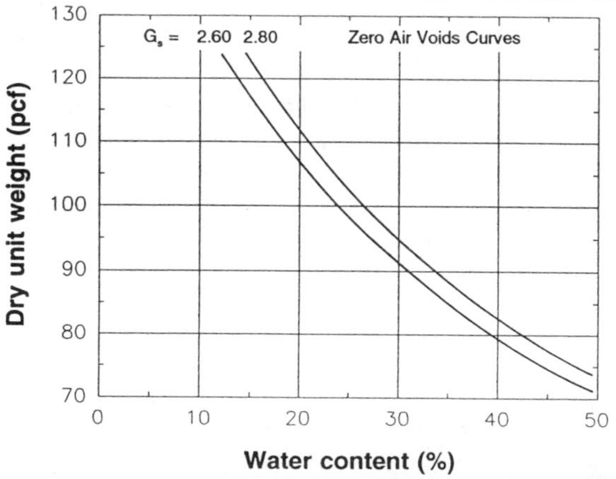

Figure A.5.2 Data form for standard compaction test.

4. The compaction curve will be established on the basis of the moisture-density data from five compaction trials: two tests with water contents dry of optimum, one at or near the optimum, and two wet of optimum. Estimate the optimum moisture content of the soil, based on plasticity data and the relationship plotted in Figure 8.16. The compaction tests are

performed such that the moisture content of the soil is progressively increased after each trial. The difference in moisture content between each trial should be roughly 2 to 3 percent.

5. Add water (mixing thoroughly) to the minus-No. 4 sample, until the first moisture content is attained (again, account for hygroscopic moisture as necessary).

6. Weigh the compaction mold and base plate to 0.01 lb. Do not include the extension collar in this weighing.

7. Assemble the extension collar and compaction mold.

8. The soil sample will be compacted in three equal lifts (layers) such that at the end of compaction of the third layer, the soil should extend approximately one-fourth in. above the top of the compaction mold. For the first layer, place the minus-No. 4 soil into the mold, until it is about half full.

9. Smooth the surface of the soil with light tamping and then begin compacting the soil with the 5.5 lb hammer. Each layer is compacted with 25 uniformly distributed blows before the next layer of soil is added.

10. After the third layer has been compacted, remove the extension collar from the compaction mold.

11. Using the steel straightedge, trim off the excess soil until the sample is even with the top of the mold. In the event that a small quantity of soil is lost from the compaction mold during removal of the collar or during the trimming process, fill the void with trimmings pressed in with moderate finger pressure.

12. Weigh the compaction mold, base plate, and compacted soil to 0.01 lb.

13. Extrude the sample from the mold and retain approximately 100 g for a moisture-content determination. The equipment used to extrude the sample is shown in Figure A.5.3. Equal portions of the sample should be obtained from each of the three lifts to ensure a representative water content.

14. Break up the extruded sample by hand, and mix with excess soil from the previous compaction test. Add water, mixing thoroughly, until the water content of the soil has been raised by 2 to 3 percent.

Figure A.5.3   Equipment used for extruding and weighing the compacted soil specimens.

15. Repeat Steps 7 through 14. Note the consistency of the soil and the total weight of the mold, collar, and moist soil throughout the five trials.

16. Once the water content samples have been dried, determine the water content and dry unit weight of the soil in each trial. This data is then plotted on the data sheet illustrated in Figure A.5.2.

17. Plot the zero-air-voids curve on the compaction curve, using Equation 8.3. Plot the zero-air-voids curve for specific gravity (G_s) values of 2.60 and 2.80.

A.6 RELATIVE DENSITY

Purpose.   The relative density (D_r) of a cohesionless soil expresses the degree of compactness with respect to the loosest and densest states achieved by standard laboratory procedures. Relative density is a widely used index parameter for estimating the strength and volume-change characteristics of cohesionless, free-draining soils.

The purpose of this laboratory exercise is to determine the maximum and minimum index densities of a dry sand. The impact method of soil compaction (presented in A.5) is not suitable for obtaining the maximum relative density of a sandy soil for engineering design purposes due to the effects of grain crushing. The standard method utilizes vibratory compaction to obtain the maximum density (100 percent relative density, or minimum void ratio) and dry pluviation to obtain minimum density (0 percent relative density, or maximum void ratio). Sample preparation and test techniques have been shown to have a significant influence on the accuracy and repeatability of these methods. The method described in this section makes use of readily available but nonstandard equipment. This exercise is intended to illustrate the concept of relative density; again, it is advisable to refer to the reference standards when performing these procedures in practice.

Reference Standards.   ASTM D4253—Maximum Index Density of Soils Using a Vibratory Table.
ASTM D4254—Minimum Index Density of Soils and Calculation of Relative Density.
EM 1110-2-1906, Appendix XII: Relative Density

Equipment (Refer to Figure A.6.1)

1. ASTM 4253 mold assembly, or compaction mold (4-in. diameter), including base plate and extension collar.

2. No. 4 sieve.

3. Vibrating table.

4. Rubber hammer.

5. Compaction hammer (if necessary).

6. Balance sensitive to 0.01 lb.

7. Surcharge weights.

8. Surcharge plate.

9. Dial gage (0.001-in. graduations) and gage holder.

10. Pouring devices (e.g., funnel, wide tubing).

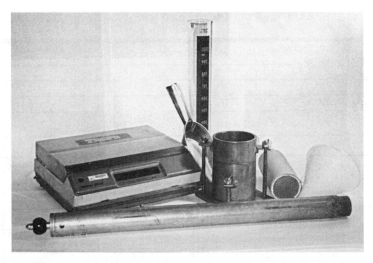

Figure A.6.1   Equipment utilized in the laboratory demonstration of relative density.

11. Steel straightedge.

12. 2000 ml graduated cylinder.

Procedure.   The following procedures are applicable for sandy soils passing the No. 4 sieve and containing less than 15 percent fines.

A.6.1   Maximum-Density Procedure (Dry Method)

1. Prepare a representative batch of the soil to be tested by breaking down soil clumps into individual particles.

2. Pass enough of the prepared soil through the No. 4 sieve to yield approximately 4 to 5 lbs of minus-No. 4 material.

3. If the mold assembly specified in ASTM D4253 is not available, use the small compaction mold described in Section A.5 for the sake of demonstration.

4. Fill the mold with the minus-No. 4 soil, leveling with the top of the mold by using the steel straightedge. Record all weights and measures on the data sheet shown in Figure A.6.2.

5. Place the extension collar on the top of the mold. Insert the surcharge plate of known thickness (± 0.001 in.) into the extension collar. Add a surcharge load of approximately 5 lbs with standard laboratory weights on the surcharge plate. If this equipment is not available, densification can be achieved by (a) vigorously tapping on the sides of the mold with a rubber-headed mallet or (b) compacting the specimen in small lifts by using a compaction hammer. The latter method is suggested for demonstration purposes only.

6. Strike the sides of the mold several times with a rubber hammer to settle the soil and to ensure even settling prior to subjecting to the vibration loading.

7. Attach the mold and base-plate assembly on a vibrating table capable of oscillating at 50 to 60 Hz.

MAXIMUM- AND MINIMUM-DENSITY TESTS

DESCRIPTION OF SOIL _____

SAMPLE NO. _____   LOCATION _____

PROJECT _____

TESTED BY _____   DATE _____

Maximum Dry Density

Test No.	Weight of Mold (lb)	Weight of Mold + Soil (lb)	Weight of Soil (lb)	Sample Height After Shaking (in.)	Volume of Sample (ft³)	Dry Unit Weight (lb/ft³)

Minimum Dry Density

Test No.	Weight of Mold (lb)	Weight of Mold + Soil (lb)	Height of Soil Sample (in.)	Volume of Sample (ft³)	Dry Unit Weight (lb/ft³)

Figure A.6.2   Data and calculations for maximum- and minimum-density tests.

8. Vibrate the mold assembly and soil specimen for approximately 10 minutes.

9. After vibrating the sample, carefully remove the mold assembly from the vibrating table and measure the distance from the top of the mold to the top of the surcharge plate. Account for the thickness of the surcharge plate and compute the volume of the sample.

10. Remove the surcharge plate from the mold and weigh the mold, base plate, and soil. Subtract the weight of the mold and base plate to determine the weight of the soil. The maximum density of the soil can be computed from the sample volume determined in Step 9 and the weight obtained in Step 10.

11. Repeat Steps 4 through 10 until consistent values (within 2 percent) of maximum index density are obtained.

A.6.2 Minimum Density Procedures

Three simple methods are described in the ASTM Standard for determining the minimum density of a free-draining cohesionless soil. All three methods are presented:

Method A   Using a funnel-pouring device or a hand scoop to place the material in a mold.
Method B   Depositing material into a mold by extracting a soil-filled tube.
Method C   Depositing material by inverting a graduated cylinder.

Method A

1. Using the oven-dried soil prepared for determining the maximum index density, fill a funnel with a spout opening of approximately 1/2 in. Place the funnel spout inside the mold at a distance of about 1/2 in. above the base of the mold and allow the sand to flow out in a steady stream. Holding the funnel in the vertical position, move the funnel in a spiral path from the outside to the center of the mold. Maintain a distance of roughly 1/2 in. between the spout and the soil surface in the mold. Continue to pour the soil until the mold has been overtopped. Avoid jarring the mold or disturbing the deposited soil.

2. Remove the extension collar and level the soil with the top of the mold by using a straightedge. It is again noted that this procedure must be carried out without disturbing the loose soil sample.

3. Once the soil has been leveled, weigh the mold, base plate, and soil. Calculate the weight of the soil. This weight is used in conjunction with the volume of the mold to determine the minimum index density.

4. Repeat Steps 1 through 3 until three consistent values (± 1 percent) of the minimum index density are obtained.

Method B   This method is similar to Method A, except that a thin-walled tube is used to deposit the soil in lieu of the funnel previously described. The tube should have the following properties: (1) a volume equal to 1.25 to 1.30 times the volume of the mold and (2) an inside diameter that is about 0.7 times the diameter of the mold.

1. Rest the tube on the base plate inside the mold. Fill the tube with the prepared soil to within 1/8 in. of the top.

2. Withdraw the tube from the mold allowing the soil to fill the mold.

3. Use the straightedge to level the sand with the top of the mold.

4. Weigh the mold, base plate, and soil. Calculate the minimum-index density from the weight of the soil and volume of the mold.

5. Repeat Steps 1 through 4 until three consistent values (± 1 percent) of the minimum index density are achieved.

Method C

1. Weigh 1000 $\pm$ 1 g of the prepared soil and pour this soil into a 2000 ml graduated cylinder.

2. Place a stopper in the top of the cylinder, tip the cylinder upside down, and then tilt it back to the vertical, upright position.

3. Record the volume occupied by the loose sand, and determine the sand's weight. Calculate the minimum-index density.

4. As with the previous methods, repeat the procedure until three values are obtained that vary by no more than 1 percent.

A.7 PERMEABILITY TESTS

Purpose.   As discussed in Chapter 5, the flow of water through soils is of importance for projects involving excavation dewatering, seepage through embankments or natural slopes, and the transport or collection of groundwater contaminants. The rate at which water flows through a soil specimen of cross-sectional area A is a function of the hydraulic gradient and the size of the interconnected void spaces between soil grains. This flow rate is expressed in Darcy's Law as

$$q = kiA,$$

where

- q = seepage quantity (l^3/t)
- k = coefficient of permeability (l/t)
- i = hydraulic gradient (l/l)

There are two types of laboratory permeability tests: constant-head tests and falling-head tests. Constant-head tests are a direct application of Darcy's Law, but for soils with a permeability of less than 10×10^{-4} cm/sec, they are too slow to be practical or accurate. For soils with permeabilities below this, the falling-head test is used.

A.7.1 Falling-Head Test

This laboratory exercise will provide experience with permeability tests and illustrate the effect of particle size, particle gradation, and void ratio on permeability coefficients.

Equipment

1. Falling-head permeameter. (Refer to Figure A.7.1.)
2. Balance sensitive to 0.1 g.
3. Scale accurate to 1 mm.
4. Stop watch.
5. Calipers sensitive to 0.1 mm.
6. Vacuum source.

Procedure

1. Obtain enough of the soil to be tested to fill the plastic cylinder of the permeameter three times. The gradation of this material should be known.
2. Measure the inside diameter of the standpipe cylinder (or burette). Also measure the inside diameter and length of the plastic cylinder that will be filled with sand. This data can be recorded on the data sheet shown in Figure A.7.2.

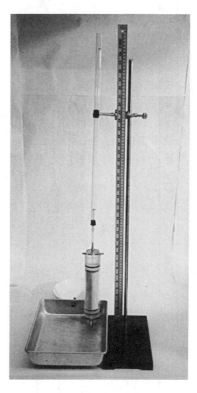

Figure A.7.1 The falling-head permeameter.

3. Assemble the base, porous stone, and cylinder of the permeameter.

4. Fill the cylinder with the sandy soil. Record the weight of soil W_s required to fill the permeameter cylinder and calculate volume of the sample V_t and the void ratio of the sample. The void ratios of the successive samples should be varied in order to demonstrate the influence of pore size on the permeability of the soil. Loose samples can be fabricated by carefully pouring the sand into the mold with a funnel or tube. Care should be taken to minimize sample disturbance during subsequent laboratory procedures prior to testing. Lower void ratios can be attained by either (1) vibrating the sample after filling the cylinder or (2) fabricating the sample in lifts and tamping each lift before adding the next layer.

5. Insert the top porous stone and the cover, and then secure the assembled cylinder with the spring.

6. Saturate the specimen. To accomplish this, first close the permeameter outlet, stand the plastic cylinder in a water reservoir and attach the standpipe to the top of the permeameter. Apply a vacuum to the top of the standpipe, and finally open the permeameter outlet and let water slowly enter the permeameter from the bottom until the system is saturated and the water level is near the top of the standpipe. Be careful not to let water enter the vacuum source.

7. Close the outlet of the permeameter. Disconnect the vacuum line from the top of the standpipe.

8. Using tape or a grease pencil, mark off the upper and lower elevations on the standpipe.

FALLING-HEAD PERMEABILITY TEST

TEST FOR ______________________________

DATE ______________ TEST BY ______________

SAMPLE NO. AND DESCRIPTION ______________________________

SAMPLE DATA

D_{10} __________ (mm)

G_s __________

W_s __________ (g)

L_s __________ (cm)

D_s __________ (cm)

A_s __________ (cm^2)

V_t __________ (cm^3)

e __________

$$e = \frac{G_s V_t \gamma_w}{W_s} - 1$$

TEST DATA

TEST NO.	h_1 (cm)	h_2 (cm)	TIME (sec)	k (cm/s)

$$k = 2.303 \frac{aL}{At} \log \left[\frac{h_1}{h_2}\right] \quad \text{(cm/s)}$$

SCHEMATIC OF APPARATUS

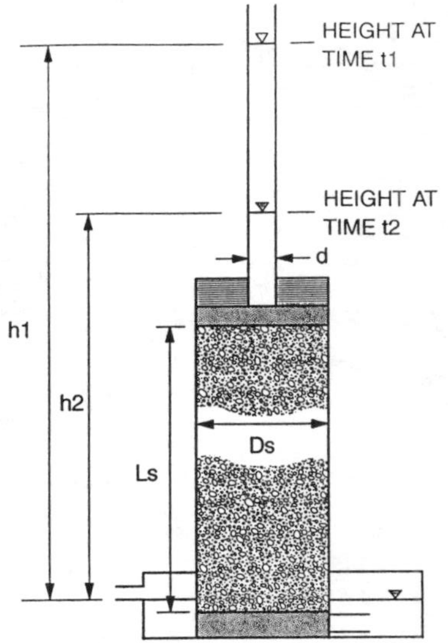

Figure A.7.2 Falling-head permeability test form.

9. Measure the elevations between the marks on the standpipe and the overflow water level. These elevations are used to determine the hydraulic gradients at the beginning and end of the test.

10. Perform the test by opening the permeameter outlet and measuring the time for the water level in the standpipe to fall from the upper mark to the lower. The value of the time interval is used in conjunction with the initial and final water elevations to compute the coefficient of permeability. (Refer to Figure A.7.2.)

11. Close the outlet before the water level reaches the bottom of the standpipe. Refill the standpipe and repeat the test until three consistent values for the coefficient of permeability are obtained.

Run additional tests as desired to demonstrate the effects of void ratio (vary the density of the test sample), particle size, and gradation (select alternate sand samples).

A.7.2 Constant-Head Test

Equipment

1. Permeameter cylinder. A schematic of the test apparatus both for this test and the falling-head test is shown in Figure A.7.3.

2. Perforated metal, or plastic disks, and circular wire screens, 35 to 100 mesh, cut for a close fit inside the permeameter.

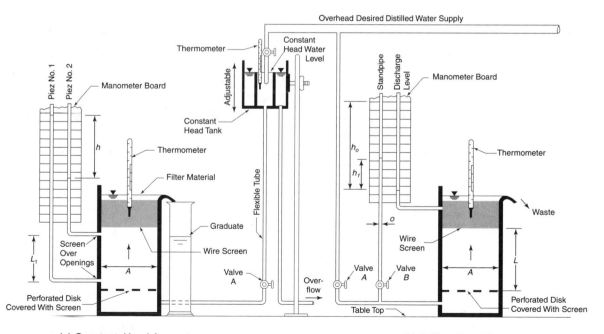

(a) Constant–Head Apparatus

(b) Falling–Head Apparatus

Figure A.7.3 Schematic diagram of constant-head and falling-head permeability apparatus. (From EM 1110-2-1906)

3. Glass tubing, rubber or plastic tubing, stoppers, screw clamps, and other hardware necessary to make connections.

4. Filter materials such as Ottawa sand, coarse sand, and gravel of various gradations.

5. Device for maintaining a constant-head water supply.

6. Deaired distilled water.

7. Manometer board.

8. Timing device.

9. Graduated cylinder, 100 ml capacity.

10. Centigrade thermometer, range of 0° to 50°C.

11. Balance.

12. Oven.

Procedure

1. Record all identifying information for the specimen on a data sheet such as is shown in Figure A.7.4.

2. Oven dry the specimen.

3. Place a wire screen, with openings small enough to retain the specimen, over a perforated disk near the bottom of the permeameter above the inlet.

4. Allow deaired distilled water to enter the water inlet of the permeameter to a height of about 1/2 in. above the bottom of the screen.

5. Mix the soil specimen thoroughly and place in the permeameter. The material should be dropped just at the water surface, keeping the water surface about 1/2 in. above the top of the soil during placement.

6. Weight the remaining soil in the container along with the container. The difference between this weight and the weight after oven drying is the weight of the specimen in the permeameter.

7. Level the top of the specimen, cover with a wire screen, and fill the remainder of the permeameter with a filter material.

8. Measure the length of the specimen and inside diameter of the permeameter to the nearest 1 mm, and record these data on the data sheet.

9. Test the specimen at the estimated natural void ratio and at a series of different void ratios. The void ratio can be varied by tapping the cylinder and causing compaction with the specimen.

10. Measure the distance between the piezometer taps.

11. Adjust the height of the constant head tank to obtain the desired hydraulic gradient. The gradient should be selected for laminar flow.

12. Open valve *A* (Figure A.7.3), and record the initial piezometer readings.

13. After allowing equilibrium conditions to be reached, measure the quantity of discharge for a given time interval. Record the quantity of flow, piezometer readings, water temperature, and time interval of the flow on the data sheet.

14. Repeat the previous step several times over a period of about one hour, and compute the coefficient of permeability corresponding to each set of measured data by using Darcy's Law.

CONSTANT–HEAD PERMEABILITY TEST

DATE _____

PROJECT _____

BORING NO. _____

Sample or Specimen No.

W_t, in grams							
Tare plus dry soil			Diameter of specimen, cm		D		
Tare			Area of specimen, sq cm		A		
Dry soil	W_s		Initial height of specimen, cm		L		
Specific gravity	G		Initial vol of spe. cc = AL		V		
Void of solids, cc = W_s + G	V_s		Initial void ratio = $(V - V_s) \div V_s$		e		

Distance between piezometer taps, cm		L_1		

Test No.		1	2	3
Height of specimen, cm	L			
Void ratio = $(AL - V_s) + V_s$	e			

		1a	1b	2a	2b	3a	3b
Reading of piez 1, cm	h_1						
Reading of piez 2, cm	h_2						
Head loss, cm = $h_1 - h_2$	h						
Quantity of flow, cc	Q						
Elapsed time, sec	t						
Water temperature, °C	T						
Viscosity correction factor [1]	R_T						
Coefficient of permeability, [2] cm/sec	k_{20}						
	Avg						

(1) Correction factor for viscosity of water at 20°C.

(2) $k_{20} = \dfrac{Q \times L \times R_T}{h \times A \times t}$.

where L = height of specimen or distance between piezometer taps if used.

Remarks _____

_____

Technician _____ Computed by _____ Checked by _____

Figure A.7.4 Constant-head permeability test. (From EM 1110-2-1906)

A.8 DIRECT-SHEAR TEST ON SAND

Purpose. The peak- and residual-shear stresses that can be mobilized by a soil are requisite parameters for the analysis of engineering projects, involving bearing capacity, earth retention and support, and slope stability. Common laboratory techniques for determining the shear resistance of soils include the direct-shear test and triaxial methods. The direct-shear test is often used to determine (1) the angle of internal friction ϕ of cohesionless soils and (2) the residual-shear strength of cohesive soils. This laboratory exercise is intended to demonstrate the effects of relative density, grain size, particle shape, and soil gradation on the angle of friction of a sandy soil as determined by the direct-shear method.

Reference Standard. ASTM D3080-72—Direct Shear Test of Soils Under Consolidated Drained Conditions.
EM 1110-2-1906, Appendix IX: Drained (*S*) Direct-Shear Test

Equipment

1. Direct-shear apparatus. (Refer to Figures A.8.1 and A.8.2.)
2. No. 4 sieve.
3. Balance sensitive to 0.1 g.
4. Spoon.
5. Tamper for compacting sand in the direct-shear box.
6. Large evaporating dish.
7. Calipers.

Procedure

1. Familiarize yourself with components of the shear box. Refer to Figure A.8.2 for photographs of two unassembled shear boxes. The boxes consist of a base plate and mold, the top half of the shear box with pins for lateral restraint, and the top cap loading block.

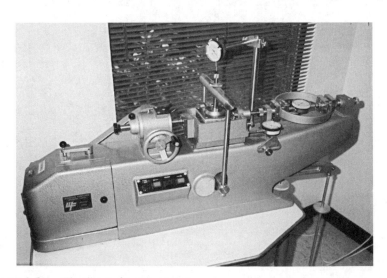

Figure A.8.1 The direct-shear apparatus.

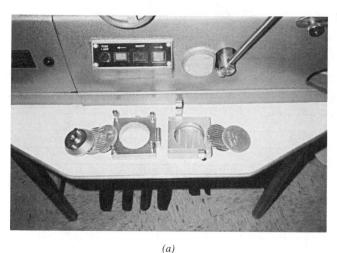

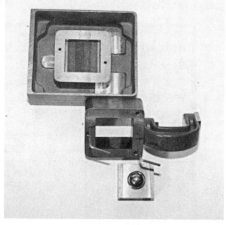

(a)   (b)

Figure A.8.2   Components of the direct-shear box: (*a*) Round sample. (*b*) Square sample.

2. Remove the shear-box assembly from the loading frame. Notice the screws on the top half of the shear box. The screws oriented vertically are used to separate the top and bottom halves of the shear box prior to testing. Back off these screws until they have been fully retracted within the top half of the shear box. The horizontally aligned screws are used to clamp the loading block to the upper half of the shear box. Back off these screws until the loading block can be removed.

3. Insert the pins into the vertical holes machined into both halves of the shear box. These pins prevent lateral displacements between upper and lower halves of the shear box during sample preparation.

4. Pass the soil to be tested through the No. 4 sieve, retain the minus-No. 4 material for testing.

5. Measure and record the interior dimensions of the shear box (length and width, or diameter and height) and the height of the loading block. Data can be recorded on the sample-data form (Figure A.8.3).

6. Fill a container of known weight with the prepared soil. Weigh the container plus the soil. Pour the sand into the shear box by using the following techniques for the preparation of either loose or dense samples. If the soil is to be tested at loose relative densities, the material should be poured into the shear box in one lift by using a funnel or tube (similar to the methods outlined in Section A.6 for determining the minimum-index density of a cohesionless soil). Dense samples can be prepared by placing the sand into the box in thin lifts and tamping or by adding soil in thin lifts and vibrating the sample with lateral blows with a rubber hammer. For the latter technique, the loading block should be inserted into the shear box and secured by moderate finger pressure during the hammer impact. After the sample has been fabricated, weigh the container and remaining sand in order to determine the weight of the specimen.

7. After leveling the surface of the sample, carefully place the loading block on the specimen. Determine the height of the sample by measuring the vertical distance between the top of

DIRECT-SHEAR TEST ON SAND

DESCRIPTION OF SOIL _____

SAMPLE NO. _____ LOCATION _____

SIZE OF SPECIMEN _____

WEIGHT OF SPECIMEN, W _____ DRY UNIT WEIGHT _____

G_s _____ VOID RATIO, e _____ NORMAL LOAD, N _____

NORMAL STRESS, σ' _____

PROVING RING CALIBRATION FACTOR _____

Horizontal Displacement (in.) (1)	Vertical Displacement (in.) (2)	No. of Div. in Proving Ring Dial Gage (3)	Shear Force, S (lb) (4)	Shear Stress, τ (lb/in.^{2}) (5)

Figure A.8.3 Example data sheet for direct-shear test.

the loading block and the top of the shear box. Once the sample height is known, the volume, unit weight, and void ratio of the specimen can be obtained.

8. Place the shear-box assembly in the direct-shear machine as shown in Figure A.8.1.

9. The sample should be tested at a range of normal stresses that correspond to the stresses anticipated in the field. Convert these estimated stresses to vertical loads based on the area of the sample on the horizontal plane. Add the appropriate load.

10. Advance the vertically oriented screws in the top half of the shear box until the upper box is elevated approximately one grain diameter above the lower half of the shear box.

11. Clamp the loading block to the upper half of the shear box by advancing the horizontally aligned screws. Back off the vertical screws so that there is now no contact between the two halves of the shear box.

12. Attach the horizontal and vertical dial gages to the shear box. During the test, measurements will be made of the horizontal and vertical displacements.

13. Remove the vertical pins from the shear box.

14. The application of the horizontal load is provided by a mechanical drive that displaces the upper half of the shear box at a constant rate. This rate should be between 0.02 and 0.1 in./min. The corresponding load is measured with a proving ring clamped between the moving-load piston and the upper half of the shear box. The deflection of the proving ring is indicated on a dial gage and converted to load. Apply the horizontal load and at 15-second intervals, record (1) the horizontal deflection, (2) vertical deformation, and (3) deflection of the proving ring, which will be used to calculate the horizontal load being sustained by the soil.

15. The load versus deformation behavior of the sample will depend on both the initial density of the sample and the normal stress at the beginning of the test. Dense samples under relatively low-confining stresses will tend to mobilize (resist) a peak horizontal load followed by a potentially significant reduction to a residual value in the horizontal load. Conversely, loose samples generally exhibit a gradual increase in the sustained horizontal load that will not appear to taper off. Continue to take readings until either (1) a maximum horizontal load has been reached and it remains constant for at least five readings or (2) the horizontal load increases to a maximum, then gradually decreases to a stable minimum value. At this load level, the residual strength of the soil is being mobilized.

16. The test data should be presented in the following plots: (1) shear stress against horizontal displacement, (2) vertical displacement against horizontal displacement, and (3) peak stress at failure against normal stress.

17. Repeating Steps 2 through 14 on samples at the same void ratio, but at two different normal stresses, will facilitate the construction of the Mohr's failure envelope for the soil.

Additional insights on the strength and stress–strain behavior of soils can be obtained by varying the sample composition (particle size and shape) to determine effects of these variables on angle of internal friction. Vary density of samples to determine its effect.

A.9 UNCONFINED COMPRESSION TEST FOR COHESIVE SOILS

Purpose.   The shear strength of a cohesive soil is primarily influenced by its void ratio, degree of saturation, and stress history. In most construction projects, saturated clay strata are subjected to rapid enough loading that the excess pore pressures which are induced do not have sufficient time to dissipate. This case is termed *undrained loading*, and the strength that is mobilized by the clay is called the *undrained shear strength (c or s)*.

The unconfined compression test is a quick, relatively simple method of determining the compressive strength for a clayey soil. An axial load is applied to a cylindrical sample of the clay and the load is gradually increased until the sample fails in shear. The axial stress at which failure occurs is the unconfined compressive strength (q_u). The unconfined compression test is a special form of the triaxial compression test in that there is no confining pressure on the sample $(p_3 = 0)$. Based on the strength concepts presented in Chapter 5, it is apparent that the unconfined compressive strength of a cohesive soil and the undrained shear strength are simply related by Equation (5.30).

The purpose of this laboratory exercise is to provide experience with strength testing, to demonstrate the stress–strain behavior and compressive strength of either partially saturated or fully saturated cohesive soils, and to illustrate how the strength of cohesive soil varies with changing water contents.

Reference Standard.   ASTM D2166-85—Compressive Strength, Unconfined, of Cohesive Soil. EM 1110-2-1906, Appendix XI: Unconfined Compression Test.

Equipment

1. Unconfined compression-testing device (complete with dial gages sensitive to 0.001 in.). Refer to Figure A.9.1.
2. Specimen trimmer (soil lathe for field samples from a sampling tube).
3. Harvard miniature compaction device and accessories (for samples compacted in the laboratory).
4. Wire saw.
5. Balance sensitive to 0.01 g.
6. Soil-drying oven.
7. Evaporating dish.

Figure A.9.1   A portable, hand-operated unconfined-compression testing machine.

8. Sample drying containers.

9. Stopwatch or timer.

10. Protractor.

11. Calipers.

Procedure

1. Familiarize yourself with the unconfined compression-test apparatus. The most common form consists of a load frame to which a proving ring and load platen are attached to the upper portion of the frame and an electric-powered motor that drives the bottom load platen is fixed to the base of the load frame. Tests are usually run at a predetermined, constant rate of axial strain (1.0 percent/min).

2. Sample height to diameter ratio should be close to 2.5. For standard laboratory equipment D = 1.4 in. and H = 3.5 in. The preparation of a specimen will depend on whether the sample to be tested is an undisturbed field sample or a specimen compacted in the laboratory.

Field-Sample Preparation

3. After extruding a portion of the field sample from the sampling tube, cut a sample approximately 5 in. long and 2 to 2.5 in. in diameter.

4. Rest the sample on its side in an appropriate sample holder (usually U-shaped to cradle the specimen) and trim the ends of the sample with a wire saw. The ends should be parallel to one another and perpendicular to the long axis of the sample. Retain the trimmings from each end of the sample for two water-content determinations.

5. Carefully place the sample upright in the soil lathe. Trim the sides of the sample by using either the wire saw or trimming knife. (Continue by following Steps 8 through 17 described next.)

Compacted-Sample Preparation

6. Extrude the compacted sample from the compaction mold (e.g., Harvard miniature compaction mold, standard-compaction mold). Sides should be smooth with no large voids.

7. Place the sample on its side in the sample holder and trim the ends of the sample with a wire saw. The ends should be parallel to one another and perpendicular to the long axis of the sample. Retain the trimmings from each end of the sample for two water-content determinations.

Testing Field or Compacted Samples

8. Once the specimen has been trimmed, measure the height, diameter (at two or three positions along the length), and weight of the sample. This procedure should be performed quickly, yet carefully, to minimize drying. Record pertinent data on the form in Figure A.9.2.

9. Place the trimmed sample at the center of the lower load platen. Move the load platen until the top platen just makes contact with the soil specimen, as shown in Figure A.9.1.

UNCONFINED-COMPRESSION TEST DATA

TEST FOR _____
TEST BY _____ DATE _____
SAMPLE IDENTIFICATION _____
PROVING RING NO. AND CONSTANT _____
SPECIMEN WEIGHT _____ SPECIMEN HEIGHT, L _____
SPECIMEN DIAMETER
    TOP _____
    CENTER _____
    BOTTOM _____
    AVERAGE _____
AVERAGE SPECIMEN AREA, A_o _____
AVERAGE LOADING RATE _____
SPECIMEN PROTECTION _____

STRAIN DIAL, ΔL (IN.)	LOAD DIAL (IN.×10⁴)	LOAD, P (LB)	AXIAL STRAIN, ϵ	AREA, A $A_o/(1-\epsilon)$ (IN.²)	P/A (LB/IN.²)	WATER CONTENT DETERMINATION
						Can no. ____
						Wt. wet ____
						Wt. dry ____
						Wt. water ____
						Tare ____
						Wt. solids ____
						w% ____
						Sketch of Failure

REMARKS:

Figure A.9.2 Unconfined-compression test data sheet.

10. Zero deformation and proving-ring dial gages.

11. Turn the machine on to begin the compression testing at an axial strain rate of 1 percent/min. Take readings of both dial gages at every 0.01 in. of sample deformation.

12. Once the compression load begins to taper off (as indicated by the proving-ring reading), the readings can be taken at 0.02-in. increments. Continue the readings until a uniform, minimum value of compression load has been reached or the sample has been subjected to 20-percent axial strain, whichever occurs first.

13. Use the protractor to measure the inclination of any failure plane in the sample, from the horizontal. Unload the sample by lowering the bottom platen.

14. Sketch the final shape of the specimen, noting the mode of failure (e.g., "bulging type," well-defined failure plane, or intermediate mode) and any discontinuities or unconformities that may exist in the sample.

15. Place the sample in an evaporating dish and use an oven to obtain the moisture content.

16. The data may be tabulated as shown in Figure A.9.2 and plotted in the form of axial stress against axial strain. Note that the axial stress must be corrected to account for bulging of the sample and, hence, increasing area of the sample during the test. This correction is provided:

$$\epsilon = \frac{\Delta L}{L}$$

$$p_1 = \frac{P}{A} = \frac{P(1 - \epsilon)}{A_o}$$

The unconfined-compression strength corresponds to the value of peak axial stress.

17. If field samples are being tested and additional material is available, repeat the test on one or two more samples at the same moisture content to demonstrate the variation in compressive strengths of similar samples. If compacted samples are being evaluated, prepare several additional specimens each at different water contents to illustrate how the strength of these samples varies with water content.

A.10 TRIAXIAL-COMPRESSION TEST ON SAND

Purpose. Various aspects of testing and interpretation of shear versus deformation data for soil have been discussed in laboratory demonstrations A.8 and A.9. In this exercise, a cohesionless soil will be tested under triaxial compression. A vacuum (negative pore pressure) applied to the sample provides the confining stress that gives the sand its stiffness and strength. This test is commonly referred to as the *Vacuum Triaxial Test.*

The triaxial test offers several advantages over the direct-shear test when evaluating the stress–strain behavior of soils. The triaxial test provides more uniform stress distribution in the specimen, sample volume changes can be measured much more accurately, and the state of stress is known throughout the duration of the test. This laboratory exercise is intended to demonstrate the effects of void ratio, grain size, particle shape, and distribution on the angle of friction and stress versus strain behavior of a sandy soil.

Figure A.10.1   Equipment required for the vacuum triaxial test.

Reference Standard: EM 1110-2-1906, Appendix X: Triaxial Compression Tests

Equipment (Refer to Figure A.10.1)

1. Triaxial compression test device.
2. Specimen preparation equipment (i.e., base plate, sample mold, rubber membrane, rubber O-rings, vacuum pump, and regulator).
3. Coarse to fine sand.
4. Balance sensitive to 0.1 g.
5. Spoon or funnel for preparing the sample.
6. Tamper for compacting sand in the sample mold.
7. Vacuum source.
8. Calipers.
9. Small level.
10. Timer.

Procedure

1. Familiarize yourself with the vacuum triaxial apparatus and associated equipment shown in Figure A.10.1.
2. Attach a vacuum line to the base plate/pedestal assembly of the triaxial apparatus.
3. Bind the rubber membrane to the sample pedestal with a rubber O-ring.
4. Being careful not to pinch the membrane, clamp the sample mold around the pedestal and membrane. Pull the membrane up, stretching it slightly, and fold the membrane over the top of the mold. This procedure should be performed in a manner that minimizes twisting or creasing of the membrane.
5. Determine the volume of the sample mold. Based on this volume and the desired void ratio of the specimen, estimate the mass of dry sand required to fabricate the sample. Note that

the top of the sample must be below the top of the mold to facilitate the placement of the top cap (and possibly a porous stone).

6. Weigh to 0.1 g the required quantity of sand. Place the sand in the mold by using either a spoon or funnel. The sample can be constructed in either one lift and vibrated to the desired volume or placed in thin lifts and lightly tamped. The choice of the appropriate method will depend on the desired void ratio. A sample-data sheet for recording sample data as well as test data is provided in Figure A.10.2.

7. Place a dry porous stone (if necessary) and top cap on the specimen. Use the bubble level to attain a horizontal upper surface.

8. Roll the membrane over the top cap and bind the membrane to the cap with a rubber O-ring.

9. Apply a vacuum to the sample. The magnitude of the vacuum (confining stress) should closely approximate the anticipated confining stress in the field. Recall that 1 in. Hg = 0.5 psi.

10. With the specimen under the imposed vacuum, remove the sample mold from the specimen. Measure the height and diameter (or circumference) of the specimen to 0.1 mm. Calculate the actual void ratio of the sample prior to testing.

11. Carefully place the specimen under the loading frame. Adjust and zero the proving-ring and axial-deformation dial gages.

12. Set the loading mechanism so that the sample will be deformed at an axial strain rate of 1 percent/min.

13. Begin the test, taking dial gage readings every 15 seconds for the first two minutes of the test and then every 30 seconds until the failure criteria are achieved.

14. Continue the test until (1) the sample has reached a peak compressive force (as indicated by the proving-ring deflection) and the values of compressive force have begun to diminish or (2) an axial strain of 15 percent is reached.

15. Once one of the failure criteria are met, stop the test and sketch the deformed shape of the specimen. Note the angle made between the failure plane (if one is evident) and the horizontal plane.

16. Remove the base plate and soil sample from the loading frame.

17. Disconnect the vacuum line and release the vacuum. Being careful not to puncture the membrane, remove the O-ring from the top cap and pour out the sand.

18. Repeating Steps 3 through 17 on samples at the same void ratio, but tested at two different normal stresses will facilitate the construction of the Mohr's failure envelope for the soil.

19. The test data should be presented in the following plots: (1) shear stress against axial strain and (2) Mohr's circle at failure on a plot of shear stress against normal stress.

A.11 CONSOLIDATION TEST

Purpose. Consolidation of a saturated soil is the result of a decrease in the volume of the void spaces in the soil. This reduction in void volume is associated with the gradual transfer of an applied load from pore water to the soil skeleton as excess pore pressures are dissipated. The amount of water that is expelled from the pore spaces depends on the magnitude of the load and on the compressibility of the soil skeleton. As the flow of water through a soil is governed by

TRIAXIAL-COMPRESSION TEST DATA

TEST FOR _____

TEST BY _____   DATE _____

SAMPLE IDENTIFICATION _____

PROVING RING NO. AND CONSTANT _____

SPECIMEN HEIGHT, L _____   SPECIMEN WEIGHT _____

SPECIMEN DIAMETER   SPECIFIC GRAVITY, G_s _____

    TOP _____   VOLUME OF SOLIDS _____

    CENTER _____   VOID RATIO _____

    BOTTOM _____

    AVERAGE _____

AVERAGE SPECIMEN AREA, A_o _____

SPECIMEN VOLUME _____   AVERAGE LOADING RATE _____

STRAIN DIAL, ΔL (IN.)	LOAD DIAL (IN.$\times 10^4$)	LOAD, P (LB)	AXIAL STRAIN, ϵ	AREA, A $A_o/(1-\epsilon)$ (IN.^{2})	P/A (LB/IN.^{2})	CONFINING PRESSURE (σ_3)
						in. Hg _____ psi _____
						Sketch of Failure

REMARKS:

Figure A.10.2   Triaxial-compression test data sheet.

the hydraulic conductivity of the material, consolidation is a time-dependent process. Consolidation results in a soil surface settlement.

Like settlement of fine-grained soils, shear strength depends on effective confining stresses. Therefore, in order to determine the strength of a soil—the intergranular pressure—the degree to which consolidation may have progressed must be known. The time-dependent change in void ratio, and shear strength, of soils undergoing consolidation is often accounted for in projects that involve staged construction. These projects take advantage of the fact that the shear strength of the soil is increasing with time.

With regard to consolidation settlements of fine-grained soils, the geotechnical consultant is commonly asked by the client to evaluate both (1) the magnitude of the ground surface settlements and (2) the pattern of settlement versus time. This laboratory exercise is intended to demonstrate the procedure for a one-dimensional consolidation test and the data reduction methods for determining the void-ratio–vertical-effective stress curve (e versus log p'), the preconsolidation pressure, p'_p, and the coefficient of consolidation, c_v.

Reference Standard. ASTM D2435-80—Standard Test Method for One-Dimensional Consolidation Properties of Soils.
EM 1110-2-1906, Appendix VIII: Consolidation Test.

Equipment (Refer to Figures A.11.1 through A.11.3)

1. Consolidation apparatus.
2. Sample extruder.

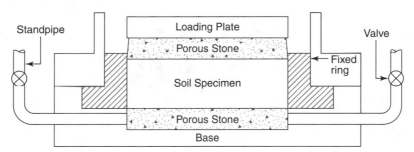

(a) Fixed-Ring Consolidometer

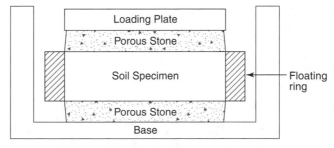

(b) Floating-Ring Consolidometer

Figure A.11.1 Schematic diagrams of fixed-ring and floating-ring consolidometers (from the U.S. Army Corps of Engineers EM 1110-2-1906).

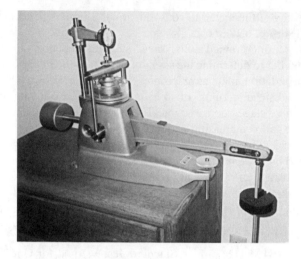

Figure A.11.2   Lever arm consolidation test loading device.

Figure A.11.3   Equipment for trimming soil specimens.

3. Specimen trimming device (vertical soil lathe).
4. Wire saw.
5. Balance sensitive to 0.1 g.
6. Timer.
7. Moisture-content cans.
8. Oven.
9. Distilled water.
10. Filter papers.

Apparatus: The consolidometer consists of a rigid base, a consolidation ring, porous stones, a rigid loading plate, and a support for a dial indicator. Two types of consolidometers are commonly used in practice: the fixed-ring and floating-ring consolidometers. These are shown in

Figure A.11.1. In addition to the different consolidometers, various types of loading devices have been developed for applying the load to the specimen. The beam and weight (lever arm) loading frame is shown in Figure A.11.2.

Procedure

1. Prepare the consolidation accessories prior to handling of the soil samples.

    (a) Determine the height, diameter, and weight of the consolidation ring.

    (b) Saturate two porous stones by either of the following methods: (1) placing them in water and boiling for 10 minutes or (2) subjecting the submerged stones to a vacuum.

    (c) Cut out of filter paper two circles with diameters slightly less than the diameter of the consolidation ring.

2. Place a saturated porous stone in the base of the consolidometer and fill the consolidometer with distilled water to a height equal with the top of the stone. Place the filter paper on the porous stone.

3. Specimen preparation is initiated either by (1) extruding the soil from the thin-walled sampling tube or by (2) obtaining an oversized sample from a block sample. It is preferable to perform all specimen preparation procedures in a humid room to prevent the evaporation of soil moisture. In addition, care should be taken to minimize sample disturbance during the trimming procedure.

4. Using a wire saw, trim the sample into a cylindrical shape with a diameter about one-half in. greater than the inside diameter of the specimen ring. Carefully trim the lower edge of the specimen until the bottom will fit exactly into the specimen ring. Two types of specimen trimming equipment are shown in Figure A.11.3.

5. Place the specimen ring on the rotating wheel (soil lathe) and the specimen on top of the consolidation ring. Begin to trim the bottom portion of the specimen, applying a light downward pressure on the sample during the trimming operation. The sample should fit snugly in the consolidation ring.

6. Retain enough soil from both the top and bottom of the specimen to determine the water content of the sample. The void ratio can be determined with the aid of Table 3.1.

7. Cut off the portion of the specimen remaining above the ring with a wire saw. Carefully trim the top and bottom surfaces so that they are flush with the ends of the consolidation ring. If an obstruction (e.g., pebble, shell) is encountered in the surface, remove it and fill the void with sample trimmings. Weigh the consolidation ring and soil specimen.

8. Position the soil specimen in the ring over the lower porous stone.

9. On top of the specimen, place a moist filter paper and the upper porous stone.

10. Place the top cap on the porous stone and place the consolidometer in the loading device.

11. Add water to the consolidometer to submerge the sample. Check this water level throughout the test to ensure that the specimen remains saturated.

12. Attach the vertical deflection dial gage to permit almost the maximum travel in the downward direction. The dial gage should be attached such that it can also allow measurement of any swelling. Zero the dial gage or record its reading at the initial height of the sample.

13. Adjust the loading device until it just makes contact with the specimen. The seating load should not exceed about 0.01 ton per sq ft (tsf).

CONSOLIDATION TEST DATA

PROJECT NAME _____   PROJECT NUMBER _____

TEST BY _____   DATE _____

SAMPLE DESCRIPTION _____

Date	Load	Time			Dial Reading (in. × 10⁻⁴)	Date	Load	Time			Dial Reading (in. × 10⁻⁴)
		Clock	Elapsed (min)	$\sqrt{t}$ (min)$^{0.5}$				Clock	Elapsed (min)	$\sqrt{t}$ (min)$^{0.5}$	

Figure A.11.4   Form for recording time versus vertical-dial reading data.

14. Apply load to the specimen such that the applied pressure is 0.25 tsf. Observe and record on the data sheet (Figure A.11.4) the deformation as determined from the dial-gage readings at the following times, t, counted from the time of the load application: 0, 0.1, 0.2, 0.5, 1.0, 2, 4, 8, 15, 30, 60, 120, 240, 480, and 1440 min. (24 hr.). In order to observe trends in the consolidation behavior of the specimen and as a check on the dial gage readings, it is advisable also to plot the deflection-versus-time data as they are obtained. Allow each load increment to remain on the specimen for a minimum of 24 hours to ensure that the primary consolidation is completed.

CONSOLIDATION TEST

Description of soil _____ Location _____
Specimen diameter _____   Initial specimen height, $H_{t(i)}$ _____
Moisture content: Beginning of test _____ (%) End of test _____ %
Weight of dry soil specimen _____ G_s _____ Height of solids, H_s _____ cm = _____

Pressure, p (ton/ft^2)	Final dial reading (in.)	Change in specimen height (in.)	Final specimen height (in.)	Height of void (in.)	Final void ratio, e	Average height during consolidation (in.)	Fitting time (sec)		c_v from $\times 10^3$ (in.^{2}/sec)	
							t_{90}	t_{50}	t_{90}	t_{50}

Figure A.11.5 Void-ratio pressure and coefficient of consolidation calculation.

15. Continue consolidation of the specimen by applying the next load increment. The following loading schedule is considered satisfactory for routine tests: 0.25, 0.5, 1.0, 2.0, 4.0, 8.0, 16.0, and 32.0 tsf.

16. After the specimen has consolidated under the maximum stress, remove the load in decrements, taking three-fourths of the load off successively for each of the first two decrements and as desired thereafter.

17. At the end of the test, remove the soil specimen from the consolidometer, remove the filter papers, and determine the moisture content of the specimen. The data reduction is performed by using the data sheet presented in Figure A.11.5.

18. From the data collected during each load increment, prepare plots of deformation, or dial reading, against the logarithm of time ($\log 10t$) and deformation against the square root of time $\sqrt{t}$. Examples of these plots are provided in Figures 5.10 and 5.11.

19. By using these plots, the *coefficient of consolidation*, c_v, can be calculated by two different methods. The first technique is illustrated in Figure 5.10 and involves the determination of the time required to reach 50-percent primary consolidation, $t50$, from the dial reading versus $\log t$ plot. Recall from Table 5.2 that a degree of consolidation (U) equal to 50 percent corresponds to a time factor, T, of approximately 0.2. The coefficient of consolidation is calculated from Equation 5.22.

    The value of c_v can also be determined from the deformation versus $\sqrt{t}$ plot (Figure 5.11). In this procedure, the time required for 90 percent consolidation is obtained, and used, in conjunction with a time factor of 0.8 to calculate c_v.

20. Based on the initial height and void ratio of the specimen as well as the deflection data obtained at the end of each load increment, determine the height and void ratio at the end of consolidation during each loading. These computations can be made on the data sheet in Figure A.11.5.

21. On Figure A.11.6, plot the final void ratio against the logarithm of pressure. Determine the preconsolidation pressure, p_c', of the sample by the approximate method that is illustrated in Figure 5.9. Also obtain the compression index, C_c, and recompression index, C_r, from the slopes of the virgin compression curve and unloading curve, respectively.

22. Plot the values of c_v, which are determined by both the deformation against $\log t$ and deformation against $\sqrt{t}$ methods in Figure A.11.6. Note that the value of c is greatly influenced by the stress history of the sample.

CONSOLIDATION TEST

	Water Content,%	Total Unit Weight, pcf	Void Ratio	Saturation, %	Height, inches	Diameter, inches	Specific Gravity	Liquid Limit, %	Plastic Limit,%
Boring No:			Sample No:				Depth, ft:		
Material:									
Initial									
Final									

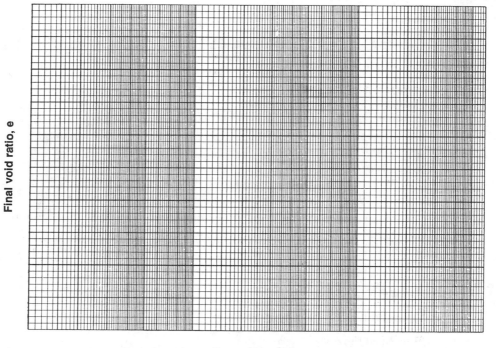

Final void ratio, e

Pressure, p (tsf)

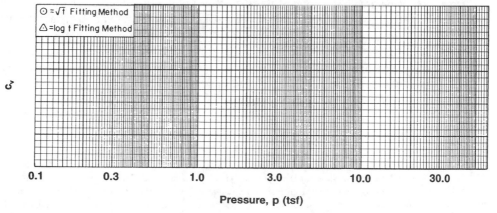

⊙ = √t̄ Fitting Method

△ = log t Fitting Method

c_v

Pressure, p (tsf)

0.1   0.3   1.0   3.0   10.0   30.0

Figure A.11.6   Consolidation test results: void ratio versus log pressure and coefficient of consolidation versus log pressure.

APPENDIX B

Pile Hammer Specifications

The following specifications are ©Copyright 2001 Pile Buck®, Inc., P. O. Box 64-3929, Vero Beach, FL, 32964, www.pilebuck.com and are reproduced with permission.[1]

B.1 AIR/STEAM HAMMERS

Air/Steam Hammer Terminology

- Rated Energy

   The rated energy of an impact hammer is the maximum energy potential per blow. For single-acting hammers, it is the product of the stroke length and ram weight. For hammers that assist the ram during its downward travel, the rated energy includes the effects of the force exerted by the motive-fluid pressure in addition to the gravitational component. In cases of inclined pile driving, the stroke value should be adjusted to account for the angle of inclination (i.e., only the vertical component of the stroke should be used in establishing hammer-potential energy). The hammer energy at impact is a function of ram weight, stroke, and overall hammer condition (i.e., efficiency).

[1]The data and commentary contained within this pile hammer specifications chart is for general information purposes only. It is provided without warranty of any kind. Pile Buck®, Inc. shall not be held responsible for any errors, omissions, or misuse of any of the enclosed information and hereby disclaims any and all liability resulting from the ability or inability to use the information contained herein. Anyone making use of this material does so at his or her risk. In no event will Pile Buck®, Inc. be held liable for any damages including lost profits, lost savings, or other incidental or consequential damages arriving from the use or inability to use the information contained within. Pile Buck®, Inc. suggests contacting the manufacturer of any listed pile driving hammer to ensure its suitability for a particular application.

- Model

  This is the model designation given by the manufacturer to each hammer. Usually, it provides some description of the hammer (e.g., VULCAN 340 model has a 3-ft stroke and 40-kip ram weight).

- Manufacturer

  The name of the manufacturing company.

- Type

  Single-acting hammers mean that the motive fluid is only used to raise the ram to the desired stroke and downwards ram movement is done by gravity. For double-acting hammers, the motive fluid is not only used to raise the ram, but it also assists during the ram fall.

- Style

  Open-style hammers are those that have visible rams; closed-style hammers enclose the ram. Generally, open-style hammers cannot be used under water, while some closed-style hammers can.

- Blows per Minute (BPM)

  The rated number of hammer impacts per minute. Reported values are under specific operating conditions and may vary depending on the actual field conditions. In most cases, it is not possible to relate the hammer rate of operation to its potential energy.

- Weight of Striking Part

  Weight of the part of the hammer that actually impacts energy to the pile. This is commonly known as the "ram." Hammer rated energy and general effectiveness is a direct function of the weight of its striking part. In some cases, this weight is indicated as part of the hammer-model designation.

- Total Weight

  The total weight of the hammer. This value is important in sizing the crane, transportation requirements, and other aspects involving the hammer.

- Hammer Length

  Total length of hammer in its normal operating configuration, excluding any accessories that may be present between the hammer and the pile head.

- Jaw Dimensions

  Dimensions of the hammer guides that interface with the leads. Also shown whether "male" or "female" type. Necessary in selecting appropriate size and style leads.

- Inlet Pressure

  The pressure (at the inlet of the hammer) of the motive fluid needed to operate the hammer at full-rated energy. In cases where direct reading is not available, pressure readings measured at the boiler or compressor must be adjusted to arrive at the pressure value at the hammer inlet.

- Inlet Size

  The number and size of hoses supplying motive fluid to the hammer.

Air/Steam Hammers

Rated Energy (Ft.–Lbs.)	Model	Manufacturer	Type	Style	BPM	Striking Weight (Lbs.)	Total Weight (lbs)	Hammer Length (Ft–In)	Jaw Dimension (In.)	Inlet Pressure (PSI)	Inlet Size (In.)
1800000	6300	VULCAN	SGL-ACT	OPEN	42	300000	575000 w/o	30'0"	22" × 144"(M)	235	2@6"
1582220	MRBS 12500	MENCK	SGL-ACT	OPEN	36	275600	540100 w/o	71'9"	CAGE	171	2@6"
1200000	200E6	CONMACO	SGL-ACT	OPEN	40	200000	490000 w/o	35'6"	CAGE	155	2@6"
1200000	6200	VULCAN	SGL-ACT	OPEN	36	200000	438218 w/o	30'1/2"	22" × 10"(M)	110	4@4"
1050000	1750E6	CONMACO	SGL-ACT	OPEN	40	175000	465000 w/o	35'6"	14 1/2" × 120"	135	2@6"
954750	MRBS 8800	MENCK	SGL-ACT	OPEN	36	194000	348300 w/o	66'3"	CAGE	150	2@6"
900000	6150	VULCAN	SGL-ACT	OPEN	41	150000	275000 w/o	35'3"	18 1/2" × 88"(M)	175	4@4"
868000	MRBS 8000	MENCK	SGL-ACT	OPEN	36	176365	330695 w/o	71'8"	CAGE	142	2@6"
868000	MRBS 7000 SL	MENCK	SGL-ACT	OPEN	36	176365	330695 w/o	71'8"	CAGE	142	2@6"
759450	MRBS 6000	MENCK	SGL-ACT	OPEN	34	132300	218300 w/o	57'7"	CAGE	171	6"
750000	5150	VULCAN	SGL-ACT	OPEN	46	150000	275000 w/o	26'3 1/2"	22" × 120"(M)	175	2@6"
750000	1500E5	CONMACO	SGL-ACT	OPEN	42	150000	283000 w/o	30'6"	14 1/2" × 120"	135	2@6"
613955	MRBS 7000	MENCK	SGL-ACT	OPEN	40	154000	303600 w/o	60'3"	CAGE	142	4@4/2@6"
542470	MRBS 5000	MENCK	SGL-ACT	OPEN	40	110200	185200 w/o	57'7"	CAGE	150	6"
512960	MRBS 3900	MENCK	SGL-ACT	OPEN	36	86900	132300 w/o	51'3"	CAGE	171	6"
510000	850E6	CONMACO	SGL-ACT	OPEN	40	85000	173600 w/o	25'2"	18 3/4" × 100"	160	2@4"
500000	5100	VULCAN	SGL-ACT	OPEN	48	100000	219000 w/o	27'4"	22 × 120"(M)	150	3@4"
499070	MRBS 4600	MENCK	SGL-ACT	OPEN	42	101410	176371 w/o	57'7"	CAGE	142	6"
350000	700E5	CONMACO	SGL-ACT	OPEN	43	70000	152000 w/o	23'2"	18 3/4" × 100"	130	2@4"
325480	MRBS 3000	MENCK	SGL-ACT	OPEN	42	66100	106000 w/o	51'3"	CAGE	142	5"
325480	MRBS 2500 SL	MENCK	SGL-ACT	OPEN	42	66100	106000 w/o	51'3"	CAGE	142	5"
300000	3100	VULCAN	SGL-ACT	OPEN	60	100000	195500 w/o	23'3"	18 3/4" × 88"(M)	130	3@4"
300000	560	VULCAN	SGL-ACT	OPEN	47	62500	134060 w/o	23'0"	18 3/4" × 88"(M)	150	3@4"
262340	MRBS 2500 S	MENCK	SGL-ACT	OPEN	40	63800	105600 w/o	51'0"	CAGE	156	2@3"

Air/Steam Hammers

Rated Energy (Ft.-Lbs.)	Model	Manufacturer	Type	Style	BPM	Striking Weight (Lbs.)	Total Weight (lbs)	Hammer Length (Ft-In)	Jaw Dimension (In.)	Inlet Pressure (PSI)	Inlet Size (In.)
226100	MRBS 2500/4	MENCK	SGL-ACT	OPEN	50	55000	96800 w/o	45'5"	CAGE	142	2@3"
226100	MRBS 2500/2	MENCK	SGL-ACT	OPEN	50	55000	74800 w/o	47'7"	CAGE	156	4.75
225000	450E5	CONMACO	SGL-ACT	OPEN	45	45000	103000 w/o	23'3"	14" × 80"	130	2@4"
204500	540	VULCAN	SGL-ACT	OPEN	48	40900	102980 w/o	22'7"	14" × 80"(M)	130	2@4"
189850	MRBS 1500 SL	MENCK	SGL-ACT	OPEN	42	38600	64600 w/o	45'0"	CAGE	142	4"
189850	MRBS-1800	MENCK	SGL-ACT	OPEN	42	38600	64600 w/o	45'0"	CAGE	142	4"
180000	360	VULCAN	SGL-ACT	OPEN	62	60000	124830 w/o	19'0"	18 3/4" × 88"(M)	130	2@4"
175000	535	VULCAN	SGL-ACT	OPEN	37	35000	61000 w/o	22'4"	10 1/4" × 54"(M)	150	3"
150000	530	VULCAN	SGL-ACT	OPEN	42	30000	57860 w/o	20'5"	10 1/4" × 54"(M)	150	3"
150000	300E5	CONMACO	SGL-ACT	OPEN	40	30000	58400 w/o	20'10"	11 1/4" × 56"	135	4"
150000	60X	RAYMOND	SGL-ACT	OPEN	60	60000	85000 w/o	22'7"	10 1/2" × 43 1/2"	165	3"
138300	MRBS 1500	MENCK	SGL-ACT	OPEN	40	33075	50715 w/o	17'2"	CAGE	145	4
135525	MRBS 1500/4	MENCK	SGL-ACT	OPEN	40	33000	58080 w/o	41'5"	CAGE	142	4"
123680	MRBS 1100	MENCK	SGL-ACT	OPEN	40	25100	37900 w/o	39'0"	CAGE	171	3"
120000	340	VULCAN	SGL-ACT	OPEN	60	40000	98180 w/o	18'7"	14" × 80"(M)	120	2@3"
100000	40X	RAYMOND	SGL-ACT	OPEN	64	40000	62000 w/o	19'1"	10 1/2" × 43 1/2"	135	3"
100000	520	VULCAN	SGL-ACT	OPEN	42	20000	47680 w/o	20'5"	11 1/4" × 37"	102	3"
100000	200E5	CONMACO	SGL-ACT	OPEN	46	20000	48000 w/o	19'1"	11 1/4" × 56"	110	4"
93340	MRBS 850	MENCK	SGL-ACT	OPEN	45	19000	30400 w/o	39'4"	CAGE	142	3"
93340	MRBS 750 SL	MENCK	SGL-ACT	OPEN	45	19000	30400 w/o	39'4"	CAGE	142	3"
92200	MRB 1000	MENCK	SGL-ACT	OPEN	45	22050	33075 w/o	15'6"	CAGE	145	3"
90000	030	VULCAN	SGL-ACT	OPEN	54	30000	53470 w/o	16'4"	11 1/4" × 37"	150	3"
90000	300	CONMACO	SGL-ACT	OPEN	52	30000	55390 w/o	16'10"	11 1/4" × 56"	150	3"
81250	8/0	RAYMOND	SGL-ACT	OPEN	40	25000	34000 w/o	19'4"	10 1/4" × 25"	150	3"
75000	30X	RAYMOND	SGL-ACT	OPEN	70	30000	52000 w/o	19'1"	10 1/2" × 43 1/2"	150	3"
67810	MRBS 750	MENCK	SGL-ACT	OPEN	40	165341	32052 w/o	38'10"	CAGE	156	3"

Air/Steam Hammers

Rated Energy (Ft.-Lbs.)	Model	Manufacturer	Type	Style	BPM	Striking Weight (Lbs.)	Total Weight (lbs)	Hammer Length (Ft-In)	Jaw Dimension (In.)	Inlet Pressure (PSI)	Inlet Size (In.)
62500	125E5	CONMACO	SGL-ACT	OPEN	41	12500	22000 w/o	18'0"	9 1/4" × 26"	100	2 1/2"
62239	MRB 600	MENCK	SGL-ACT	OPEN	45	14884	20947 w/o	14'1"	CAGE	145	3
60000	512	VULCAN	SGL-ACT	OPEN	41	12000	23480 w/o	18'5"	9 1/4" × 26"	100	2 1/2"
60000	200	CONMACO	SGL-ACT	OPEN	55	20000	44560 w/o	15'0"	11 1/4" × 56"	110	3"
60000	020	VULCAN	SGL-ACT	OPEN	59	20000	41670 w/o	14'8"	11 1/4" × 37"	120	3"
57500	115E5	CONMACO	SGL-ACT	OPEN	42	11500	21000 w/o	17'9"	9 1/4" × 26"	100	2 1/2"
56875	5/0	RAYMOND	SGL-ACT	OPEN	44	17500	26450 w/o	16'9"	10 1/4" × 25"	150	3"
50000	100E5	CONMACO	SGL-ACT	OPEN	47	10000	19500 w/o	17'9"	9 1/4" × 26"	100	2 1/2"
50000	510	VULCAN	SGL-ACT	OPEN	41	10000	21480 w/o	18'5"	9 1/4" × 26"	83	2 1/2"
50000	200-C	VULCAN	DIFFER	OPEN	95	20000	39000 w/o	13'11"	11 1/4" × 37"	142	4"
48750	160	CONMACO	SGL-ACT	OPEN	50	16250	33200 w/o	13'10"	11 1/4" × 42"	100	3"
48750	150-C	RAYMOND	DIFFER	OPEN	95–105	15000	32500 w/o	15'9"	10' × 27 1/2"	120	3"
48750	016	VULCAN	SGL-ACT	OPEN	58	16250	30250 w/o	13'11"	11 1/4" × 32"	120	3"
48750	4/0	RAYMOND	SGL-ACT	OPEN	46	15000	23800 w/o	16'1"	10 1/4" × 25"	120	2 1/2"
46100	MRB 500	MENCK	SGL-ACT	OPEN	50	11025	15655 w/o	13'9"	CAGE	145	3"
44000	MS-500	MKT	SGL-ACT	OPEN	40–50	11000	15500 w/c	15'1"	8 1/2" × 26"	115	3"
42000	014	VULCAN	SGL-ACT	OPEN	59	14000	27500 w/o	13'11"	11 1/4" × 32"	110	3"
42000	140	CONMACO	SGL-ACT	OPEN	55	14000	30750 w/o	13'10"	11 1/4" × 42"	100	3"
40600	3/0	RAYMOND	SGL-ACT	OPEN	50	12500	21000 w/o	15'7"	10 1/4" × 25"	120	2 1/2"
40000	508	VULCAN	SGL-ACT	OPEN	41	8000	19480 w/o	18'5"	9 1/4" × 26"	65	2 1/2"
40000	80E5	CONMACO	SGL-ACT	OPEN	47	8000	17500 w/o	17'9"	9 1/4" × 26"	80	2 1/2"
37500	S-14	MKT	SGL-ACT	CLOSED	60	14000	31700 w/c	13'7"	×36"	100	3
37375	115	CONMACO	SGL-ACT	OPEN	52	11500	20830 w/o	14'2"	9 1/4" × 32"	100	2 1/2"
36000	140-C	VULCAN	DIFFER	OPEN	101	14000	27984 w/o	12'3"	11 1/4" × 32"	140	3"
32885	100-C	VULCAN	DIFFER	OPEN	103	10000	22200 w/o	14'0"	9 1/4" × 26"	140	2 1/2"
32500	65E5	CONMACO	SGL-ACT	OPEN	50	6500	12500 w/o	16'10"	8 1/4" × 20"	95	2 1/2"
32500	S-10	MKT	SGL-ACT	CLOSED	55	10000	22380 w/c	14'1"	×30"	80	2 1/2"
32500	2/0	RAYMOND	SGL-ACT	OPEN	50	10000	18550 w/o	15'0"	10 1/4" × 25"	110	2"
32500	100	CONMACO	SGL-ACT	OPEN	55	10000	19280 w/o	14'2"	9 1/4" × 32"	100	2 1/2"
32500	010	VULCAN	SGL-ACT	OPEN	50	10000	18780 w/o	15'0"	9 1/4" × 26"	105	2 1/2"
32500	506	VULCAN	SGL-ACT	OPEN	46	6500	13025 w/o	17'5"	8 1/4" × 20"	120	2"
30800	MS-350	MKT	SGL-ACT	OPEN	40–50	7716	10500 w/c	15'1"	8 1/2" × 26"	105	2 1/2"
26000	85-C	VULCAN	DIFFER	OPEN	111	8525	19020 w/o	12'7"	9 1/4" × 26"	128	2 1/2"
26000	08	VULCAN	SGL-ACT	OPEN	50	8000	16750 w/o	14'10"	9 1/4" × 26"	83	2 1/2"
26000	80	CONMACO	SGL-ACT	OPEN	56	8000	17280 w/o	14'2"	9 1/4" × 32"	80	2 1/2"
26000	S-8	MKT	SGL-ACT	CLOSED	55	8000	18300 w/c	14'4"	×26	80	2 1/2"

Air/Steam Hammers

Rated Energy (Ft.-Lbs.)	Model	Manufacturer	Type	Style	BPM	Striking Weight (Lbs.)	Total Weight (lbs)	Hammer Length (Ft-In)	Jaw Dimension (In.)	Inlet Pressure (PSI)	Inlet Size (In.)
25000	505	VULCAN	SGL-ACT	OPEN	46	5000	11800 w/o	17'5"	8 1/4" × 20"	100	2"
25000	50E5	CONMACO	SGL-ACT	OPEN	48	5000	11000 w/o	16'10"	8 1/4" × 20"	70	2"
24450	80-C	RAYMOND	DIFFER	OPEN	95–105	8000	17885 w/o	12'2"	10 1/4" × 25"	120	2 1/2"
24450	80-C	VULCAN	DIFFER	OPEN	109	8000	17885 w/o	12'7"	9 1/4" × 26"	120	2 1/2"
24375	0	RAYMOND	SGL-ACT	OPEN	50	7500	16000 w/o	15'0"	10 1/4" × 25"	110	2"
24375	0	VULCAN	SGL-ACT	OPEN	50	7500	16250 w/o	15'0"	9 1/4" × 26"	80	2 1/2"
24000	C-826	MKT	COMPOUND	CLOSED	85–95	8000	17750 w/c	12'2"	×26	125	2 1/2"
19500	65E3	CONMACO	SGL-ACT	OPEN	61	6500	12100 w/o	12'10"	8 1/4" × 20	100	2"
19500	06	VULCAN	SGL-ACT	OPEN	60	6500	11200 w/o	12'9"	8 1/4" × 20"	100	2"
19500	1-S	RAYMOND	SGL-ACT	OPEN	58	6500	12500 w/o	12'9"	7 1/2" × 28 1/4"	100	1 1/2"
19500	65-C	RAYMOND	DIFFER	OPEN	110	6500	14675 w/o	11'8"	9 1/4" × 19"	120	2"
19200	65-C	VULCAN	DIFFER	OPEN	117	6500	14886 w/o	12'1"	8 1/4" × 20"	150	2"
19150	11B3	MKT	DBL-ACT	CLOSED	95	5000	14000 w/o	11'2"	8 1/2" × 26"	100	2 1/2"
16250	S-5	MKT	SGL-ACT	CLOSED	60	5000	12460 w/c	13'3"	×24"	80	2"
16000	C-5(STM)	MKT	DBL-ACT	CLOSED	100–110	5000	11880 w/c	—	×26"	100	2 1/2"
15100	50C	VULCAN	DIFFER	OPEN	117	5000	11782 w/o	11'0"	8 1/4" × 20"	120	2"
15000	1	RAYMOND	SGL-ACT	OPEN	60	5000	11000 w/o	12'9"	7 1/2" × 28 1/4"	80	1 1/2"
15000	50E3	CONMACO	SGL-ACT	OPEN	64	5000	10600 w/o	12'10"	8 1/4" × 20"	80	2"
15000	1	VULCAN	SGL-ACT	OPEN	60	5000	9700 w/o	12'9"	8 1/4" × 20"	80	2"
14200	C-5(AIR)	MKT	COMPOUND	CLOSED	100–110	5000	11880 w/c	8'9"	×26	100	2 1/2"
13100	10B3	MKT	DBL-ACT	CLOSED	105	3000	10850 w/o	9'2"	8 1/2" × 24"	100	2 1/2"
8750	9B3	MKT	DBL-ACT	CLOSED	145	1600	7000 w/o	8'4"	8 1/2" × 20	100	2"
7260	2	VULCAN	SGL-ACT	OPEN	70	3000	6700 w/o	11'7"	7 1/4" × 19"	80	1 1/2"
7260	30-C	VULCAN	DIFFER	OPEN	133	3000	7036 w/o	8'11"	7 1/4" × 19"	120	1 1/2"
5000	PM 252	PILE EQUIP.	SGL-ACT	AIR	55	2500	3600 w/o	8'	8" × 21"	120	1"
4150	#7	MKT	DBL-ACT	CLOSED	225	800	5000 w/o	6'1"	6 1/2" × 21	100	1 1/2"
4000	DGH-900	VULCAN	DIFFER	CLOSED	328	900	5000 w/o	6'9"	VARIES	78	1 1/2"
4000	PM 202	PILE EQUIP.	SGL-ACT	AIR	55	2000	3100 w/o	8'	8" × 21"	120	1"
2500	#6	MKT	DBL-ACT	CLOSED	275	400	2900 w/o	5'3"	—	100	1"
2000	PM 102	PILE EQUIP.	SGL-ACT	AIR	55	1000	2100 w/o	8'	8" × 21"	120	1"
1000	#5	MKT	DBL-ACT	CLOSED	300	200	1500 w/o	4'7"	6" × 11"	100	1 1/2"
386	DGH-100	VULCAN	DIFFER	CLOSED	303	100	786 w/o	4'2"	4 1/4" × 8 3/4"	60	1"
—	#1	MKT	DBL-ACT	CLOSED	500	21	145 w/c	3'3"	—	100	3/4"
—	#3	MKT	DBL-ACT	CLOSED	400	68	675 w/c	4'5"	—	100	1"
—	#2	MKT	DBL-ACT	CLOSED	500	48	343 w/c	2'5"	3 1/4" × 8 1/4"	100	3/4"

B.2   DIESEL HAMMERS

Diesel Hammer Terminology

- Energy Range

  The operating range of hammer-potential energy. For single-acting hammers, the rated energy is essentially the product of ram weight and stroke; for double-acting hammers, the force resulting from "bounce chamber pressure" is added to the gravitational component. Some manufacturers may include the effects of the explosive force to the hammer-potential energy. For inclined pile driving, only the vertical component of the stroke should be used in computing hammer-potential energy.

- Model

  This is the model name designation given by the manufacturer to each hammer. Usually, it provides some description of the hammer [e.g., Delmag D30 hammer has a ram weight of 3000 kg (equivalent to 6600 lbs)].

- Manufacturer

  The name of the manufacturing company.

- Type

  Single-acting hammers are open ended at the top, while double-acting hammers are closed ended. Single-acting hammers allow the ram to travel outside the cylinder, makes it visible for inspection of the stroke. Double-acting hammers utilize a bounce chamber for increasing the hammer rate of operation. The ram is not visible in a double-acting hammer.

- Blows per Minute

  The range of the hammer operating rate per minute. For single-acting hammers, the rate can be empirically correlated to the stroke. The hammer rate depends on many factors, including, but not limited to, the hammer, pile, as well as soil conditions.

- Weight of Striking Part

  Weight of the part of the hammer that actually impacts energy to the pile. This is commonly known as the *ram* or *piston*. Hammer-rated energy and general effectiveness is a direct function of the weight of its striking part. In some cases, this weight is indicated as part of the hammer-model designation.

- Total Weight

  The total weight of the hammer. This value is important in sizing the crane, transportation requirements, and other aspects involving the hammer.

- Hammer Length

  Total length of hammer in its normal operating configuration, excluding any accessories that may be present between the hammer and the pile head.

- Maximum Stroke

  Maximum attainable stroke. Values obtained under favorable controlled conditions. Strokes under common-field conditions vary, depending on the mechanical condition of the hammer, cushion and pile elastic effects, soil resistance, and general hammer-cushion-pile-soil dynamic compatibility.

- Jaw Dimensions

  Dimensions of the hammer guides that interface with the leads. All diesel hammers have "female" type jaws, and most have provisions for changeable guides.

- Fuel Consumption

  Amount of fuel (diesel) per hour that a hammer might consume. Actual amount is subject to operating variations. For proper hammer function, the appropriate type of fuel must be used.

Diesel Hammers

Energy Range (Ft. Lb.)	Model	Manufacturer	Type	BPM	Striking Weight (Lbs.)	Total Weight (Lbs.)	Total Length (Ft.-In.)	Max. Stroke (Ft.-In.)	Jaw Dim.	Fuel Usage (GPH)
500000–315000	D200-42	DELMAG	SINGLE	36–52	44065	118350 w/o	27'	11'2"	60"	10.60
350000–179478	120-13	APE	SINGLE	34–44	27612	47994 w/c	20'4"	12'8"	36"	9.43
300000	100-13	APE	SINGLE	34–45	23612	44894 w/c	20'4"	12'8"	36"	7.93
300000–157443	D100-13	DELMAG	SINGLE	34–45	23612	43703 w/o	20'4"	12'8"	36"	7.93
225000–126190	D80-23	DELMAG	SINGLE	36–45	19500	36085 w/o	20'4"	11'6"	36"	6.60
225000–129870	80-23	APE	SINGLE	36–45	19500	37275 w/c	20'4"	11'6"	36"	6.60
212400–126192	I-80	ICE	SINGLE	35–45	17700	47664 w/c	13.50'	20.4'	42"	6.6
210000–80000	205S	ICE	SINGLE	40–55	20000	33500 w/o	17'11"	8'6"	32"	4.2
209400	B-6505C	BERMINGHAMMER	SINGLE	39–60	22046	42946 w/o	23'4"	9'5"	37.5"	3
202400–79200	B-6505	BERMINGHAMMER	SINGLE	35–60	17600	38500 w/o	23'4"	11'6"	37.5"	3
198300	G4	BERMINGHAMMER	SINGLE	36–60	16805	33000 w/o	20'9"	11'9"	37.5"	3
19347–347225	HMC15000	HMC	SINGLE	37–50	33069	72552	31.5'	10.5'	—	0
183750–60000	120S-15	ICE	SINGLE	41–55	15000	26600 w/o	19'11"	8'10"	32"	4.2
16500–78960	62-22	APE	SINGLE	36–50	17000	27077 w/c	19'4"	11'3"	32"	5.28
16500–78960	D62-22	DELMAG	SINGLE	36–50	14600	26173 w/o	19'4"	11'4"	32"	5.28
16500–78960	I-62	ICE	SINGLE	35–50	14600	35964 w/c	14.25'	19.7'	32"	5.3
15732C–231483	HMC10000	HMC	SINGLE	37–50	22046	48501	32.2'	10.5'	—	0
15686C–61380	B-6005	BERMINGHAMMER	SINGLE	35–60	13640	34500 w/o	23'4"	11'6"	37.5"	3
14960–88000	MH80B	MITSUBISHI	SINGLE	42–60	17600	43600 w/o	19'6"	14'10"	42"	8–12
141000–63360	MB70	MITSUBISHI	SINGLE	38–60	15840	46000 w/o	19'6"	8'11"	42"	7–10
135100–79500	MH72B	MITSUBISHI	SINGLE	38–60	15900	44000 w/o	19'6"	14'10"	42"	7–10
127500–90000	DE-150/110C	MKT	SINGLE	40–50	15000	32150 w/c	19'11"	10'	32"	7
127500–90000	DE-150/110	MKT	SINGLE	40–50	15000	32150 w/c	19'10"	10'9"	32"	7
120000–48000	120S	ICE	SINGLE	38–55	12000	23800 w/o	19'10"	10'	32"	4.2
117000–62500	D55	DELMAG	SINGLE	36–47	12128	26300 w/o	17'9"	9'8"	32"	5.50
107770–52250	46-32	APE	SINGLE	37–53	10143	19580 w/c	17'4"	10'6"	30"	4.23
107700–52260	I-46	ICE	SINGLE	36–53	10145	27355 w/c	12.12'	17.3'	32"	4.2
107177–52260	D46-32	DELMAG	SINGLE	37–53	10143	18720 w/o	17'4"	10'7"	30"	4.23
105800–41400	B-5505	BERMINGHAMMER	SINGLE	36–60	9200	21400 w/o	21'8"	11'9"	32"	2.2
105000–48400	D46-23	DELMAG	SINGLE	37–53	10143	19900 w/o	17'3"	10'5"	30"	3.3
104101–203700	HMC8800	HMC	SINGLE	37–50	19400	37037	27.4'	10.5'	—	0
100000–40000	200S	ICE	SINGLE	53–70	20000	33.600 w/o	17'	5'0"	32"	4
100000–40000	100S	ICE	SINGLE	38–55	10000	18800 w/o	19'	10'	32"	3.5
95321–173618	HMC7500	HMC	SINGLE	37–50	16535	34171	25.5'	10.5'	—	0
93500–66000	DE-150/110	MKT	SINGLE	40–50	11000	27150 w/c	17'10"	10'9"	32"	5.7
93500–66000	DE-150/110C	MKT	SINGLE	40–50	11000	28150 w/c	19'11"	10'	32"	5.7

Diesel Hammers

Energy Range (Ft. Lb.)	Model	Manufacturer	Type	BPM	Striking Weight (Lbs.)	Total Weight (Lbs.)	Total Length (Ft.-In.)	Max. Stroke (Ft.-In.)	Jaw Dim.	Fuel Usage (GPH)
92752-39000	K45	KOBE	SINGLE	39-60	9920	25300 w/o	18'6"	9'2"	36"	4.5-5.5
92000	B-5505C	BERMINGHAMMER	SINGLE	42-60	11500	23700 w/o	21'8"	8'	32"	2.2
90540-40900	I-36	ICE	SINGLE	35-53	7940	23738 w/c	12.12'	17.3'	32"	3.0
90000-36000	90S	ICE	SINGLE	38-55	9000	16800 w/o	17'3"	10'0"	29"	4
87400-34200	B-5005	BERMINGHAMMER	SINGLE	35-60	7600	19800 w/o	21'8"	11'6"	26"	2.2
87000-43000	D44	DELMAG	SINGLE	37-56	9500	22300 w/o	15'11"	9'2"	32"	5.5
85400-50200	MH45	MITSUBISHI	SINGLE	42-60	10500	24600 w/o	17'11"	11'5"	36"	4.6
84000-37840	M43	MITSUBISHI	SINGLE	40-80	9460	22660 w/o	16'3"	8'10"	37"	4-6
83880-40900	36-32	APE	SINGLE	36-52	7938	17375 w/c	17'4"	10'6"	30"	3.04
83880-40900	D36-32	DELMAG	SINGLE	36-53	7938	16515 w/o	17'4"	10'7"	30"	3.04
83100-38000	D36-23	DELMAG	SINGLE	37-53	7938	17700 w/o	17'4"	10'5"	30"	3
80000-32000	80S	ICE	SINGLE	38-55	8000	15400 w/o	18'11"	10'	26"	2.9
79500	J44	IHI	SINGLE	42-70	9720	21500 w/o	14'10"	8'2"	37"	6.86
79000	K42	KOBE	SINGLE	39-60	9260	23000 w/o	18'6"	8'6"	36"	4.5-5.5
75900-29700	B-4505	BERMINGHAMMER	SINGLE	36-60	6600	16100 w/o	19'1"	11'9"	26"	2.1
72340-131943	HMC5700	HMC	SINGLE	37-50	12566	23809	24.8'	10.5'	—	0
72182-31700	K35	KOBE	SINGLE	39-60	7720	19100 w/o	17'8"	9'2"	30"	3-4
71700-35383	I-30	ICE	SINGLE	35-53	6615	16434 w/c	12.14'	16.1'	26"	2.6
70000-42000	DE-70/50OC	MKT	SINGLE	40-50	7000	16285 w/c	17'11"	11'6"	26"	3.3
70000-28000	70S	ICE	SINGLE	38-55	7000	14100 w/o	16'8"	10'	26"	2.10
70000-36100	1070	ICE	DOUBLE	64-68	10000	32500 w/o	17'10"	7'	30"	3.5
69898-35383	D30-32	DELMAG	SINGLE	36-52	6615	12855 w/o	17'3"	10'7"	26"	2.64
69898-35383	30-32	APE	SINGLE	36-52	6615	14600 w/o	17'3"	10'6"	26"	2.64
66100-33700	D30-23	DELMAG	SINGLE	38-54	6615	13150 w/o	17'2"	10'	26"	1.7
65600-38600	MH35	MITSUBISHI	SINGLE	42-60	7720	18500 w/o	17'3"	11'	32"	3.4-5
64600-29040	M33	MITSUBISHI	SINGLE	40-60	7260	16940 w/o	13'2"	8'	32"	3.4-5
64000-36000	B-4505C	BERMINGHAMMER	SINGLE	42-60	8000	17500 w/o	19'1"	8'	26"	2.1
63500	J35	IHI	SINGLE	72-70	7730	16900 w/o	14'6"	8'3"	32"	4.76
63000-34650	3000	F.E.C.	SINGLE	40-60	6600	13200 w/c	17'	10'6"	26"	2.8-4.2
60100	K32	KOBE	SINGLE	39-60	7050	16250 w/o	17'8"	8'6"	30"	3-4
60000-26000	60S	ICE	SINGLE	41-59	7000	13900 w/o	17'	8'7"	26"	2.1
59500-42000	DE-70/50B	MKT	SINGLE	40-50	7000	16035 w/c	16'11"	8'6"	26"	3.3
58248-29486	D25-32	DELMAG	SINGLE	37-52	5513	11752 w/o	17'3"	10'7"	26"	2.11
57882-29480	25-32	APE	SINGLE	36-52	5512	13500 w/c	17'3"	10'6"	26"	2.11
57500-22500	B-4005	BERMINGHAMMER	SINGLE	36-60	5000	14500 w/o	19'1"	11'9"	26"	2.1
54250-23800	D30	DELMAG	SINGLE	39-60	6615	12300 w/o	14'3"	8'2"	26"	2.9
54000-22500	B-400	BERMINGHAMMER	SINGLE	38-60	5000	16000 w/c	19'4"	10'10"	26"	1.5
52250-115742	HMC5000	HMC	SINGLE	37-50	11023	22266	23.3'	10.5'	—	0

Diesel Hammers

Energy Range (Ft.-Lb.)	Model	Manufacturer	Type	BPM	Striking Weight (Lbs.)	Total Weight (Lbs.)	Total Length (Ft.-In.)	Max. Stroke (Ft.-In.)	Jaw Dim.	Fuel Usage (GPH)
51520–23500	JK25	KOBE	SINGLE	39–60	5510	13200 w/o	17'6"	9'3"	26"	2.5–3
50000–25100	660	ICE	DOUBLE	84–88	7564	24480 w/o	17'4"	6'7"	30"	3.25
50000–30000	DE-70/50C	MKT	SINGLE	40–50	5000	14285 w/c	17'11"	11'6"	26"	3.3
50000–27500	2500	F.E.C.	SINGLE	40–60	5500	12100 w/c	17'	10'6"	20"	2.8–4.2
48500–24500	D22-23	DELMAG	SINGLE	38–52	4850	11400 w/o	17'2"	10'	26"	1.6
46900–27550	MH25	MITSUBISHI	SINGLE	42–60	5510	13200 w/o	16'8"	10'9"	26"	2.4–4
46000–18000	B-3505	BERMINGHAMMER	SINGLE	36–60	4000	10500 w/o	18'6"	11'9"	26"	1.4
45020–104181	HMC4500	HMC	SINGLE	37–50	9922	18662	24.1'	10.5'	—	0
45000–20240	M23	MITSUBISHI	SINGLE	42–60	5060	11220 w/o	14'1"	8'10"	26"	2.4–3.7
44800	DE50C	BSP	SINGLE	42–54	4980	10300 w/o	14'4"	9'	26"	2.68
43225–21510	I-19	ICE	SINGLE	35–53	4015	11034 w/c	12.14'	15.4'	26"	1.5
42800–20500	D19-32	DELMAG	SINGLE	37–53	4190	9650 w/o	15'6"	10'3"	20"	1.45
42800–20500	D19-42	DELMAG	SINGLE	37–53	4190	9650 w/o	15'6"	10'2"	20"	1.45
42800–20500	19–42	APE	SINGLE	37–53	4190	9600 w/o	15'6"	10'2"	20"	1.45
42500–30000	DE-70/50B	MKT	SINGLE	40–50	5000	14035 w/c	16'11"	6'6"	26"	3.3
42500–30000	DA-55C	MKT	SINGLE	40–50	5000	17635 w/c	17'4"	10'6"	26"	2.7
42000–25200	DE-42/35	MKT	SINGLE	40–50	4200	10250 w/c	16'7"	10'0"	26"	3
42000–16000	42S	ICE	SINGLE	37–55	4088	7610 w/o	16'1"	10'3"	20"	1.3
41300	K22	KOBE	SINGLE	39–60	4850	11300 w/o	17'4"	8'6"	26"	2.5–3
40888–81028	HMC3500	HMC	SINGLE	37–50	7717	15057	24.1'	10.5'	—	0
40500–16875	B-300	BERMINGHAMMER	SINGLE	38–60	3750	9892 w/c	18'4"	10'10"	26"	1.2
40200–18871	D16-32	DELMAG	SINGLE	36–52	3528	7386 w/o	15'6"	11'5"	20"	1.45
40200–18871	16-32	APE	SINGLE	36–52	3528	8850 w/c	15'6"	11'5"	20"	1.45
40000	B-3505C	BERMINGHAMMER	SINGLE	42–60	5000	11500 w/o	18'6"	8"	26"	1.4
40000–16000	40S	ICE	SINGLE	38–55	4000	7500 w/o	15'9"	10'0"	20"	1.2
40000–25400	640	ICE	DOUBLE	74–77	6000	14460 w/o	15'7"	6'8"	26"	3
40000–24000	DE-33/30/20C	MKT	SINGLE	40–50	4000	9400 w/c	15'11"	10'6"	26"	3
39700	D22	DELMAG	SINGLE	42–60	4850	11200 w/o	14'2"	8'2"	26"	2.9
39100–12000	J22	IHI	SINGLE	42–70	4850	10800 w/o	14'	10'0"	26"	3.2
38200–31200	DA-55C	MKT	DOUBLE	78–82	5000	17635 w/c	17'4"	—	26"	3
36000–24000	DE-40	MKT	SINGLE	40–50	4000	11275 w/c	14'0"	10'6"	20"	3
35000–21000	DE-42/35	MKT	SINGLE	40–50	3500	9550 w/c	16'7"	10'0"	26"	2
34400–13500	B-3005	BERMINGHAMMER	SINGLE	36–60	3000	9500 w/o	18'6"	11'9"	26"	1.4
34000–24000	DA-45	MKT	SINGLE	40–50	4000	15525 w/c	15'1"	10'6"	26"	2.5
33000–19800	DE-33/30/20C	MKT	SINGLE	40–50	3300	8700 w/o	5'11"	10'6"	20"	2
32000–11000	32S	ICE	SINGLE	41–60	3000	5725 w/o	15'3"	8'8"	20"	.80
31200–15660	D12-32	DELMAG	SINGLE	36–52	2820	5730 w/o	15'6"	11'1"	20"	.95
30700–18500	DA-45	MKT	DOUBLE	78–82	4000	15525 w/c	15'1"	—	26"	2.8
3000–17000	520	ICE	DOUBLE	80–84	5070	13400 w/o	13'7"	5'11"	26"	1.35
29456–57876	HMC2500	HMC	SINGLE	37–50	5512	11772	23.6'	10.5'	—	0
29400–13500	B-225	BERMINGHAMMER	SINGLE	39–60	3000	9142 w/c	18'4"	9'10"	26"	1.2

Diesel Hammers

Energy Range (Ft. Lb.)	Model	Manufacturer	Type	BPM	Striking Weight (Lbs.)	Total Weight (Lbs.)	Total Length (Ft.-In.)	Max. Stroke (Ft.-In.)	Jaw Dim.	Fuel Usage (GPH)
28100–16550	MH15	MITSUBISHI	SINGLE	42–60	3310	8400 w/o	16'1"	10'3"	26"	1.3–2
28050–19800	DE-33/30/20B	MKT	SINGLE	40–50	3300	8700 w/c	15'11"	8'6"	20"	2
28000–16800	DE-33/30/20C	MKT	SINGLE	40–50	2800	8200 w/c	15'11"	10'6"	20"	2
27190	D15	DELMAG	SINGLE	40–60	3300	6615 w/o	13'11"	8'2"	20"	1.75
27100–14900	1500	F.E.C.	SINGLE	40–60	3300	7225 w/c	15'8"	10'12"	20"	1.5–2.3
27000	DE30C	BSP	SINGLE	42–54	3000	7600 w/o	14'2"	9'	26"	1.7
26300–17700	520	LINKBELT	DOUBLE	80–84	5070	12545 w/o	13'6"	5'3"	26"	1.35
26000–11800	M14S	MITSUBISHI	SINGLE	42–60	2970	7260 w/o	13'6"	8'9"	26"	1.3–2.2
25428–13200	K13	KOBE	SINGLE	40–60	2870	7800 w/o	16'8"	8'6"	26"	.75–2
25200–16800	DE-30	MKT	SINGLE	40–50	2800	9075 w/c	15'0"	10'6"	20"	2
23800–16800	DE-33/30/20B	MKT	SINGLE	40–50	2800	8200 w/c	15'11"	8'6"	20"	2
23800–16800	DA-35C	MKT	SINGLE	40–50	2800	11750 w/c	17'0"	10'6"	20"	1.7
23000	B-23	BERMINGHAMMER	DOUBLE	82	2800	9940 w/o	20'10"	4'6"	26"	1.9
22500–9000	30S	ICE	SINGLE	44–67	3000	6250 w/o	12'4"	7'6"	20"	.80
22500	D12	DELMAG	SINGLE	42–60	2750	6050 w/o	13'11"	8'2"	20"	1.75
22500–9000	422	ICE	DOUBLE	76–82	4000	9750 w/o	13'11"	5'8"	22"	1
22500–12375	1200	F.E.C.	SINGLE	40–60	2750	6540 w/c	15'6"	10'10"	20"	1.5–2.3
21506–43985	HMC1900	HMC	SINGLE	37–50	4189	8774	23.4'	10.5'	—	0
21000–15600	DA-35C	MKT	DOUBLE	78–82	2800	11750 w/c	17'0"	—	20"	2.7
20400–9000	B-200	BERMINGHAMMER	SINGLE	38–58	2000	6940 w/c	15'	10'2"	—	1.2
20000–12000	DE-33/30/20C	MKT	SINGLE	40–50	2000	7400 w/c	15'11"	10'6"	20"	2
18901–34724	HMC1500	HMC	SINGLE	37–50	3307	7275	23.3'	10.5'	—	0
18200	B-2005	BERMINGHAMMER	SINGLE	36–60	2000	6600 w/o	16'8"	9'	26"	1
18100–7700	440	ICE	DOUBLE	88–92	4000	9840 w/o	13'6"	4'8"	20"	1.16
18000–12000	DE-20	MKT	SINGLE	40–50	2000	6325 w/c	13'3"	9'6"	20"	1.6
18000–7500	312	LINKBELT	DOUBLE	100–105	3857	10375 w/o	10'9"	4'8"	26"	1.1
18000–9434	D8-22	DELMAG	SINGLE	38–52	1762	4000 w/o	15'5"	10'3"	20"	1
18000–8435	8-22	APE	SINGLE	38–52	1764	4220 w/c	15'5"	10'2"	20"	1
17000–12000	DE-33/30/20B	MKT	DOUBLE	40–50	2000	7400 w/o	15'11"	8'6"	20"	2
10500–6300	D6-32	DELMAG	SINGLE	39–52	1322	3570 w/o	12'6"	7'11"	20"	.70
9350–6600	DA-15C	MKT	SINGLE	40–50	1100	5700 w/c	13'11"	10'6"	20"	1
9150	D5	DELMAG	SINGLE	42–60	1100	2730 w/o	12'6"	9'4"	20"	1.32
8800–6600	DE-10	MKT	SINGLE	40–50	1100	3518 w/c	12'2"	9'	20"	.90
8800–6600	DA-15C	MKT	DOUBLE	86–92	1100	5700 w/c	13'11"	—	20"	1.8
8100–4060	180	ICE	DOUBLE	90–95	1725	4645 w/o	11'3"	4'9"	20"	.65
3630–1125	D4	DELMAG	SINGLE	50–60	836	1360 w/o	7'9"	4'4"	—	.21
1815–868	D2	DELMAG	SINGLE	60–70	484	792 w/o	6'9"	3'8"	—	.13

B.3   HYDRAULIC IMPACT HAMMERS

Hydraulic Impact Hammer Terminology

- Rated Energy

  The rated energy of an impact hammer is the maximum-energy potential per blow. For single-acting hammers, it is the product of the stroke length and ram weight. For hammers that provide assistance to the ram during its downward travel, the rated energy includes the effects of the force exerted by the motive-fluid pressure in addition to the gravitational component. In cases of inclined pile driving, the stroke value should be adjusted to account for the angle of inclination (i.e., only the vertical component of the stroke should be used in establishing hammer-potential energy). The hammer energy at impact is a function of ram weight, stroke, and overall hammer condition (i.e., efficiency).

- Model

  This is the model designation given by the manufacturer to each hammer. Usually, it provides some description of the hammer. (e.g., VULCAN 340 model has a 3-ft stroke and 40-kip ram weight).

- Manufacturer

  The name of the manufacturing company.

- Type

  Single-acting hammers mean that the motive fluid is only used to raise the ram to the desired stroke and downwards ram movement is done by gravity. For double-acting hammers, the motive fluid is not only used to raise the ram, but it also assists during the ram fall.

- Style

  Open-style hammers are those which have visible rams; closed-style hammers enclose the ram. Generally, open-style hammers cannot be used under water, while some closed-style hammers can.

- Blows per Minute

  The rated number of hammer impacts per minute. Reported values are under specific operating conditions and may vary depending on the actual field conditions. In most cases, it is not possible to relate the hammer rate of operation to its potential energy.

- Weight of Striking Part

  Weight of the part of the hammer that actually impacts energy to the pile. This is commonly known as the "ram." Hammer-rated energy and general effectiveness is a direct function of the weight of its striking part. In some cases, this weight is indicated as part of the hammer-model designation.

- Total Weight

  The total weight of the hammer. This value is important in sizing the crane, transportation requirements, and other aspects involving the hammer.

- Hammer Length

  Total length of hammer in its normal operating configuration, excluding any accessories that may be present between the hammer and the pile head.

- Jaw Dimensions

  Dimensions of the hammer guides that interface with the leads. Also shown whether "male" or "female" type. Necessary in selecting appropriate size and style leads.

- Inlet Pressure

  The pressure (at the inlet of the hammer) of the motive fluid needed to operate the hammer at full-rated energy. In cases where direct reading is not available, pressure readings measured at the boiler or compressor must be adjusted to arrive at the pressure value at the hammer inlet.

- Inlet Size

  The number and size of hoses supplying motive fluid to the hammer.

Rated Energy (Ft–Lbs)	Model	Manufacture	Type	Style	BPM	Striking Weight (Lbs.)	Total Weight (Lbs.)	Hammer Length (Ft.–In.)	Hydraulic Pressure (PSI)	Hydraulic Flow (GPM)	Width between Leads (In.)
2434000	MHU 3000T	MENCK	DBL-ACT	CLOSED	40	396800	683550 w/o	66'7"	3335	1268	NO LEADS
2212600	MHU 3000	MENCK	DBL-ACT	CLOSED	30	370377	662145 w/o	58'10"	3343	1450	NO LEADS
1703700	MHU 2100T	MENCK	DBL-ACT	CLOSED	40	277830	453050 w/o	52'2"	3335	845	NO LEADS
1695800	S-2300	IHC	ASSISTED	CLOSED	29–80	253530	573190 w/o	59'2"	4300	1188	NO LEADS
1549000	MHU 2100	MENCK	DBL-ACT	CLOSED	25	256839	397714 w/o	50'3"	3556	845	NO LEADS
1379200	MHU 1700T	MENCK	DBL-ACT	CLOSED	45	224910	396900 w/o	55'5"	3335	845	NO LEADS
1327100	S-1800	IHC	ASSISTED	CLOSED	40–80	165340	440920 w/o	51'11"	4300	1188	NO LEADS
1254000	MHU 1700	MENCK	DBL-ACT	CLOSED	30	207235	335103 w/o	45'6"	3414	845	NO LEADS
885000	S-1200	IHC	ASSISTED	CLOSED	32–80	132250	286,600 w/o	43'6"	4300	791	NO LEADS
811300	MHU 1000T	MENCK	DBL-ACT	CLOSED	40	132300	242550 w/o	53'2"	3480	423	NO LEADS
738000	S-1000	IHC	ASSISTED	CLOSED	40–80	110200	257900 w/o	41'11"	4300	739	NO LEADS
737600	MHU 1000	MENCK	DBL-ACT	CLOSED	30	126986	239202 w/o	36'3"	2134	726	NO LEADS
689600	MHU 850T	MENCK	DBL-ACT	CLOSED	45	112445	209475 w/o	42'4"	3335	423	NO LEADS
590000	S-800	IHC	ASSISTED	CLOSED	40–80	92600	211600 w/o	38'7"	4000	739	NO LEADS
590000	MHU 800S	MENCK	DBL-ACT	CLOSED	50	99200	176370 w/o	41'4"	3770	726	NO LEADS
567900	MHU 700T	MENCK	DBL-ACT	CLOSED	50	52920	167580 w/o	41'4"	3190	383	NO LEADS
486800	MHU 600T	MENCK	DBL-ACT	CLOSED	50	79380	145530 w/o	45'9"	3190	330	NO LEADS
443000	S-600	IHC	ASSISTED	CLOSED	36–100	66100	138900 w/o	36'1"	4300	369	NO LEADS
442500	MHU 600	MENCK	DBL-ACT	CLOSED	42	76721	151017 w/o	37'6"	3343	396	NO LEADS
405600	MHU 500T	MENCK	DBL-ACT	CLOSED	55	66150	119070 w/o	39'9"	3335	303	NO LEADS
400000	HI 400U	APE	FREE	CLOSED	30–60	80000	152500	39'3"	4500	250	74"
390582	HH45	BSP	FREE	OPEN	33	99200	138940 w/o	28'10"	3988	265	VARIABLE
368600	S-500	IHC	ASSISTED	CLOSED	45–100	55110	125660 w/o	33'5"	4300	396	NO LEADS
347520	V400A-40	TWINWOOD	FREE	CLOSED	20	89082	125244 w/o	24'11"	4260	120	—
347185	HH40	BSP	FREE	OPEN	37	88180	126834 w/o	27'6"	3843	265	VARIABLE
324500	MHU 400T	MENCK	DBL-ACT	CLOSED	55	52920	97020 w/o	35'5"	2190	264	NO LEADS
320000	320	APE	ASSISTED	BOTH	40–80	80000	165000 w/c	25'	3900	304	VARIABLE
305786	HH35	BSP	FREE	OPEN	40	77160	114730 w/o	26'1"	3626	250	VARIABLE
295000	S-400	IHC	ASSISTED	CLOSED	50–100	44090	108000 w/o	30'10"	3600	396	NO LEADS
295000	MHU 400	MENCK	DBL-ACT	CLOSED	42	51147	115963 w/o	32'5"	2347	370	NO LEADS
282100	SGH-3013	BRUCE	FREE	OPEN	26	66139	116850 w/o	33'2"	3917	160	84"
260640	V400A-30	TWINWOOD	FREE	CLOSED	32	67032	102753 w/o	22'4"	3970	120	—
260388	HH30	BSP	FREE	OPEN	43	66150	102530 w/o	24'8"	3408	265	VARIABLE
260387	SGH-30	BRUCE	FREE	OPEN	26	66139	116845 w/o	32'2"	3917	145	—

Hydraulic Impact Hammers

Rated Energy (Ft–Lbs)	Model	Manufacture	Type	Style	BPM	Striking Weight (Lbs.)	Total Weight (Lbs)	Hammer Length (Ft.–In.)	Hydraulic Pressure (PSI)	Hydraulic Flow (GPM)	Width between Leads (In.)
225888	V400A-26	TWINWOOD	FREE	CLOSED	34	58212	93933 w/o	22'4"	3970	120	—
221200	MHU 300	MENCK	DBL-ACT	CLOSED	42	36376	66139 w/o	23'9"	3343	198	NO LEADS
219000	MHU 270T	MENCK	DBL-ACT	CLOSED	50	35280	66150 w/o	30'6"	3625	145	NO LEADS
216990	HH 25	BSP	FREE	OPEN	43	55120	90425 w/o	23'4"	3263	—	VARIABLE
206400	S-280	IHC	ASSISTED	CLOSED	45–100	30000	63930 w/o	33'5"	4300	198	37"
184000	SC-250	IHC	ASSISTED	CLOSED	30–80	40800	68300 w/o	21'3"	4300	198	52"
176325	3505	HPSI	FREE	CLOSED	30–60	35265	50690 w/o	27'2"	3850	90	57"
175906	HH16	BSP	FREE	CLOSED	30	35300	50177 w/o	27'0"	4133	98	57"
173760	V400A-20	TWINWOOD	FREE	CLOSED	36	44982	80703 w/o	22'4"	3820	120	—
173592	SGH-20	BRUCE	FREE	OPEN	30	44092	80028 w/o	28'8"	3917	108	—
173592	HH 20	BSP	FREE	OPEN	36	44100	71698 w/o	22'4"	2900	211	VARIABLE
173592	DKH 20	PILEMER	FREE	OPEN	40–100	44092	99208 w/o	22'4"	4000	120	78"
173592	NH-150B	NISSHA	DOUBLE	—	20–52	33075	73868 w/o	23'3"	4266	109	23"
162300	MHU 220	MENCK	DBL-ACT	CLOSED	36	25357	49384 w/o	25'10"	3556	116	NO LEADS
162300	MHU 200T	MENCK	DBL-ACT	CLOSED	60	26460	55125 w/o	27'0"	3335	145	NO LEADS
159600	HHK 18A	JUNTTAN	ASSISTED	CLOSED	40–100	39800	61712 w/o	22'9"	3333	211	VARIABLE
156232	SGH-18	BRUCE	FREE	OPEN	30	39683	61729 w/o	24'11"	3627	108	—
154800	MHF 10–20	MENCK	FREEFALL	OPEN	60	44100	61740 w/o	20'2"	4495	118	55"
153918	HH 14	BSP	FREE	CLOSED	31	30850	45040 w/o	25'2"	3770	99	57"
149856	HHK 16A	JUNTTAN	ASSISTED	CLOSED	40–100	35282	57334 w/o	26'3"	3043	211	VARIABLE
149723	NH-115B	NISSHA	DOUBLE	—	23–62	25357	60638 w/o	23'6"	4266	109	23"
147500	SC-200	IHC	ASSISTED	CLOSED	38–90	30000	55110 w/o	18'6"	4000	199	52"
147500	S-200	IHC	ASSISTED	CLOSED	45–100	22050	54010 w/o	29'3"	3600	198	37"
141043	NH-150S	NISSHA	SINGLE	—	40–80	33075	53022 w/o	27'2"	4053	182	23"
139008	V160B-16	TWINWOOD	FREE	CLOSED	28	42336	54305 w/o	20'	4120	76	—
138873	DKH 16	PILEMER	FREE	OPEN	40–100	35274	66800 w/o	20'9"	4000	100	64"
136873	SGH-16	BRUCE	FREE	OPEN	32	35274	59084 w/o	21'1"	3627	93	—
124135	HHK 14A	JUNTTAN	ASSISTED	CLOSED	40–100	30900	39100 w/o	24'7"	2753	211	VARIABLE
124135	HHK 14	JUNTTAN	FREE	OPEN	30–100	30900	41876 w/o	24'4"	2680	158	VARIABLE
121632	V160B-14	TWINWOOD	FREE	CLOSED	30	37044	49895 w/o	20'	3820	76	—
121511	SGH-14	BRUCE	FREE	OPEN	35	30865	49163 w/o	20'11"	3627	85	57"
116100	MHF 10–15	MENCK	FREEFALL	OPEN	80	33075	50715 w/o	20'2"	3480	118	56"
112845	DKH 13	PILMER	FREE	OPEN	40–100	28660	46945 w/o	20'7"	4000	85	58"
111000	S-150	IHC	ASSISTED	CLOSED	50–100	15400	34800 w/o	28'8"	4300	108	30"

Hydraulic Impact Hammers

Rated Energy (Ft–Lbs)	Model	Manufacture	Type	Style	BPM	Striking Weight (Lbs.)	Total Weight (Lbs)	Hammer Length (Ft.–In.)	Hydraulic Pressure (PSI)	Hydraulic Flow (GPM)	Width between Leads (In.)
110600	SC-150	IHC	ASSISTED	CLOSED	38–90	24250	42300 w/o	20'11"	4200	108	40"
110000	275	ICE	FREE	OPEN	40–60	27500	47650 w/o	21'11"	3500	130	36"
110000	149	HMC (J & M)	FREE	OPEN	40–75	27500	47650 w/o	21'11"	5000	87	36"
106402	HHK 12A	JUNTTAN	ASSISTED	CLOSED	40–100	26490	34600 w/o	2'11"	3188	171	VARIABLE
104256	V160B-12	TWINWOOD	FREE	CLOSED	34	31752	45485 w/o	20'	3830	76	—
104155	NH-100	NISSHA	DOUBLE	—	27–62	22050	49613 w/o	22'7"	2488	64	23"
104153	SGH-12	BRUCE	FREE	OPEN	38	26455	41006 w/o	20'6"	3627	85	57"
100000	2500	HPSI	FREE	OPEN	40–60	25000	33500 w/o	20'8"	4900	65	27"
95475	HH11(HD)	BSP	FREE	OPEN	30	24250	29431 w/o	22'7"	4351	50	27"
88500	MHF 5–12	MENCK	FREEFALL	OPEN	60	26460	36162 w/o	18'1"	4640	84	36"
88000	119	HMC (J & M)	FREE	OPEN	40–75	22000	35000 w/o	20'3"	5000	87	32"
88000	220	ICE	FREE	OPEN	45–60	22000	35000 w/o	20'3"	5000	87	32"
88000	S-120	IHC	ASSISTED	CLOSED	50–100	13400	31500 w/o	26'1"	4000	108	30"
86880	V160B-10	TWINWOOD	FREE	CLOSED	37	26460	41075 w/o	20'	3530	76	—
86880	V100D-10	TWINWOOD	FREE	CLOSED	28	22270	30826 w/o	20'4"	4116	50	13"
86798	DKH 10	PILEMER	FREE	OPEN	40–100	22043	36585 w/o	19'	4000	71	58"
86794	SGH-10	BRUCE	FREE	OPEN	38	22046	33731 w/o	19'5"	3627	69	57"
81100	SC-110	IHC	ASSISTED	CLOSED	38–100	17420	33700 w/o	18'4"	3700	108	40"
81100	MHU 100T	MENCK	DBL-ACT	CLOSED	60	13230	33075 w/o	21'4"	3770	84	NO LEADS
80000	2000	HPSI	FREE	OPEN	32–60	20000	24800 w/o	21'2"	4600	51	27"
79801	HHK 9A	JUNTTAN	ASSISTED	CLOSED	40–100	19846	29768 w/o	24'2"	3043	118	VARIABLE
79801	HHU 9A	JUNTTAN	ASSISTED	CLOSED	40–100	19846	23400 w/o	24'2"	3043	118	32"
78192	V100D-9	TWINWOOD	FREE	CLOSED	30	20066	28621 w/o	20'	3970	50	13"
78166	HH9S	BSP	FREE	OPEN	32	19850	28802 w/o	20'9"	3843	50	27"
73700	MHF 5-10	MENCK	FREEFALL	OPEN	80	22050	31752 w/o	18'1"	4640	66	35"
69504	V160B-8	TWINWOOD	FREE	CLOSED	42	21168	36665 w/o	20'	3530	76	—
69504	V100D-8	TWINWOOD	FREE	CLOSED	32	17860	26416 w/o	20'4"	3820	50	13"
66000	S-90	IHC	ASSISTED	CLOSED	50–125	9900	20300 w/o	25'10"	4000	66	25"
64808	NH-70	NISSHA	DOUBLE	—	30–72	15435	31532 w/o	15'8"	2488	58	13"
64000	160	ICE	FREE	OPEN	40–60	16000	28600 w/o	21'	5000	67	32"
64000	86	HMC (J & M)	FREE	OPEN	40–75	16000	28600 w/o	21'0"	5000	67	32"
62068	HHK 7	JUNTTAN	FREE	OPEN	30–100	15430	21378 w/o	21'2"	2680	79	VARIABLE
62068	HHU 7A	JUNTTAN	ASSISTED	CLOSED	40–100	15430	23800 w/o	21'8"	2753	118	32"

Hydraulic Impact Hammers

Rated Energy (Ft–Lbs)	Model	Manufacture	Type	Style	BPM	Striking Weight (Lbs.)	Total Weight (Lbs)	Hammer Length (Ft.–In.)	Hydraulic Pressure (PSI)	Hydraulic Flow (GPM)	Width between Leads (In.)
62068	HHK 7A	JUNTTAN	ASSISTED	CLOSED	40–100	15430	24250 w/o	21'8"	2753	118	VARIABLE
61740	HH7S	BSP	FREE	OPEN	40	15500	20350 w/o	20'9"	3940	50	27"
60816	V100D-7	TWINWOOD	FREE	CLOSED	34	15656	24211 w/o	20'4"	3675	50	13"
60757	DKH 7	PILEMER	FREE	OPEN	40–100	15430	29315 w/o	18'	4000	61	44"
60757	NH-70S	NISSHA	SINGLE	—	38–80	15435	25357 w/o	18'5"	4053	59	13"
60756	SGH-7	BRUCE	FREE	OPEN	40	15432	24251 w/o	18'7"	3337	50	33"
59000	MHF 5-8	MENCK	FREEFALL	OPEN	80	17640	27342 w/o	18'1"	4060	66	36"
55000	SC-75	IHC	ASSISTED	CLOSED	45–100	12600	21600 w/o	20'4"	3900	66	30"
53196	HHK 6A	JUNTTAN	ASSISTED	CLOSED	40–100	13230	18630 w/o	20'6"	2608	95	VARIABLE
53196	HHK 6	JUNTTAN	FREE	OPEN	30–100	13230	18954 w/o	16' B	2318	79	VARAIBLE
52128	V100D-6	TWINWOOD	FREE	CLOSED	36	13450	21551 w/o	20'4"	3380	50	13"
52078	HH6LD	BSP	FREE	CLOSED	40–90	13230	19621 w/o	15'5"	3480	53	32"
52000	S-70	IHC	ASSISTED	CLOSED	50–125	7700	18300 w/o	23'4"	3300	66	25"
52000	10.6a LOW	APE	ASSISTED	C/O	40–80	17000	21000 w/o	10'6"	3800	90	26"
51600	MHF 3-7	MENCK	FREEFALL	OPEN	80	15435	21388 w/o	17'4"	4640	39	32"
50000	1250	HPSI	FREE	OPEN	40–60	12500	18000 w/o	19'2"	4250	45	21"
47800	SC-65	IHC	ASSISTED	CLOSED	50–100	10100	16300 w/o	19'10"	4000	52	26"
47000-											
18100	HPH 6500	DAWSON	DBL-ACT	CLOSED	80–130	10250	21000 w/o	17'	3500	67	32"
46000	115	ICE	FREE	OPEN	40–60	11500	17850 w/o	19'10"	2500	100	26"
46000	62	HMC (J & M)	FREE	OPEN	40–75	11500	18000 w/o	19'11"	2500	95	26"
44334	HHU5A	JUNTTAN	ASSISTED	CLOSED	40–100	11020	16800 w/o	15'3"	3043	95	26"
44334	HHK 5A	JUNTTAN	ASSISTED	CLOSED	40–100	11020	17260 w/o	15'3"	3043	95	VARIABLE
44334	HHK 5	JUNTTAN	FREE	OPEN	30–100	11020	16309 w/o	18'9"	2174	79	VARAIBLE
44200	MHF 3-6	MENCK	FREEFALL	OPEN	80	13230	19183 w/o	17'4"	4495	39	32"
44200	SC-60	IHC	ASSISTED	CLOSED	50–100	13200	20900 w/o	19'9"	3400	58	30"
44000	12a LOW	APE	ASSISTED	C/O	40–80	11000	16500 w/o	12'3"	3800	70	26"
43976	NH-40	NISSHA	DOUBLE	—	33–80	8820	21609 w/o	15'2"	2488	58	13"
43440	V100D-5	TWINWOOD	FREE	CLOSED	38	11245	19801 w/o	20'4"	3090	50	13"
43398	HH 5S	BSP	FREE	OPEN	42	11020	15984 w/o	18'10"	3045	48	27"
43398	DKH5	PILEMER	FREE	OPEN	40–100	11020	22049 w/o	17'	3600	50	44"
43398	HH5LD	BSP	FREE	CLOSED	42–90	11020	15652 w/o	14'7"	3625	48	32"
43397	SGH-5	BRUCE	FREE	OPEN	40	11023	19842 w/o	17'1"	3337	45	33"

Hydraulic Impact Hammers

Rated Energy (Ft–Lbs)	Model	Manufacture	Type	Style	BPM	Striking Weight (Lbs.)	Total Weight (Lbs)	Hammer Length (Ft.–In.)	Hydraulic Pressure (PSI)	Hydraulic Flow (GPM)	Width between Leads (In.)
40000	1000	HPSI	FREE	OPEN	40–70	10000	13100 w/o	20'	4000	37	21"
37000	SC-50	IHC	ASSISTED	CLOSED	50–100	7300	13000 w/o	17'3"	4200	52	26"
36800	MHF 3-5	MENCK	FREEFALL	OPEN	80	11025	16978 w/o	17'4"	4060	39	32"
35464	HHK 4	JUNTTAN	FREE	OPEN	30–100	8820	13885 w/o	17'6"	2028	79	VARIABLE
35464	HHK 4A	JUNTTAN	ASSISTED	CLOSED	40–100	8820	15876 w/o	18'4"	2173	95	VARIABLE
34752	V20B-4	TWINWOOD	FREE	CLOSED	42	9040	13671 w/o	17'8"	2940	26	13"
34752	V100D-4	TWINWOOD	FREE	CLOSED	40	9040	17596 w/o	20'4"	2650	50	13"
34718	DKH 4	PILEMER	FREE	OPEN	40–100	8816	19830 w/o	17'	3300	50	44"
34718	SGH-4	BRUCE	FREE	OPEN	42	8819	17637 w/o	17'1"	3337	45	33"
29500	MHF 3-4	MENCK	FREEFALL	OPEN	85	8820	14773 w/o	17'4"	3625	39	32"
29000	SC-40	IHC	ASSISTED	CLOSED	50–100	5600	11200 w/o	19'5"	3100	44	23"
26598	HHK 3A	JUNTTAN	ASSISTED	CLOSED	40–100	6600	13230 w/o	16'10"	1594	95	VARIABLE
26598	HHK 3	JUNTTAN	FREE	OPEN	30–100	6600	13230 w/o	16'2"	1884	79	VARIABLE
26064	V20B-3	TWINWOOD	FREE	CLOSED	44	6836	11465 w/o	17'8"	2350	26	13"
26039	HH 3S	BSP	FREE	OPEN	46	6615	11574 w/o	17'1"	2320	40	27"
26038	SGH-3	BRUCE	FREE	OPEN	42	6614	15432 w/o	17'1"	3337	45	33"
26000	S-35	IHC	ASSISTED	CLOSED	60–125	6610	15700 w/o	18'4"	3100	44	25"
24500	8a LOW	APE	ASSISTED	C/O	40–80	11000	12200 w/o	8'	3800	70	26"
23100	NH-20	NISSHA	DOUBLE	—	30–90	4410	14815 w/o	12'4"	2275	34	13"
22100	SC-30	IHC	ASSISTED	CLOSED	50–100	3750	9040 w/o	16'7"	2900	44	23"
21000	28	HMC (J & M)	FREE	OPEN	50–100	7000	13400 w/o	15'5"	2500	33	20"
21000	70	ICE	FREE	OPEN	50–100	7000	13400 w/o	15'5"	2500	33	20"
20400	7.5a LOW	APE	ASSISTED	C/O	40–80	12000	11500 w/o	7'6"	3800	70	26"
17376	V20B-2	TWINWOOD	FREE	CLOSED	46	4630	9261 w/o	17'8"	2060	26	13"
17360– 13750–	HPH 2400	DAWSON	DBL-ACT	CLOSED	80–120	4189	13227 w/c	14'7"	3300	40	26"
7074	HPH 1800	DAWSON	DBL-ACT	CLOSED	80–120	3300	8800 w/c	12'9"	3300	28	21"
7235 13381	HHI-5	BSP	ASSISTED	CLOSED	80	3300	8342 w/o	12'5"	1885	45	21"
8680–4640	HPH 1200	DAWSON	DBL-ACT	CLOSED	80–120	2300	6614 w/c	12'6"	3300	20	21"

B.4   VIBRATORY HAMMERS[2]

Hydraulic Vibratory Driver/Extractor Terminology

- Model
  Manufacturer's designation of the equipment. (Usually has some relationship to the eccentric moment or dynamic force.)
- Manufacturer
  The name of the manufacturing company.
- Frequency
  Frequency is the measurement of the rotational speed of the eccentrics and may be referred to as vibrations per minute (VPM) or cycles per minute (CPM).
- Eccentric Moment
  Eccentric moment is the distance from the center of rotation to center of gravity of each eccentric, times their total weight. For example if an eccentric weight weighs 500 pounds and the distance from the center of rotation to the center of gravity is one inch (500 × one inch = 500 inch pounds), then the eccentric moment is 500 inch pounds total.
- Amplitude
  Amplitude is the measurement of the distance the vibratory pile driver/extractor is traveling up and down, while it is vibrating in the "free-hanging position" without a pile attached. *Note*: Technically, amplitude is equal to half the distance of total travel. However, the pile-driving industry in the United States has adopted this term to mean the full travel up and down.
- Maximum Hydraulic Horsepower
  $HP = ((Flow\ (GPM) \times Pressure\ (PSI))/1714) \times Efficiency\ \%$
- Engine Horsepower
  This is the engine manufacturer's rated horsepower output of the engine. Some manufacturers have dual ratings: continuous and intermittent.
- Maximum Line Pull
  This is a measurement in tons of the maximum crane-line pull that can be exerted on the vibro suppressor housing while in the act of extracting piles.
- Pipe Clamp Cylinder Force
  Pile clamp force is a term used to describe the force exerted by the hydraulic clamp cylinder. *Note*: Some pile clamping devices use a lever arm system that adds to the clamping force. In these cases, the moment arm ratio must be considered (sometimes referred to as *Force of Clamp Cylinder*.)
- Suspended Weight
  This is a measurement of the total weight of the vibro while it is hanging from the crane line. Also included is the weight of the clamp assembly and 50 percent of the hose bundle weight.
- Shipping Weight
  This is the total weight of all the equipment, including the vibratory pile driver/extractor, power unit, hose bundle, and clamp assembly.
- Height
  Total height of the vibratory driver/extractor from the top of lifting ball to the bottom of pile guide.
- Width
  Total width of the vibratory driver/extractor, at its widest point.
- Length
  Total length (left to right) of the vibratory driver/extractor, including hoses.
- Throat Width
  Widest point of the vibratory (at the center area) that could be capable of falling between two adjoining sheets.

---

[2]To come

Vibratory Hammers

Manufacturer	Model	Freq. (VPM)	Eccentric Moment	Amplitude (In.)	Max. Hyd. H.P.	Engine H.P.	Max. Pull Extract. (Tons)	Pile Clamp Force (Tons)	Susp. Weight Lbs. (w/Clamp)	Ship Weight (Lbs.)	Height (w/clamp) (Ft.–In.)	Width (Thick.)	Length (Ft/In)	Throat Width (In.)
EXCAVATOR														
TRAMAC	428B	2800–3100	348	.35	113	147	14	56	2815	3015	4'7"	3'8"	2'	3.15"
HMC/MOVAX	SP-50	3000	650	.38	95	NA	15	60	4070	4070	6'8"	4'4"	3'1"	—
PTC	7HF4	2300	550	.60	141	N/A	12	40	2550	2550	4'11"	2'	2'6"	13"
PTC	7PHF4	2300	550	.70	141	N/A	12	40	2750	2750	4'5"	2'	3'5"	15"
TRAMAC	625SH	2750	522	.51	133	173	14	56	3250	3450	4'7"	6'	27.5"	3.15"
HMC/MOVAX	SP40	3000	500	.38	N/A	N/A	3–4	60	3936	4086	6'8"	5'6"	3'1"	—
MKT	V-2B	1800	440	.75	44	85	16	16	2700	2900	3'9"	10"	3'9"	12"
DAWSON	EMV300A	2400	400	.58	195	70	16	40	2000	2200	4'	2'	3'4"	6"
DAWSON	EMV300	2400	400	.58	195	70	8.8	40	1890	2100	3'7"	1'10"	3'4"	6"
HPSI	65E	2200	650	.75	120	175	15	50	3150	3150	5'5"	2'1"	3'7"	12"
TRAMAC	428SH	2800–3100	348	.35	113	147	14	56	3210	3410	4'8	6'	2'4"	3.15"
DAWSON	EMV400	2460	545	.55	272	107	16	48	2632	2800	4'1"	6'	4'7"	6"
TRAMAC	328SH	2800–3000	278	.35	73	93	11	39	3035	3235	4'7"	6'	2'2"	3.15"
TRAMAC	328B	2800–3000	278	.35	73	93	11	39	2400	2600	4'5"	3'8"	1'9	3.15"
PTC	3HF3	2800	260	.50	80	N/A	6	20	2070	2120	4'5"	1'7"	2'4"	11"
ICE	23E	1600	230	.58	23	N/A	13	20	1040	1040	4'10"	1'4"	2'2"	10.5"
HMC (J & M)	3X	2.200	230	.58	75	NA	13	20	1040	1040	4'10"	1'4"	2'2"	10.5
HPSI	20	2200	200	.625	60	80	12	25	1000	3500	5'5"	3'	1'10"	10"
TRAMAC	230SH	3000–3500	191	.24	60	80	11	39	3015	3215	4'7"	6'	2'2"	3.15"
TRAMAC	230B	3000–3500	191	.24	60	80	11	39	2375	2575	4'5"	3'8"	1'9"	3.15"
DAWSON	EMV70	3000	60	.134	94	16	3	33	1150	1300	3'2"	1'2"	2'7"	6"
HPSI	40E	2200	400	.875	60	80	15	30	3150	3150	5'5"	2'1"	3'7"	12"
HMC (J & M)	13X	1600	1100	.75	146	NA	45	50	2005	2005	6'5"	1'4"	4'2"	16"
TRAMAC	625B	2750	522	.51	133	173	14	56	2860	3060	4'7"	3'8"	2'	3.15"
HMC/MOVAX	SP50	3000	700	.38	N/A	N/A	3–4	60	3950	4100	6'8"	5'6"	3'1"	—
APE	100E	100–1800	2200	.89	340	N/A	50	125	6250–10000	10000	5'8"	1'2"	7'	14"
APE	50E	400–2000	1300	.875	188	N/A	50	125	5500	5800	4'7"	1'2"	4'4"	12"
ICE	216E	1600	1100	1.02	146	N/A	45	50	2005	2005	6'6"	1'4"	4'2"	16"
PTC	13H2	1750	1100	.75	141	N/A	12	55	3750	3750	5'6"	2'1"	3'1"	13"
PTC	13PH2	1750	1100	.80	141	N/A	12	55	4180	4180	4'6"	2'2"	4'1"	15"
HMC/MOVAX	SP60	3000	900	.38	N/A	N/A	3–4	60	3966	4116	6'8"	5'6"	3'1"	—
PTC	10HF5	2100	900	.60	141	N/A	12	55	3700	3700	5'6"	2'1"	3'1"	13"
PTC	10PHF5	2100	900	.60	141	N/A	12	55	4135	4135	4'6"	2'1"	4'2"	15"
HPSI	80E	2200	800	.75	120	175	15	50	3150	4100	5'5"	2'6	4'6"	12"
APE	20E	2000	750	.875	87	N/A	20	50	3600	3600	3'10	1'	3'1"	12"

Vibratory Hammers

Manufacturer	Model	Freq. (VPM)	Eccentric Moment	Amplitude (In.)	Max. Hyd. H.P.	Engine H.P.	Max. Pull Extract. (Tons)	Pile Clamp Force (Tons)	Susp. Weight Lbs. (w/Clamp)	Ship Weight (Lbs.)	Height (w/clamp) (Ft.-In.)	Width (Thick.)	Length (Ft/In.)	Throat Width (In.)
MKT	V-5E	1700	1300	1.0	135	185	30	62	6200	6400	6'6"	1'1"	6'6"	13"
HIGH		—	—									—		
PVE	2520	2000	2200	.99	375	389	45	124	12980	25080	7'10"	2'4"	7'5"	13.8"
PVE	2307	2300	570	.66	122	130	20	62	3828	12028	5'6"	1'6"	3'7"	12.4"
PTC	13HF3	2300	1150	.90	163	270	34	120	7260	17930	6'10"	2'	5'9"	12"
PTC	30HF3A	2300	2350	.90	397	522	45	240	13530	30180	8'11"	2'6"	7'3"	13"
PTC	23HF3A	2300	2000	.80	250	385	45	170	11330	23100	8'11"	2'6"	7'3"	13"
PVE	2323	2300	2000	.91	375	533	45	124	12210	24970	7'10"	2'4"	7'5"	13.8"
PTC	15HF3	2300	1300	.90	227	385	34	120	7480	19250	6'10"	2'4"	5'9"	12"
PVE	2315	2300	1300	.70	288	318	34	96	7480	18040	6'7"	2'3"	5'7"	13.8"
PVE	1420	2000	1250	.82	240	260	34	96	6120	14920	6'	2'	4'4"	13.8"
VULCAN	1400	2400	600	.33	123	175	25	50	4350	11850	8'8"	1'11"	3'8"	12"
APE	3HF	3000	30	.375	9	14	5	20	285	400	3'	7"	1'6"	5"
APE	6HF	3000	60	.50	9	14	10	20	600	800	3'	7"	2'1"	5"
DELMAG/TUNKERS	HVB10	2100	191	.31	43	131	7/13	27	2000	7835	3'3"	—	3'4"	14.56"
VULCAN	400A	2400	200	.375	42	58	15	15	1250	5100	4'4"	1'5"	2'-3/4"	10"
DELMAG/TUNKERS	HVB16	2100	278	.39	64	131	7/13	27	2000	7835	3'3"	—	3'4"	14.56"
PTC	7HF3	2300	550	.60	125	167	22	55	4642	12162	6'9"	1'11"	3'8"	12"
PVE	2330	2300	2400	.79	429	600	45	169	16046	28806	9'1"	2'8"	7'5"	13.8"
PTC	46HF3	2300	4000	.50	607	684	67	2@150	29700	52860	12'4"	3'11"	7'	31"
DELMAG/TUNKERS	HVB24	2100	434	.55	86	131	7/13	27	2100	7935	3'3"	—	3'6"	14.56"
RING		—	—									—		
PTC	13HT3	1700	1100	0.20	145	N/A	45	60	8800	13200	6'10"	4'6"	6'5"	14"/18"
PTC	50HT1	1500	4400	0.55	378	515	45	150	15400	19850	9'3"	4'10"	7'2"	14"/18"
IHC	FV-50R	1500	4300	0.6	328	430	44	280	13900	15400	6'1"	5'	7'6"	4'4"
PTC	50HTV	1500	0-4400	0.49	429	515	45	150	17600	22000	10'6"	6'3"	6'11"	14"/18"
PTC	40HT1	1500	3500	0.45	350	379	45	150	15200	19650	9'3"	4'10"	7'2"	14"/18"
PTC	30HT1	1700	2600	0.45	268	379	45	75	11900	16300	7'10"	4'6"	5'10"	14"/18"
PTC	30HTV	1700	0-2200	0.43	296	379	45	75	11000	15500	8'4"	6'2"	6'8"	14"/18"
IHC	FV-60R	1500	5200	0.7	328	441	88	280	14300	15900	6'1"	5'	7'6"	4'4"
PTC	13HFT3	2300	1100	0.20	145	N/A	45	72	8800	13200	6'10"	4'6"	6'5"	14"/18"
STANDARD		—	—									—		
HPSI	200	1600	2000	.875	204	300	45	150	8600	19850	5'3"	1'8"	7'11"	14"

Vibratory Hammers

Manufacturer	Model	Freq. (VPM)	Eccentric Moment	Amplitude (In.)	Max. Hyd. H.P.	Max. Engine H.P.	Max. Pull Extract. (Tons)	Pile Clamp Force (Tons)	Susp. Weight Lbs. (w/Clamp)	Ship Weight (Lbs.)	Height (w/clamp) (Ft.–In.)	Width (Thick.)	Length (Ft/In)	Throat Width (In.)
MKT	V-17	1700	2200	.75	254	325	60	70	12000	31500	10'0"	1'1"	8'6"	13"
FOSTER	4000	1400	4000	.31–1.25	299	445	45	100/200	18800	29960	6'4"	1'10"	9'8"	12"
PVE	23M	1650	2200	.75	261	318	45	96	11000	21560	7'11"	2'	6'1"	13.8"
APE	100	2000	2200	.89	217	260	50	125	6250–10000	17500	5'8"	1'2"	7'	14"
HMC (J & M)	25–220	1250	2200	1.03	192	220	40	80	5500	15000	7'8"	1'7"	5'3"	14.25"
H & M	H-3400	1200	3400	—	375	460	30	100	10000	20700	10'9"	—	5'6"	12"
DELMAG/ TUNKERS	HVB60.05	1500/1000	2065/ 3879	.67– 1.18	251/ 191	312	27.5	91	7050	18850	10'2"/ 5'3"	—	5'7"	11.81"
PVE	38M	1700	3300	.89	394	533	57	124	13530	26290	8'10"	2'4"	7'5"	13.8"
PTC	25H1A	1750	2000	.85	181	270	45	120	9660	20330	7'11"	2'2"	8'2"	13"
MKT	V-16	1750	1800	.47	161	210	40	75	9250	24000	11'2"	1'2"	6'6"	14"
FOSTER	1800	1600	1800	.32–.75	—	220	30	90/120	11000	24375	6'8"	1'9"	9'0"	12"
FOSTER	1700	1400	1740	.31–.75	165	220	30	100/120	12900	19930	6'3"	1'10"	7'	12"
H & M	H-1700	1200	1700	.50–.75	200	225	30	75	7000	13000	8'11"	1'	5'	12"
PTC	18H2	1600	1560	1.00	146	173	30	55	6070	13555	7'	2'	5'2"	13"
DELMAG/ TUNKERS	HVB70.01	1630	2065	.74	290	312	28/55	120	7050	17850	10'2"/5'3"	—	5'7"	11.81"
HMC (J & M)	26–335	1600	2200	1.02	279	335	40	125	9550	20885	6'11"	1'5"	7'11"	14"
PTC	30H1A	1700	2600	.90	262	385	45	120	10760	22530	7'11"	2'5"	8'2"	13"
APE	150T	1800	2600	.95	525	350	80	125	8700– 12000	22400	7'1"	1'7"	7'4"	14"
HPSI	260	1600	2600	.80	280	335	60	200	10750	23000	8'4"	1'10"	7'11"	14"
HPSI	250	1600	2500	.875	280	335	45	200	8600	20250	5'3"	1'8"	7'11"	14"
MKT	V-20B	1700	2400	.75	265	325	60	75	10750	22000	10'0"	1'1"	8'6"	13"
DELMAG/ TUNKERS	HVB130.01	1600	3991	.79	547	625	55/110	154	12600	31300	13'3"/5'9"	—	7'0"	17.32"
APE	150	1800	2200	.875	340	350	80	125	8500–12000	22000	7'1"	1'7"	7'4"	14"
PVE	25M	1700	2600	.75	365	389	45	96	12276	24376	7'10"	2'4"	7'5"	13.8"
ICE	416L	1600	2200	1.02	264	325	40	125	9550	20885	7'0"	1'5"	7'11"	17"
APE	50	2000	1300	.875	197	260	50	125	6000–10000	17000	5'8"	1'2"	4'4"	14"
H & M	H2700	1200	2700	.875	285	335	30	75	9300	19000	12'2"	2'	5'6"	12"
DELMAG/ TUNKERS	HVB100.01	1600	2950	.70	359	503	55	154	10580	23600	13'/5'5"	—	7'0"	14.74"
MKT	V-20	1650	3000	.66	310	350	60	75	12500	31500	12'2"	1'2"	8'0"	14"

Vibratory Hammers

Manufacturer	Model	Freq. (VPM)	Eccentric Moment	Amplitude (In.)	Max. Hyd. H.P.	Engine H.P.	Max. Pull Extract. (Tons)	Pile Clamp Force (Tons)	Susp. Weight Lbs. (w/Clamp)	Ship Weight (Lbs.)	Height (w/clamp) (Ft.-In.)	Width (Thick.)	Length (Ft/In)	Throat Width (In.)
HPSI	300	1600	3000	.875	335	400	60	200	10750	23000	8'4"	1'10	7'11"	14"
VULCAN	2300A	1600	2300	.75"	237	360	48	87	8550	16550	9'7"	2'4"	5'	14"
HMC (J & M)	13-200	1600	1100	.75	146	200	40	50	5096	14744	5'9"	1'4"	4'2"	12"
MKT	V-2A	1800	400	.75	44	70	8	16	2400	4400	3'8"	1'0"	3'8"	12"
MKT	V-2B	1800	440	.75	44	70	16	16	2700	5525	3'9"	1'0"	3'7"	12"
PVE	7M	1800	520	.64	86	133	20	62	3872	12072	5'10"	2'	4'1"	13.8"
PTC	7H5	2000	550	1	63	109	13.5	40	2375	5980	4'10"	1'11"	2'3"	12"
FOSTER	4030	1500	4000/3000	.31-1.25	300	450	45	100/200	17600	32500	6'5"	1'11"	9'10"	12"
MKT	V-14	1500	1500	.32	140	210	40	75	10000	29500	11'1"	1'2"	5'3"	14"
HMC (J & M)	115-800	1250	10000	1.3	760	800	150	240	28590	53181	13'9"	3'5"	7'11"	32"
HMC (J & M)	3-28	1600	230	.58	23	28	13	20	1800	3600	4'2"	1'4"	2'2"	10.5"
ICE	216	1600	1100	.94	146	175	40	50	5096	15346	5'9"	1'4"	4'2"	16"
H & M	H-75E	1800	—	.25-.50	75	210	15	40	2400	—	4'	—	3'1"	14"
H & M	H-75B	1900	—	.25-.50	105	210	15	62	4000	9000	5'7"	—	3'1"	14"
MKT	V-5B	1800	1100	.75	118	175	30	62	7200	11200	7'8"	1'1"	6'4"	13"
FOSTER	1050	1500	1050	.25-1	82	175	32	96	6990	15880	7'6"	1'5"	5'1"	14"
MKT	V-5	1450	1000	.50	59	175	30	62	5200	9200	9'4"	1'2"	6'10"	14"
DELMAG/TUNKERS	HVB30	1800	712	.79	112	158	7/13	33	2100	7935	3'4"	—	3'6"	14.56"
HPSI	130	1600	1300	.80	137	175	30	50	7050	15000	6'6"	1'7"	7'3"	14"
HPSI	150	1600	1500	.875	175	210	30	50	7050	15000	6'6"	1'7"	7'3	14"
ICE	11-23	1900	1100	.75	176	220	40	85	5250	15650	6'7"	1'4"	4'2"	17"
DELMAG/TUNKERS	HVB40.01	1500	1346	.67	178	253	28	77	5700	16100	9'8"/4'11"	—	5'3"	11.81"
FOSTER	1000	1600	1000	.25-.75	—	185	20	90	6094	10500	5'11"	1'8"	5'7"	12"
MKT	V-5E	1700	1300	1	135	185	30	62	6200	11000	6'6"	1'1"	6'6"	13"
H & M	H-150	1700	—	.50-.75	150	225	15	60	4300	10200	6'3"	—	3'6"	14"
H & M	H-50	2000	230	.25-.50	50	90	10	30	1700	5800	4'9"	—	2'11"	10"
HPSI	130L	1600	1300	.80	137	175	30	50	4250	13000	5'3"	1'7"	5'11"	14"
ICE	23	1600	230	.58	23	28	13	20	1440	3373	4'2"	1'4"	2'2"	10.5"
PTC	15H1	1650	1300	1	146	167	22	55	5590	13110	6'1"	2'3"	5'5"	12"
VULCAN	1150A	1600	1150	.75	131	155	32	48	6300	13800	7'11"	2'1"	5'	14"
HMC (J & M)	13H-220	1900	1100	.75	192	220	40	80	5100	14800	6'8"	1'6"	4'10"	13.25"
APE	6	1600	60	.50	9	14	10	20	600	800	3'	7"	2'1"	5"
H & M	H-25	2000	85	—	20	25	5	12	950	1700	3'3"	—	—	9"

Vibratory Hammers

Manufacturer	Model	Freq. (VPM)	Eccentric Moment	Amplitude (In.)	Max. Hyd. H.P.	Engine H.P.	Max. Pull Extract. (Tons)	Pile Clamp Force (Tons)	Susp. Weight Lbs. (w/Clamp)	Ship Weight (Lbs.)	Height (w/clamp) (Ft.-In.)	Width (Thick.)	Length (Ft/In)	Throat Width (In.)
HPSI	150L	1600	1500	.875	175	210	30	50	4250	13000	5'3"	1'7"	5'11"	14"
MKT	V-5C	1700	1300	1	135	185	30	62	7200	12030	7'7"	1'1"	6'4"	13"
APE	400B	1400	13000	1.5	791	990	250	4@125	38–52000	78000	11'	2'2"	10'	26"
ICE	66–65	1300	6600	1.23	595	650	80	125	17400	40080	11'5"	1'10"	8'1"	14.25"
HMC (J & M)	76–740	1300	6600	1.23	610	740	80	125	15075	40000	11'5"	1'10"	8'1"	14.25"
ICE	66–80	1600	6600	1.23	760	800	80	196	19644	41995	11'5"	1'10"	8'1"	14.25"
HMC (J & M)	76–800	1600	6600	1.23	760	800	80	196	15550	42000	11'5"	1'10"	8'1"	14.25"
HPSI	750	1400	7500	1	600	860	100	200	19500	42000	8'9"	1'10"	8'6"	14"
DELMAG/ TUNKERS	HVB260.01	1600	7983	.86	1079	1153	110	287	22000	44000	14'9"/6'	—	7'0"	18.1"
APE	3	1600	30	.375	9	14	5	20	225	400	3'	7"	1'6"	5"
HPSI	400	1600	4000	1.12	408	505	75	200	16000	30000	8'9"	1'10	8'6"	14"
HPSI	1000	1400	10000	.75	600	860	150	4@150	39000	78000	8'7"	1'10"	11'	20"
PTC	75HL	1500	6500	1.10	475	507	67	240	25285	42565	8'6"	2'4"	11'4"	15"
HPSI	1200	1200	12000	.75	850	1200	150	4@150	44340	92000	8'7"	5'8"	11'	40"
ICE	1412-B	1250	10000	1.3	760	800	150	200	28590	53181	17'1"	3'5"	7'11"	32"
MKT	V-140	1400	14000	1	1400	1800	150	300	42000	96000	14'6"	4'0"	11'10"	48"
PTC	175HD	1400	15600	1	830	4065	202	4@150	55850	79750	16'8"	4'11"	7'7"	31"
HPSI	1600	1400	16000	1	1200	1600	150	4@150	55000	107000	8'7"	5'11"	11'	40"
PVE	200M	1400	17600	1.14	1420	1650	176	676	52976	94776	11'	4'5"	10'4"	27.6"
APE	600B	1400	19500	1.75	1000	1400	250	4@125	42–58000	80000	7'	2'2"	13'9"	26"
HMC (J & M)	230–800	1250	20000	1.3	1520	1600	300	480	71914	121029	5'4"	8'	8'10"	64"
PTC	240HD	1380	21000	1	1500	1823	202	4@280	76680	113500	10'3"	6'6"	13'4"	55"
HPSI	2000	1400	21500	1.1	1500	1600	150	4@150	58000	110000	10'8"	5'11"	11'	40"
IHC	FV-450	950	39100	1.6	1415	1606	110	8@67	94800	99200	14'3"	7'10"	7'10"	7'10"
PVE	110M	1350	9600	1.19	610	798	136	225	30800	45100	10'7"	3'2"	8'8"	26.4"
ICE	612	1200	4400	1	254	325	40	125	12275	23500	8'4"	2'5"	7'11"	29"
PTC	100HD	1400	10400	1.10	613	684	135	2@150	35920	58720	13'4"	4'11"	7'7"	31"
FOSTER	4200	900–1500	4166	.32–1	436	503	80	196	16300	30950	7'6"	2'	9'2"	14"
APE	300	1400	6500	1.375	525	625	150	2@125	18–22000	40000	9'10"	2'2"	9'3"	14"
ICE	44–30	1600	4400	1.17	264	325	80	125	13650	24985	8'2"	1'10"	8'1"	14.25"
HMC (J & M)	51–335	1200	4400	1.17	279	335	80	125	13650	26000	8'2"	1'10"	8'1"	14.25"
APE	200A	1400	4400	1.125	525	350	150	125	16–18500	26000	8'	1'7"	8'5"	14"
ICE	815	1600	4400	1	452	505	50	125	14750	30500	9'1"	2'4"	7'7"	28"
ICE	44–50	1600	4400	1.17	452	505	80	125	13650	30738	8'2"	1'10"	8'1"	14.25"

Vibratory Hammers

Manufacturer	Model	Freq. (VPM)	Eccentric Moment	Amplitude (In.)	Max. Hyd. H.P.	Engine H.P.	Max. Pull Extract. (Tons)	Pile Clamp Force (Tons)	Susp. Weight Lbs. (w/Clamp)	Ship Weight (Lbs.)	Height (w/clamp) (Ft.–In.)	Width (Thick.)	Length (Ft/In.)	Throat Width (In.)
PTC	50HL	1600	4400	1.18	414	522	80	170	14650	31950	8'11"	2'3"	8'4"	15"
HMC (J & M)	51–535	1500	4400	1.17	452	535	80	125	13650	30738	8'2"	1'10"	8'1"	14.25"
MKT	V-30	1700	4400	1	510	600	80	100	15500	32500	11'1"	1'1"	8'6"	13"
PTC	60HD	1650	5200	1	414	522	67	2@120	19150	36450	11'4"	2'5"	7'9"	12"
FOSTER	4150	900–1500	4166	.32–1.25	—	503	55	145/200	16495	31995	6'6"	2'3"	8'0"	12"
APE	200B	1800	4400	1.125	525	625	150	125	16–18500	35000	8'	1'7"	8'5"	14"
APE	200T	1800	5200	1.135	525	625	150	125	16–18500	36000	8'	1'7"	8'5"	14"
HPSI	500	1600	5000	1.125	600	700	75	200	16500	35000	8'9"	1'10"	8'6"	14"
MKT	V-36	1600	5000	.75	550	650	80	100	18800	36300	13'1"	1'2"	12'0"	14"
MKT	V-35	1700	4800	1	550	650	80	100	15500	32650	11'4"	1'1"	8'0"	13"
VULCAN	4600A	1600	4600	1–1.375	—	560	80	177	14000	24600	10'11"	2'6"	7'2"	14"
PVE	52M	1700	4500	.95	508	750	57	169	17094	31394	9'8"	2'4"	8'3"	14.2"
HPSI	450	1600	4500	1	555	600	75	200	16500	33000	8'9"	1'10"	8'6	14"
HMC (J & M)	51–740	1700	4400	1.17	610	740	80	125	13650	38931	8'2"	1'10"	8'1"	14.25"
ICE	44–65	1650	4400	1.17	560	650	80	200	13650	37500	8'2"	1'10"	8'1"	14.25"
VARIABLE		—	—		—	—								—
PTC	15HFVS	2300	0–1300	0–0.60	272	385	34	120	8760	20620	8'2"	2'7"	5'10"	13"
HPSI	VM30	2200	0–2500	0–0.75	455	505	60	200	9500	19700	6'5"	1'10"	7'1"	14"
PTC	60HFV	2300	0–4800	0–0.70	578	684	67	2@240	31400	54700	12'1"	4'7"	7'4"	31"
PVE	50VM	1800	0–4400	0–0.83	494	600	45	169	18194	32494	9'11"	2'6"	7'7"	12.6"
PVE	40VM	2000	0–3500	0–0.79	680	750	45	169	16434	30734	9'11"	2'6"	7'7"	12.6"
PVE	2335VM	2300	0–3100	0–0.60	787	1200	57	225	18480	33000	10'4"	2'4"	6'7"	13.4"
PTC	34HFV	2300	0–3000	0–0.70	578	684	45	240	20130	43430	10'7"	3'1"	7'7"	15"
PVE	2332VM	2300	0–2800	0–0.62	680	750	57	225	17600	31900	10'4"	2'4"	6'7"	13.4"
PTC	60HVW	1680	0–4800	0–0.70	449	507	67	2@120	22850	39200	10'12"	3'7"	7'3"	31"
PVE	2330VM	2300	0–2400	0–0.62	429	600	45	169	12474	25234	9'3"	2'10"	7'5"	13.8"
PTC	30HFV	2300	0–2400	0–0.70	417	522	45	240	17030	33780	10'7"	2'10"	7'3"	13"
PVE	2025VM	2000	0–2200	0–0.60	332	533	45	124	10890	24210	9'2"	2'4"	6'1"	13.8"
PVE	2323VM	2300	0–2000	0–0.60	384	533	45	124	10450	23210	9'2"	2'4"	6'1"	13.8"
PVE	23HFV	2300	0–2000	0–0.60	302	385	45	170	14650	26510	9'2"	2'10"	7'3"	13"
PVE	2316VM	2300	0–1400	0–0.67	336	389	34	124	8250	20350	7'9"	2'1"	5'5"	12.5"
HPSI	VM15	2200	0–1300	0–0.50	236	250	30	150	7500	16000	6'5"	1'10"	7'1"	14"
PTC	15HFV	2300	0–1300	0–0.70	190	270	34	120	8350	19120	8'2"	2'7"	5'10"	13"
PTC	10HFV	2300	0–700	0–0.40	129	167	22	55	5390	11100	5'4"	2'7"	5'1"	12"
PTC	17HFV	2300	0–1500	0–0.70	314	385	34	120	11360	23220	8'10"	2'8"	6'3"	15"

Nomenclature

Conventional units in the English and SI systems are shown with terms where they apply. Where no units are shown, the terms are dimensionless.

A_p	Area of the tip of the pile or cross-sectional area of the pile (ft^3 or m^3)
A_s	Area of the side of the pile (ft^3 or m^3)
B	Footing width (ft or m)
C_c	Coefficient of curvature
C_c	Compression index
C_u	Uniformity coefficient
D_f	Depth of footing (ft or m)
D_n	Maximum particle size in finer nth percentage (mm)
D_r	Relative density (%)
D_{10}	Effective size of soil particles (mm)
E_p	Modulus of elasticity of the pile material (psi, psf, ksf or MPa)
E_r	Rated striking energy of a pile hammer (ft-lbs or kJ)
F_D	Pile length factor, cohesive soils
F_L	Pile length factor, cohesionless soils
F_{SD}	Skin friction factor, cohesionless soils
FS	Factor of safety
FS_c	Factor of safety with respect to cohesion
FS_f	Factor of safety with respect to friction
G_s	Specific gravity of the soil solids
H	Longest path taken by water seeping from the soils as a result of application of the consolidating pressure increment, D_p (in., cm, or ft)
H	Height of water surface above datum (ft or m)
H_c	Critical height of slope (ft or m)

H_w	Height of water surface in well (ft or m)
K	Coefficient of lateral earth pressure
K_A	Active earth pressure coefficient
K_o	Coefficient of lateral earth pressure at rest
L, L_p	Length of pile or drilled shaft (ft or m)
LL	Liquid limit (%)
N	Standard penetration resistance (blows/ft)
N_{60}	standard penetration resistance, corrected to a hammer of 60% efficiency (blows/ft)
N	Number of taps of liquid-limit device needed to close standard groove
N_c	Bearing capacity factor for cohesion
N_q	Bearing capacity factor for surcharge
N_s	Stability number
N_γ	Bearing capacity factor for soil weight
N_ϕ	Rankine passive earth pressure coefficient, level backfill
P	Wheel load (kips, lb, or kN)
P_l	Plasticity index (%)
PL	Plastic limit (%)
Q	Total flow volume (ft^3 or m^3)
Q_a	Allowable load capacity of a deep foundation (tons, kips, or MN)
Q_d	Ultimate load capacity of deep foundation (tons, kips, or MN)
Q_p	Point load capacity of deep foundation (tons, kips, or MN)
Q_s	Friction load capacity of deep foundation (tons, kips, or MN)
R	Radius of influence of well (ft or m)
R_w	Radius of well (ft or m)
S	Degree of saturation (%)
SF	Shrinkage factor (%)
SL	Shrinkage limit (%)
T_s	Surface tension (g/cm)
T_u	Time factor for consolidation analysis
V_A	Volume of air in soil (cm^3, ft^3, or m^3)
V_S	Solid volume in soil (cm^3, ft^3, or m^3)
V_t	Total soil volume (cm^3, ft^3, or m^3)
V_{tcut}	Total soil volume of the cut soil (ft^3 or m^3)
V_{tfill}	Total soil volume of the fill soil (ft^3 or m^3)
V_V	Total soil void volume (cm^3, ft^3, or m^3)
V_W	Volume of water in soil (cm^3, ft^3, or m^3)
W	Pile hammer ram weight (kips, lb, or kN)
W_S	Weight of soil solids (g, lb, or kN)
W_{scut}	Weight of soil solids in the cut soil (lb, or kN)
W_{sfill}	Weight of soil solids in the fill soil (lb or kN)
W_t	Total weight of soil volume (g, lb or kN)
W_{tcut}	Total weight of the cut soil (lb or kN)
W_{tfill}	Total weight of the fill soil (lb or kN)
W_W	Weight of water in soil (g or lb)
W_{wcut}	Weight of water in the cut soil (lb or kN)
W_{wfill}	Weight of water in the fill soil (lb or kN)

c	Cohesion (tsf, ksf, psf, or kPa)
c	Constant for *Engineering News* pile driving formula (in.)
c_c	Critical cohesion (tsf, ksf, or kPa)
c_v	Coefficient of consolidation (cm^2/sec)
$c_{req'd}$	Cohesion required for stable slope (tsf, ksf, psf or kPa)
d	Diameter of capillary (cm)
d_p	Diameter of tire contact area (in. or mm)
d_s	Width of loaded area on subgrade (ft, in. or m)
e	Void ratio
e_l	Initial void ratio
$e_{(max)}$	Maximum void ratio determined by standard test method
$e_{(min)}$	Minimum void ratio determined by standard test method
f_s	Unit side friction for deep foundation (tsf, ksf, psf, or kPa)
h	Height of drop of pile hammer (ft or m)
h_l	Thickness of compressible stratum (ft or m)
i	Hydraulic gradient
k	Permeability coefficient (cm/sec)
m	Slope of elastic deflection line (lb/ft, kips/ft, MN/m)
n	Porosity (%)
p	Total stress (ksf, psf, or kPa)
p	Tire pressure (psi or Pa)
p'	Effective stress (ksf, psf, or kPa)
p_{1c}	Preconsolidation pressure (ksf, psf, or kPa)
p_h	Horizontal earth pressure (ksf, psf, or kPa)
p_o	Vertical effective stress (ksf, psf, or kPa)
p_v	Total vertical stress or vertical earth pressure (ksf, psf, or kPa)
p_1'	Initial effective stress (ksf, psf, or kPa)
p_2'	Effective stress after increment, D_p (ksf, psf, or kPa)
q	Flow rate (water; ft^3/sec or m^3/sec)
q_a	Allowable bearing pressure (ksf, psf, or kPa)
q_d	Ultimate bearing capacity (ksf, psf, or kPa)
q_p	Net unit point resistance for deep foundation (tsf, ksf, psf, or kPa)
q_p	Tire contact stress on pavement (psi or Pa)
q_s	Pressure on subgrade (psf or kPa)
q_u	Unconfined compressive strength (tsf, ksf, pcf, or kPa)
s	Penetration of the pile resulting from one hammer blow (in. or mm)
s	Settlement (in. or mm)
s	Shearing strength (tsf, ksf, psf or kPa)
$s_{(max)}$	Maximum shear stress (tsf, ksf, psf, or kPa)
t	Pavement thickness (in. or mm)
t_u	Time for a given percentage of consolidation to occur (min, mo, or yr)
u	Pore water pressure (ksf, psf, or kPa)
v	Apparent velocity of flow (fps or m/s)
w	Water content (%)
z_c	Height of capillary rise (cm or ft)
α	Unslope angle (°)

α	Meniscus contact angle (°) or adhesion factor for clay soils in piles
β	Slope angle from the horizontal (°)
γ, γ_t	Density or unit weight (pcf or kN/m^3)
γ_d	Dry density or dry unit weight (pcf or kN/m^3)
γ_{dcut}	Dry unit weight of the cut soil (pcf or kN/m^3)
γ_{dfill}	Dry unit weight of the fill soil (pcf or kN/m^3)
$\gamma_{d(max)}$	Maximum dry density determined by standard test method (pcf or kN/m^3)
$\gamma_{d(min)}$	Minimum dry density determined by standard test method (pcf or kN/m^3)
$\gamma_{d(zav)}$	Dry density for saturated soil at specified water content (pcf or kN/m^3)
γ_s	Density or unit weight of solid material (pcf or kN/m^3)
γ_{tcut}	Unit weight of the cut soil (pcf or kN/m^3)
γ_{tfill}	Unit weight of the fill soil (pcf or kN/m^3)
γ_w	Density or unit weight of water (pcf or kN/m^3)
Δ_h	Settlement (in. or mm)
Δ_p	Effective stress increment (ksf, psf or kPa)
ϕ	Angle of internal friction (°)
$\phi_{req'd}$	Angle of internal friction required for stable slope (°)
ϕ'_s	Approximate angle of internal friction for slope with seepage (°)

UNIFIED CLASSIFICATION SYMBOLS

G	gravel
S	sand
M	silt
C	clay
O	organic
Pt	peat
W	well-graded soil
P	poorly graded soil
L	low compressibility or plasticity
H	high compressibility or plasticity
NP	nonplastic

Problem Solutions

CHAPTER 3

1. (a) 113.6 lb/ft^3
   (b) 100 lb/ft^3
   (c) 0.63
   (d) 39 percent
   (e) 59 percent
   (e) 124 lb/ft^3

2. (a) 105.5 lb/ft^3
   (b) 68.6 lb/ft^3

3. 0.26 tons = 0.52 kips

4. (a) 103.4 lb/ft^3
   (b) 37.2 percent
   (c) 0.592
   (d) 71.5 percent

5. 114 percent

6. 20 gal/yd^3

7. 104.9 lb/ft^3, 73.07 lb/ft^3, 2.38

CHAPTER 4

1. OL

2. SM

3.

Soil No.	Unified	AASHTO
2	SC	A-2-6 (1)
4	MH	A-7-5 (17)
6	CL	A-6 (17)

CHAPTER 5

1.

Depth (ft)	Total, p_v (lb/ft^2)	Water, u (lb/ft^2)	Effective, p_v' (lb/ft^2)
0 (surface)	0	0	0
10 (water table)	1100	0	1100
32 (stratum change)	3828	1373	2455
50 (bedrock)	5628	2496	3132

2. 624 psf

3. 49 ft

4. $k = 1.0 \times 10^{-2}$ cm/sec

5. $k = 3.0 \times 10^{-2}$ cm/sec

6. Improve by including an observation well.

7. $C_c = 0.63$ and $C_c' = 0.09$

8. $c_v = 1.2 \times 10^{-4}$ cm^2/sec

9. Undisturbed $q_u = 0.50$ tsf $= 1$ ksf and remolded $q_u = 0.30$ tsf $= 0.60$ ksf

10. $\phi = 34°$

11. $q_u = 345$ psf

CHAPTER 8

1. 89.9 pcf, 27.8 percent

2. An error was made in the test.

4. $\gamma_d = \dfrac{\gamma_w}{w + \dfrac{1}{G_s}}$

5. Specifications not met.

6. (a) $D_r = 59$ percent

   (b) $D_r = 29$ percent

   (c) $D_r = 117$ percent

7. 15 percent

8. $\gamma_{d-\#4} = 124.1$ lb/ft^3

   $w_{-\#4} = 7.9$ percent

CHAPTER 9

1. (a) Dewatering method: Wellpoints; initial spacing is 12 ft.

   (b) Dewatering method: Cofferdam.

2. (a) Dewatering method: Sump pump.

   (b) Dewatering method: Caisson.

   (c) Do not plan operation without consultation of specialty subcontractor.

CHAPTER 10

1. FS $= 3.4$. Cut can be made safely, and normal work can progress without danger.

2. Trench will have to be structurally supported.

3.

Depth (ft)	p'_v (lb/ft^2)	p'_h (lb/ft^2)	u (lb/ft^2)	p_h (lb/ft^2)
0	0	0	0	0
10	1100	440–550	0	440–550
32	2455	982–1228	1373	2355–2601
50	3132	1253–1566	2496	3749–4062

5. Support method: Soldier piles and horizontal lagging.

6. Support method: Vertical sheeting with tiebacks.

CHAPTER 11

1. For 5-ft footing, estimated settlement $= 1.72$ in.

   For 10-ft footing, estimated settlement $= 2.42$ in.

2. FS $= 5.554$

3. $P = 17$ kips

4. $P = 15$ kips

5. $Q_u = 83$ kips; minimum hammer energy $= 8,300$ ft-lbs; hammer selections include Vulcan or Raymond #1, or Vulcan 50C.

6. Length estimate $= 86$ ft.; minimum hammer energy $= 24,000$ ft-lbs; hammer selections include Vulcan 08, 80C or 505.

7. Select a single-acting hammer with energy of 30,000 to 40,000 ft-lb, such as MKT S-10, Conmaco 100, Raymond 2/0, or MKT S-14.

8. Static capacity calculations show that you should continue driving piling. Consider a load test to demonstrate acceptance.

9. Belled piers at 26 ft

CHAPTER 12

1. 39 inches.

2. Compaction improves capacity by about 50%.

3. Yes, since the minimum thickness required is 6.3 inches.

4. Minimum thickness is 31 inches without geotextile, 17″ with geotextile.

Index

354